Synchrotron Radiation, Cultural Heritage, Biomineralization

SCIENCES

Physics of Condensed Matter,
Field Directors – Roland Pellenq and Pierre Levitz

Cultural and Industrial Heritage Materials,
Subject Head – Philippe Sciau

Synchrotron Radiation, Cultural Heritage, Biomineralization

Coordinated by
Catherine Dejoie
Pauline Martinetto
Nobumichi Tamura

iSTE

WILEY

First published 2024 in Great Britain and the United States by ISTE Ltd and John Wiley & Sons, Inc.

ISTE Ltd
27-37 St George's Road
London SW19 4EU
UK

www.iste.co.uk

John Wiley & Sons, Inc.
111 River Street
Hoboken, NJ 07030
USA

www.wiley.com

Library of Congress Control Number: 2024943302

British Library Cataloguing-in-Publication Data
A CIP record for this book is available from the British Library
ISBN 978-1-78945-199-3

ERC code:
PE4 Physical and Analytical Chemical Sciences
PE4_14 Radiation and Nuclear chemistry
SH5 Cultures and Cultural Production
SH5_8 Cultural studies, cultural identities and memories, cultural heritage

Contents

Preface

Catherine DEJOIE[1], Pauline MARTINETTO[2] and Nobumichi TAMURA[3]

[1]*European Synchrotron Radiation Facility, Grenoble, France*
[2]*Institut Néel CNRS/UGA, Grenoble, France*
[3]*Advanced Light Source, Lawrence Berkeley National Lab, USA*

The last significant book focusing substantially on the use of synchrotron radiation to study cultural heritage materials was published more than two decades ago[1]. Therefore, we strongly believe that the time is ripe for a new publication, considering the remarkable progress made in synchrotron techniques since then. These advancements have greatly broadened the range of applications, encompassing diverse materials. Hence, we have decided to curate contributions from several experts in the field to create a fresh and updated volume on the utilization of synchrotron radiation in studying cultural heritage materials, fossils and biominerals. Divided into three parts, this volume covers various contributions that showcase the impact of synchrotron-based techniques in these fields.

The first part begins with an introduction to synchrotron radiation and facilities, highlighting their increasing importance in cultural heritage and related research. The second part delves into the application of synchrotron radiation in examining cultural and paleontological artifacts. Techniques such as full-field X-ray absorption spectroscopy imaging and X-ray diffraction-computed tomography are explored, shedding light on various materials and offering valuable insights.

1 Creagh, D.C. and Bradley, D.A. (2000). *Radiation in Art and Archeometry*. Elsevier, Amsterdam.

Synchrotron Radiation, Cultural Heritage, Biomineralization,
coordinated by Catherine DEJOIE, Pauline MARTINETTO and Nobumichi TAMURA.

The third part focuses on specific materials significant to Cultural Heritage. Ceramics, which played a crucial role in ancient civilizations, are studied to understand glaze and slip compositions using synchrotron techniques. Iron-based materials are also investigated, revealing insights into their corrosion and long-term evolution. The study of Roman concrete using synchrotron-based techniques showcases the material's exceptional resilience and provides inspiration for modern, sustainable materials.

Finally, the book explores biomineralization, where living organisms produce minerals, impacting materials like bones and ivory artifacts. Synchrotron methods aid in understanding the mineralization process and contribute to preservation and conservation strategies.

We hope this book captivates readers, offering insights into the diverse scientific studies at synchrotron facilities worldwide. It showcases significant contributions in cultural heritage, fossils and biominerals, highlighting the significant contributions made by synchrotron-based investigations in these fields. By exploring the pages of this book, we aim to provide readers with a glimpse of the vast array of applications and breakthroughs made possible by synchrotron technology. Our aspiration is to inspire future researchers, encouraging them to explore the endless possibilities of synchrotron technology and contribute to knowledge in these fields. May this book serve as a catalyst for transformative research ahead.

September 2024

1

Introduction to Synchrotron Radiation: Application to the Study of Cultural Heritage Materials and Biominerals

Catherine DEJOIE[1], Pauline MARTINETTO[2] and Nobumichi TAMURA[3]

[1] *European Synchrotron Radiation Facility, Grenoble, France*

[2] *Institut Néel CNRS/UGA, Grenoble, France*

[3] *Advanced Light Source, Lawrence Berkeley National Lab, USA*

1.1. Introduction

Cultural Heritage materials are often complex and heterogeneous with a hierarchical architecture spanning from the nanometer range to the macroscopic. An ancient ceramic is usually made of a clay body, on which a surface decoration is apposed (clay patterning, slip, glaze, etc.). The chemical and structural composition of the raw materials and the firing technique (temperature and atmosphere control) affect the end quality and intrinsic properties of the product for its everyday use. An ancient iron axe preserves traces of the manufacturing process in its inner structure, hinting at the know-how of the craftsman who manufactured it. The passing of time, storage conditions and conservation intervention are assessed by the presence of corrosion products and/or passivation layers at the surface of the axe. Cultural Heritage does not only refer to manufactured objects, but also encompasses the tools and techniques developed by ancient societies. In such a context, Roman concrete has shown exceptional resilience over time, as evidenced by Roman monuments still standing today after two millennia.

Synchrotron Radiation, Cultural Heritage, Biomineralization,
coordinated by Catherine DEJOIE, Pauline MARTINETTO and Nobumichi TAMURA.

Finally, preservation and conservation of our Cultural Heritage is of societal concern, a legacy from the past to be transmitted to future generations.

Cultural Heritage is at the junction of several disciplines, such as history, art, archaeology, materials science, chemistry, physics, geology and biology. The reasons for studying Cultural Heritage materials are diverse: knowledge of ancient societies, evolution of practices, development of trading exchanges, material properties, artifact dating, artwork preservation and restoration, etc. Today, the study of Cultural Heritage materials often requires scientific collaborations across multiple disciplines, necessitating combined approaches and promoting interactions between different scientific communities. The rarity, fragility and complexity of Cultural Heritage materials makes them challenging to study, justifying the use of the most advanced scientific tools such as synchrotron radiation facilities to decipher the secrets hidden in their structure.

Synchrotron radiation is the electromagnetic radiation emitted when charge particles travelling at relativistic velocities are radially accelerated. One of the most spectacular manifestations of synchrotron radiation is the Crab Nebula (and associated pulsar, a 30.2 Hz spinning neutron star), remnant of a supernova observed in 1054 by Chinese and Japanese astronomers (Figure 1.1a) (Burbidge 1957; Caroff and Scargle 1969; Bychkov 1973). The strong magnetic field produced by the pulsar bends the electron path, thus generating synchrotron radiation.

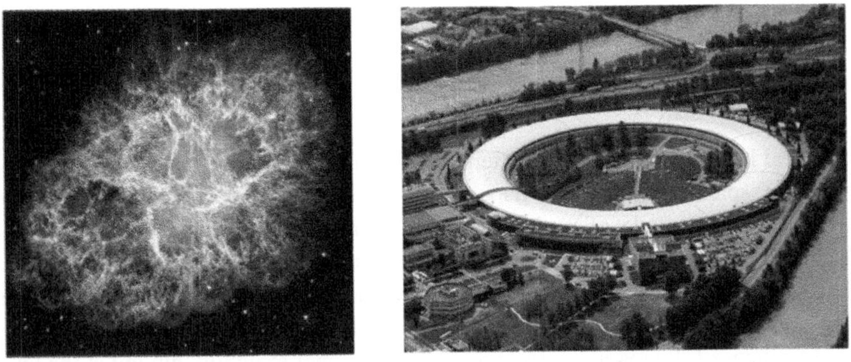

a b

Figure 1.1. *a) Crab nebula mosaic image, taken by NASA Hubble Space Telescope (credit: NASA, ESA). b) View of the European Synchrotron Radiation Facility (ESRF), Grenoble, France (credit: ESRF/D. Morel). For a color version of this figure, see www.iste.co.uk/dejoie/synchrotron.zip*

Part of the theory around synchrotron radiation was formulated at the end of the 19th century (Liénard 1898). The first particle accelerators emerged in the first half of the 20th century, and synchrotron radiation was observed for the first time at General Electric in 1947 (Goward and Barnes 1946; Elder et al. 1947). In the beginning, such radiation, causing the particles to lose energy, was mainly seen as a nuisance in high-energy electron accelerators (Blewett 1998). Nevertheless, the possibility to produce X-rays of unprecedented brilliance was soon recognized. Synchrotron beams were first used in a parasitic way by scientists at particle accelerators. Today, more than 50 fully dedicated synchrotron sources exist, distributed all over the world (Figure 1.1b)[1].

Pioneering work using synchrotron radiation for the study of Cultural Heritage materials was performed in the 1990s at both the Synchrotron Radiation Source (SRS) at the Daresbury Laboratory in the UK (Pantos 2005) and at the European Synchrotron Radiation Facility (ESRF) in France (Walter et al. 1999). The use of synchrotron radiation for the study of Cultural Heritage materials is today more common. Over the last 20 years, the potential of such an approach, the main relevant synchrotron methods and their applications to Cultural Heritage material studies have been extensively reviewed (for example, Bertrand et al. 2012; Dejoie et al. 2018b; Cotte et al. 2019; Janssens and Cotte 2020). The same synchrotron tools used in Cultural Heritage materials are also relevant to other fields such as paleontology and biomineralization. Like in Cultural Heritage, seashells and other biominerals have a complex and multiscale hierarchical structure, the properties of which we have only started to understand. Millions of years of evolution have led nature to create materials that are unique in their properties and a new branch of science, biomimetics, which seeks to understand how nature does it and how we can create new materials with that knowledge.

The objective of this book is to show some recent applications of synchrotron radiation in the field of Heritage Science through a series of contributions about ancient ceramics, the corrosion of iron-based materials, the concrete used in Roman monuments and in the related field of biomineralization. In this introductory chapter, a few elements concerning synchrotron radiation will be briefly described, before introducing the contents of the book in more detail.

1.2. What is synchrotron radiation?

Synchrotron radiation is emitted when charged particles travelling at relativistic speed are accelerated in a curved trajectory. In a synchrotron facility, electrons are

1 See: https://lightsources.org/lightsources-of-the-world/.

circulated in a storage ring near the speed of light. They are guided by magnetic fields coming from bending magnets that cause deflection of their trajectory in the horizontal plane, generating synchrotron radiation tangentially to the electron orbit. The storage ring has a polygonal shape, made of a series of cells, alternating bending magnets and straight sections, with insertion devices (wigglers or undulators) sources occupying the straight sections (Figure 1.2a). Bending magnets generate a continuous polychromatic X-ray spectrum with maximum intensity at a bending magnet critical energy. Insertion devices are made of arrays of magnets that provide a sinusoidal magnetic field, thus causing the trajectory of the electrons to oscillate, and, in so doing, to emit synchrotron radiation at each trajectory bend. In an insertion device, each emission of synchrotron radiation either adds up to produce a photon energy spectrum similar to a bending magnet but brighter (wiggler) or interfere constructively at certain energies, resulting in a series of radiation peaks (called harmonics) an order of magnitude brighter than a bending magnet or wiggler (undulators). On each turn in the storage ring, radio frequency (RF) cavities, which contain electromagnetic fields oscillating at radio frequencies, restore the energy lost by the electrons as they circulate and emit synchrotron radiation (Figure 1.2a). Beamlines where synchrotron radiation is used (mainly X-ray photons, and also IR and gamma rays) are constructed tangentially from both bending magnets and straight sections. Additional information on synchrotron radiation can be found in Margaritondo (1988), Als-Nielsen and McMorrow (2001), Kim (2001), Fitch (2019) or Hwu and Margaritondo (2021).

The principle attributes of synchrotron radiation can be defined as follows:

– high brightness, so a highly collimated and intense beam emitted from a small source size, delivering high flux of photons to the sample. This brightness parameter will be discussed further in the next paragraphs;

– a tunable range of wavelengths, extending from infrared, to soft and hard X-ray regimes, depending on the synchrotron facility. A specific wavelength can be chosen, or variable energy used, for example, for spectroscopy;

– an X-ray beam with a certain degree of coherence that can be exploited for specific experiments, for example, ptychography;

– a polarized source, as the synchrotron radiation is normally linearly polarized in the plane of the synchrotron orbit;

– a pulsed source, as the electrons do not circulate individually in the storage ring but are confined in bunches. The distribution of the bunches allows the time structure to be exploited for specific experiments.

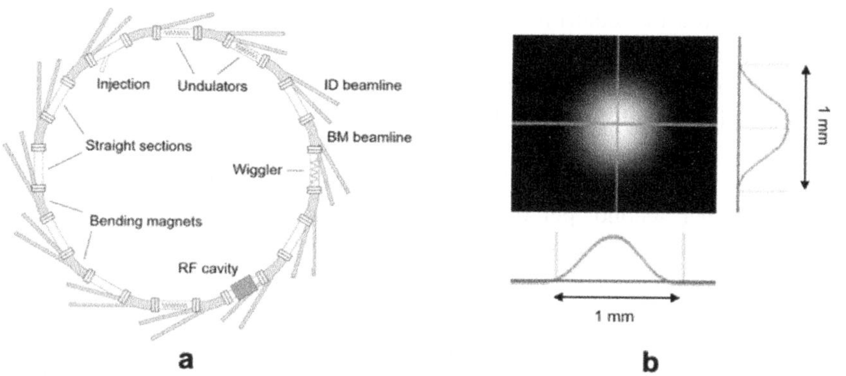

Figure 1.2. *a) Schematic representation of a synchrotron storage ring. Adapted from Fitch 2019. b) X-ray beam profile obtained at the ID22 beamline (ESRF) on September 2020. As a result of the EBS (Extremely Brilliant Source, see below) upgrade, the beam is quasi-symmetric. For a color version of this figure, see www.iste.co.uk/dejoie/synchrotron.zip*

Two parameters are often highlighted when discussing storage rings and beamline performance: the energy of the electrons in the storage ring and the spectral brightness (or brilliance). These two parameters will be discussed further in the next few paragraphs.

The energy E_e of the electrons circulating at a speed v is given by:

$$E_e = \frac{m_e c^2}{\sqrt{1 - v^2/c^2}} = \gamma m_e c^2$$

where m_e is the mass of the electron at rest ($m_e = 9.10938356 \times 10^{-31}$kg), and γ is the factor by which the mass of an electron increases because of its relativistic speed. After conversion of m_e in eV, the electron rest mass becomes 5.10998946×10^5 eV, and then the γ factor can be expressed as a function of the energy (in GeV), with

$$\gamma \approx 1957 E_e \text{ [GeV]}$$

Thus, for a 1.9-GeV (e.g. the Advanced Light Source (ALS), Berkeley, USA, optimized for the use of soft to medium energy X-rays) and a 6-GeV machine (e.g. the ESRF, Grenoble, France, optimized to access harder energy X-rays), γ takes a value of 3,718 and 11,742 respectively. The mass of an electron with an energy of 6 GeV is then roughly 10,000 times heavier than at rest. Expressed in atomic mass units, the mass of a 1.9-GeV electron and a 6-GeV electron is then 2.04 (comparable to the atomic mass of deuterium!) and 6.44 (almost a lithium atom!) respectively. Knowing the circumference and beam current of the synchrotron facility, the electron lap time, the number of laps per second or the expected number of electrons in the storage ring can also be calculated.

The second parameter, the brightness of a beamline, is defined as the number of X-ray photons per second, per 0.1% bandpass, per $mrad^2$, per mm^2, where 0.1% bandpass corresponds to a $\delta\lambda/\lambda$ of 0.001, $mrad^2$ is the solid-angle emission of the X-rays from the source and mm^2 is the cross-sectional area of the source. As a result, a high-brightness source produces (at a given energy) a lot of photons per second, into a narrow solid angle, with a small source size. The source size plays an important role and is strongly influenced by the electron beam properties, in particular its emittance. The lower the emittance, the smaller the source size, and thus the higher the resulting brightness. Lowering the horizontal emittance of the electron beam is the main challenge today, and a number of synchrotron facilities are currently undergoing or planning an upgrade to tackle this (e.g. Raimondi 2016; Sajaev 2019). In order to achieve this, the ESRF, the first among the third-generation facilities to achieve such an upgrade, has implemented a new ring lattice, the Hybrid Multi-bend Achromat (HMBA), based on an increase and a new arrangement of magnets within a cell. As a result, the horizontal divergence of the beam was reduced, and a more symmetric X-ray beam, of increased brightness, was obtained (Figure 1.2b). The new source, called EBS (Extremely Brilliant Source) became available to users in August 2020.

The X-ray source and the optical elements used to condition the X-ray beam differ from beamline to beamline and are chosen and tuned to best fit specific types of techniques and experiments (e.g. diffraction, spectroscopy, imaging, etc.). Bending magnet and wiggler sources provide a broad spectral range of X-ray photons and are more suited for polychromatic applications and techniques, where energy tunability is desirable (spectroscopy). An undulator source provides a sinusoidal magnetic field, so by choosing the period of the magnetic array and by varying the strength of the field, both wavelength distribution and X-ray beam divergence can be controlled. As well as its bending magnet or insertion device source, a beamline will incorporate a number of optical elements such as beam-size defining slits, a monochromator to select the wavelength, and focusing elements such as Kirkpatrick–Baez (KB) mirrors and compound refractive lenses (CRL). To illustrate this, the layout of two X-ray diffraction beamlines will be described in the next two paragraphs. The first beamline is dedicated to microdiffraction, thus providing high-spatial resolution for structural studies, while the second one is dedicated to high-angular resolution powder diffraction, for which the use of a larger beam of low divergence is favored.

Beamline 12.3.2 (ALS, Berkeley) is built on a superconducting magnet (or superbend) source, thus providing a continuous range of energy in the 5–24 keV range (Figure 1.3a). A grazing incidence toroidal mirror relays the source to the entrance of the experimental hutch. Two operation modes for single-crystal or powder diffraction measurements are provided, using either the full polychromatic (white) beam delivered by the superconducting magnet source or a monochromatic beam. A four-bounce monochromator allows for fast-switching between the two modes, while accurately maintaining the beam position on the sample. The X-ray beam is focused down to 0.5µm×0.5µm in polychromatic mode and 5µm×2µm in monochromatic mode using a pair of KB mirrors. Laue (white beam) microdiffraction has proven to be a successful approach to map at the micron scale crystal orientation and crystal distortion in polycrystalline and composite materials, including Cultural Heritage materials (Tamura et al. 2003; Kunz et al. 2009; Chen et al. 2016).

The ID22 beamline of the ESRF is built on an in-vacuum undulator source with a 26 mm magnetic period. The corresponding spectrum is shown in Figure 1.3b. The polychromatic beam delivered by the undulator source arrives untouched onto a cryogenically cooled channel-cut Si 111 monochromator, and, after passing a series of slits, a 1 mm × 1 mm highly monochromatic beam ($\Delta\lambda/\lambda \approx 10^{-4}$), of low divergence, is delivered to the sample. Aluminum-CRL can be inserted in the monochromatic beam as desired, for a resulting focusing down to 50 µm × 50 µm. The high-angular resolution of the powder pattern is ensured by the presence of analyzer crystals (e.g. perfect Si 111 crystals) intercepting the diffracted beam (Hodeau et al. 1998). As a result, the instrumental contribution to the full width at half maximum of a diffraction peak, measured on the NIST standard 640c Si 111 peak, is very small (< 0.0025° (2θ) at 31 keV); and this is exploited for a wide range of powder diffraction experiments (Fitch 2004; Dejoie et al. 2018a; Fitch et al. 2023).

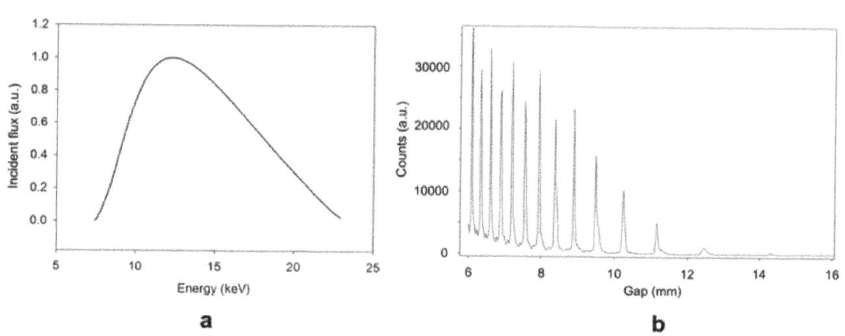

Figure 1.3. *a) Incident flux delivered by the Superbend source of the 12.3.2 beamline at ALS (extracted using the reverse method, Dejoie et al. 2011). b) Spectrum delivered by the U26 undulator of the ID22 beamline, ESRF (70 keV incident energy). For a color version of this figure, see www.iste.co.uk/dejoie/synchrotron.zip*

Synchrotron radiation facilities are user facilities, usually publicly funded. To gain access to a particular beamline and instrument, scientists from external academic or industrial laboratories submit scientific proposals. Proposals are peer-reviewed and accepted on the basis of their scientific merit. Users do not need to be experts in synchrotron radiation methods, as support is provided by the facility and the scientific and technical staff of the beamlines. Access to carry out proprietary research can also be arranged. Users can also form consortia around similar scientific interest and techniques and apply for beam time as is. This procedure is currently under implementation at the ESRF for Materials Science users (Streamline project) and will also benefit to the Cultural Heritage community through a dedicated BAG (Block Allocation Group) (Cotte et al. 2022).

1.3. Synchrotron radiation and Cultural Heritage

Owing to their non-standard shapes, their unique nature, their heterogeneity and sometimes multiscale and composite architecture, Cultural Heritage materials present some analytical challenges that synchrotron-based techniques may help to overcome. The intense beams provided by synchrotron facilities allow the study of samples via a series of X-ray based techniques, such as X-ray diffraction and scattering, X-ray absorption and emission spectroscopy, tomography and other imaging methods. The reasons to study Cultural Heritage materials can be diverse, but, most of the time, four main types of information are exploited:

– the chemical composition of the sample, using X-ray fluorescence (XRF);

– the oxidation state of the relevant chemical elements, using X-ray absorption spectroscopy (XAS);

– the structural composition of the sample using X-ray powder diffraction scattering (XRD) techniques, to retrieve information about the crystalline phases, and, in some cases, about the non-crystalline content;

– the general organization of the sample using two-dimensional and/or three-dimensional chemical (XRF, XAS) and structural (XRD) mapping. Most of the time, the spatial resolution that can be achieved is intrinsically linked to the size of the incident X-ray beam, so micro- to nano-size beams are favored.

Most of the examples described in this book will exploit one, or several of these four types of information, also combined with results obtained from lab-based techniques. In some cases, average information about the material is favored and macroprobes are used; while in others, information at a more local scale is necessary, thus requiring the use of micro- or nano-probes. It is often necessary to combine several approaches and results obtained from independent and complementary probes when materials related to Cultural Heritage are studied.

This book dedicated to Cultural Heritage and synchrotron radiation is divided into three parts. The first part is composed of a single contribution, covering the development of the use of synchrotron radiation for the study of Cultural Heritage materials, with the description of a few pertinent examples in the field of ancient ceramics, mural and oil paintings, rock art, cosmetics, manuscripts and paleontology.

The second, composed of two contributions, is oriented toward synchrotron method developments and the applications of synchrotron mapping techniques to Cultural Heritage materials and biomineralization. One unique feature provided by synchrotron facilities is the ability to generate high-brilliance micro- to nano-beams. These beams enable rapid chemical and structural mapping of relatively large regions within just a few minutes. This capability sets synchrotron facilities apart, offering researchers a powerful tool to explore and analyze materials with exceptional precision and efficiency. We choose to focus on two main techniques exploiting absorption and diffraction respectively, both of which show increasing interest in all areas of materials science. First, the full-field X-ray absorption spectroscopy imaging technique will be described, and a few applications in relation to Cultural Heritage will be presented, in the first contribution. The second technique we choose to highlight is X-ray diffraction – computed tomography (XRD-CT). The basic principle will be explained, before showing its potential for Cultural Heritage studies, in the second contribution.

The third part of the book is composed of four contributions, oriented toward the study of four different types of material for which synchrotron-based techniques have had significant impact. The first type of material is ceramics. Ceramics are among the first examples in history of a large-scale development of crafted objects, for everyday use, decoration or trading. The manufacturing process involves a series of steps, sometimes approaching the industrial scale, as in the case of *Terra Sigillata* ceramics (Sciau et al. 2006) produced by the thousands in Roman times, or the Chinese Jian ceramics, fired in giant dragon kilns during the Song dynasty (Dejoie et al. 2014a). Ancient ceramics are found on mostly all continents, each civilization having developed its own specificities in terms of shape, decoration or firing technique. Ceramics can be seen as cultural markers, and, owing to their resilience toward underground storage and time, a lot of sherds are available today for analytical studies. In this context, a review on the studies carried out on the glaze and slips of ancient ceramics, for which synchrotron radiation was used, is presented.

A different kind of material is the topic of the second contribution, also widely used in ancient times: iron-based materials. Metallic objects have been widely manufactured by ancient societies, to be used as tools, jewelry or weapons. Metallic objects have also been used as support materials, for example, as reinforcement in monuments. Iron-based materials are subject to corrosion, and through this contribution, the concept of long-term evolution of a material through the ages will be introduced. The study of corrosion involves knowledge of electrochemistry, and synchrotron radiation can be used to follow specific reactions, in situ. This illustrates how the study of Cultural Heritage materials is not simply limited to an ancient artifact, but can also lead to interesting research on a specific topic such as corrosion.

The concept of "support material" with great Cultural Heritage significance will be at the center of the third contribution, through a study of Roman concrete. Due to its specific properties, this material allowed quite complex monuments to be built and to survive over two millennia. In this contribution, a set of synchrotron-based techniques including X-ray fluorescence, X-ray powder microdiffraction and X-ray Laue microdiffraction was applied to the study of ancient concrete from Imperial Rome. The close examination of chemical and structural compositions of these concretes using synchrotron microprobe techniques allowed us to understand the chemical mechanisms behind the extraordinary resilience and longevity of these materials. Cultural Heritage materials can provide information about the long-lasting resistance of a material, something that cannot be reproduced easily in a laboratory. As a result, a material such as Roman concrete can also be attractive for modern applications, and this is an example showing how our past can be a source of inspiration to design new, long-lasting materials, following an archaeomimetic approach (Dejoie et al. 2014b). Specifically, roman concrete is currently actively studied by several researchers because of their durability and low carbon footprint, when compared to modern technologies based on Portland cement.

The last contribution opens the path toward a different field: biomineralization. Biomineralization, at the frontier of biology and materials science, refers to the process by which living organisms produce minerals. In the field of Cultural Heritage, materials obtained from a biomineralization process involve, among others, bones or ivory-based artefacts. The 3D structure of these materials is related to their origin and can serve as fingerprint of a particular species (e.g. mammoth tusk vs. modern elephant tusk), and their properties can be influenced by external factors, such as heating, weathering or burial conditions (Reiche and Chalmin 2008; Albéric et al. 2017). Biomineralization can sometimes lead to the long-term degradation of the artifact, but it can also be used as a strategy for conservation, for example, to preserve and reinforce stone-based materials (Marjadi 2016; Marvasi et al. 2020). Biomineralization leads to the formation of complex mineral architectures, with properties resulting from a multiscale organization. Understanding the mineralization process requires this organization to be probed from the nanoscale to the macroscale, and synchrotron-based techniques can be used to provide such multiscale information. The contribution will focus on the biomineralization process in sea urchins' spines, and how synchrotron techniques can help in deciphering calcium carbonate formation and stabilization.

1.4. Conclusion

The use of synchrotron radiation X-rays has strongly influenced many areas in materials science, including the field of Cultural Heritage materials and biominerals. Today, some efforts are still required to make synchrotron facilities a more routine solution for Culture Heritage and related fields communities. The first problem is practical: to access a synchrotron facility, a researcher has to submit a proposal which will be peer-reviewed. Not all proposals get funded and many beamlines are oversubscribed, so beamtime is not guaranteed. On the contrary, new access procedures such as the one mentioned earlier (BAG proposal at the ESRF) should ease and guarantee more regular access to beamtime to the Cultural Heritage community. There is also a lag time of up to six months or more between a proposal being submitted/accepted and the measurements being done, which may create some logistic problems (will the specimen be available for the beamtime? Are the measurements time sensitive?). Moreover, cultural heritage specimens are often unique and precious. Would the results justify the insurance cost of travelling the artifacts to the synchrotron radiation facility? Beam damage to the sample resulting from even short exposure to very intense X-rays is also a concern that needs to be carefully weighted in. Progress in lab-based X-ray technologies also made synchrotron radiation facilities not always the necessary tool for cultural heritage samples. Sometimes, long measurements in a readily accessible lab machine is preferable to very short measurements which may (or may not) happen in six months' time, especially when spatial or temporal resolution is not a critical issue! Cheap tabletop bright coherent X-ray source provided by a new generation of laser-driven plasma accelerators (Kneip et al. 2010) may also one day be a good alternative to synchrotron radiation facilities. For many researchers, in-situ measurements at the museum or directly on the excavation site may be preferred, and it is expected that portable X-ray fluorescence imaging or X-ray diffraction devices (Cuevas and Gravie 2011; Nakai and Abe 2012; Cuevas et al. 2015; Castaing et al. 2016) is becoming more commonplace. For the time being, however, synchrotron radiation continues to offer some sizeable advantages when it comes to spatial resolution and elemental sensitivity, compared to even the most efficient lab sources.

In this book, after an overview on the development of the use of synchrotron radiation for Cultural Heritage studies, we focus on two mapping strategies, through either X-ray absorption or X-ray diffraction and illustrate the enhanced capabilities provided by synchrotron radiation, surpassing what is generally possible with conventional lab-based techniques. Then, an overview of different studies carried out on a series of specific materials is presented, with a final chapter on the close-related field of biomineralization, highlighting how diverse the applications of synchrotron radiation for such complex and heterogeneous materials can be.

1.5. Acknowledgments

The Advanced Light Source at the Lawrence Berkeley National Laboratory is supported by the Office of Science, Office of Basic Energy Sciences, Scientific User Facilities Division of the US Department of Energy under contract No. DE-AC02-05CH11231. The project Streamline has received funding from the European Union's Horizon 2020 research and innovation program under the INFRADEV grant agreement No. 870313. The authors also thank A. N. Fitch for his advice.

1.6. References

Albéric, M., Dean, M.N., Gourrier, A., Wagermaier, W., Dunlop, J.W.C., Staude, A., Fratzl, P., Reiche, I. (2017). Relation between the macroscopic pattern of elephant ivory and its three-dimensional micro-tubular network. *PLoS ONE, Public Library of Science*, 12(1), 0166671.

Als-Nielsen, J. and McMorrow, D. (2001). *Elements of Modern X-ray Physics*. Wiley, New York.

Bertrand, L., Robinet, L., Thoury, M., Janssens, K., Cohen, S.X., Schoder, S. (2012). Cultural heritage and archaeology materials studied by synchrotron spectroscopy and imaging. *Applied Physics A*, 106, 377–396.

Blewett, J.P. (1998). Synchrotron radiation – early history. *Journal of Synchrotron Radiation*, 5, 135–139.

Burbidge, G. (1957). *The Crab Nebula a Cosmic Synchrotron*. Astronomical Society of the Pacific, San Francisco.

Bychkov, K.V. (1973). Synchrotron radiation of the Crab Nebula. *Soviet Astronomy*, 17, 163–168.

Caroff, L.J. and Scargle, J.D. (1969). Coherent synchrotron emission in the Crab Nebula. *Nature*, 225, 168.

Castaing, J., Dubus, M., Gianoncelli, A., Moignard, B., Walter, P. (2016). Development of a portable X-ray diffraction/X-ray fluorescence device for non-destructive analysis of works of art. *Techne*, 43, 79–83.

Chen, X., Dejoie, C., Jiang, T., Ku, C.S., Tamura, N. (2016). Quantitative microstructural imaging by scanning Laue X-ray micro- and nano-diffraction. *MRS Bulletin*, 41, 445–453.

Cotte, M., Autran, P.O., Berruyer, C., Dejoie, C., Susini, J., Tafforeau, P. (2019). Cultural and natural heritage at the ESRF looking back and to the future. *Synchrotron Radiation News*, 32, 34–40.

Cotte, M., Gonzalez, V., Vanmeert, F., Monico, L., Dejoie, C., Burghammer, M., Huder, L., de Nolf, W., Fisher, S., Fazlic I. et al. (2022). The "historical materials BAG": A new facilitated access to synchrotron X-ray diffraction analyses for cultural heritage materials at the European synchrotron radiation facility. *Molecules*, 27, 1997.

Cuevas, A.M. and Gravie, H.P. (2011). Portable energy dispersive X-ray fluorescence and X-ray diffraction and radiography system for archaeometry. *Nuclear Instruments and Methods in Physics Research Section A: Accelerators, Spectrometers, Detectors and Associated Equipment*, 633(1), 72–78.

Cuevas, A.M., Bernardini, F., Gianoncelli, A., Tuniz, C. (2015). Energy dispersive X-ray diffraction and fluorescence portable system for cultural heritage applications. *X-Ray Spectrometry*, 44, 105–115.

Dejoie, C., Kunz, M., Tamura, N., Bousige, C., Chen, K., Teat, S., Beavers, C., Baerlocher, C. (2011). Determining the energy-dependent X-ray flux variation of a synchrotron beamline using Laue diffraction pattern. *Journal of Applied Crystallography*, 44, 177–183.

Dejoie, C., Sciau, P., Li, W., Noe, L., Mehta, A., Chen, K., Luo, H., Kunz, M., Tamura, N., Liu, Z. (2014a). Learning from the past: Rare e-Fe2O3 in the ancient black-glazed Jian (Tenmoku) wares. *Scientific Reports*, 4, 4941.

Dejoie, C., Martinetto, P., Tamura, N., Kunz, M., Porcher, F., Bordat, P., Brown, R., Dooryhee, E., Anne, M., McCusker, L.B. (2014b). Crystal structure of an indigo@silicalite hybrid related to the ancient Maya Blue pigment. *Journal of Physical Chemistry C*, 118(48), 28032–28042.

Dejoie, C., Coduri, M., Petitdemange, S., Giacobbe, C., Covacci, E., Grimaldi, O., Autran, P.O., Mogodi, M.W., Sisak-Jung, D., Fitch, A.N. (2018a). Combining a nine-crystal multi-analyser X-ray diffraction stage with a two-dimensional detector for high-resolution powder. *Journal of Applied Crystallography*, 51(6), 1721–1733.

Dejoie, C., Autran, P.O., Bordet, P., Fitch, A.N., Martinetto, P., Sciau, P., Tamura, N., Wright, J. (2018b). X-ray diffraction and heterogeneous materials: An adaptive crystallography approach. *Comptes Rendus Physiques*, 19, 553–560.

Elder, F.R., Gurewitsch, A.M., Langmuir, R.V., Pollock, H.C. (1947). Radiation from electrons in a synchrotron. *Physical Review*, 71, 829–830.

Fitch, A.N. (2004). The high-resolution powder diffraction beam line at ESRF. *Journal of Research of the National Institute of Standards and Technology*, 109, 133–142.

Fitch, A.N. (2019). Synchrotron radiation and powder diffraction. In *International Tables for Crystallography, Volume H*, Gilmore, C.J., Kaduk, J.A., Schenk, H. (eds). Wiley, New York.

Fitch, A.N, Dejoie, C., Covacci, E., Confalonieri, G., Grendal, O., Claustre, L., Guillou, P., Kieffer, J., de Nolf, W., Petitdemange, S. et al. (2023). ID22 – The high-resolution powder diffraction beamline at ESRF. *Journal of Synchrotron Radiation*, 30, 1003–1012.

Goward, F.K. and Barnes, D.E. (1946). Experimental 8 MeV. Synchrotron for electron acceleration. *Nature*, 4012, 413.

Hodeau, J.L., Bordet, P., Anne, M., Prat, A., Fitch, A.N., Dooryhee, E., Vaughan, G., Freund, A. (1998). Nine crystal multianalyser stage for high-resolution powder diffraction between 6 and 40 keV. *Conference Proceedings of SPIE*, 3448, 353–361.

Hwu, Y. and Margaritondo, G. (2021). Synchrotron radiation and X-ray free-electron lasers (X-FELs) explained to all users, active and potential. *Journal of Synchrotron Radiation*, 28, 1014–1029.

Janssens, K. and Cotte, M. (2020). Using synchrotron radiation for characterization of Cultural Heritage materials. In *Synchrotron Light Sources and Free-Electron Lasers*, 2nd edition, Jaeschke, E.J., Khan, S., Schneider, J.R., Hastings, J.B. (eds). Springer, Cham.

Kim, K.J. (2001). Characteristics of synchrotron radiation. In *X-Ray Data Booklet*, Thompson, A.C. and Vaughan, D. (eds). Lawrence Berkeley National Laboratory, California.

Kneip, S., McGuffey, C., Martins, J.L., Martins, S.F., Bellei, C., Chvykov, V., Dollar, F., Fonseca, R., Huntington, C., Kalintchenko, G. et al. (2010). Bright spatially coherent synchrotron X-rays from a table-top source. *Nature Physics*, 6, 980–983.

Kunz, M., Tamura, N., Chen, K., MacDowell, A.A., Celestre, R.S., Church, M.M., Fakra, S., Domning, E.E., Glossinger, J.M., Kirschman, J.L. et al. (2009). A dedicated superbend x-ray microdiffraction beamline for materials, geo-, and environmental sciences at the advanced light source. *Review of Scientific Instrument*, 80, 035108.

Liénard, A. (1898). Champ électrique et magnétique produit par une charge électrique concentrée en un point et animée d'un mouvement quelconque. *L'Eclairage Electrique*, 16, 5.

Margaritondo, G. (1988). *Introduction to Synchrotron Radiation*. Oxford University Press, Oxford.

Marjadi, D.S. (2016). Conservation and restoration of cultural heritage: A biotechnological approach. *Advances in Applied Science Research*, 7(4), 159–167.

Marvasi, M., Mastromei, G., Perito, B. (2020). Bacterial calcium carbonate mineralization in situ strategies for conservation of stone artworks: From cell components to microbial community. *Frontiers in Microbiology*, 11, 1386.

Nakai, I. and Abe, Y. (2012). Portable X-ray powder diffractometer for the analysis of art and archaeological materials. *Applied Physics A*, 106, 279–293.

Pantos, E. (2005). Synchrotron radiation in archaeological and cultural heritage science. In *X-Rays for Archaeology*, Uda, M., Demortier, G., Nakai, I. (eds). Springer, Dordrecht.

Raimondi, P. (2016). ESRF-EBS: The extremely brilliant source project. *Synchrotron Radiation News*, 29, 8–15.

Reiche, I. and Chalmin, E. (2008). Synchrotron radiation and cultural heritage: Combined XANES/XRF study at Mn K-edge of blue, grey or black coloured palaeontological and archaeological bone material. *Journal of Analytical Atomic Spectrometry*, 23, 799–806.

Sajaev, V. (2019). Commissioning simulations for the Argonne Advanced Photon Source upgrade lattice. *Physical Review Accelerators and Beams*, 22, 040102.

Sciau, P., Relaix, S., Roucau, C., Kihn, Y., Chabanne, D. (2006). Microstructural and microchemical characterization of Roman period Terra Sigillata slips from archeological sites in southern France. *Journal of the American Ceramic Society*, 89(3), 1053–1058.

Tamura, N., McDowell, A.A., Spolenak, R., Valek, B.C., Bravman, J.C., Brown, W.L., Celestre, R.S., Padmore, H.A., Batterman, B.W., Patel, J.R. (2003). Scanning X-ray microdiffraction with submicrometer white beam for strain/stress and orientation mapping in thin films. *Journal of Synchrotron Radiation*, 10, 137–143.

Walter, P., Martinetto, P., Tsoucaris, G., Brniaux R., Lefebvre, M.A., Richard, G., Talabot, J., Dooryhee, E. (1999). Making make-up in ancient Egypt. *Nature*, 397, 483–484.

2

Development of the Use of Synchrotron Radiation for the Study of Cultural Heritage Materials

Nobumichi TAMURA[1], Catherine DEJOIE[2] and Pauline MARTINETTO[3]

[1] *Advanced Light Source, Lawrence Berkeley National Lab, USA*

[2] *European Synchrotron Radiation Facility, Grenoble, France*

[3] *Institut Néel CNRS/UGA, Grenoble, France*

The late 2000s saw an explosion in the use of synchrotron radiation to investigate cultural heritage materials. High brightness, collimation and energy tunability are some of the main features that made synchrotron appealing to the community of archeologists, historians and paleontologists. They enable researchers to examine small amounts of material with unprecedented chemical and structural details. From paint samples, to fabrics, metalwork to pigments and cosmetics, a vast array of artifacts have been analyzed using synchrotron radiation, providing insights into ancient cultures and know-how, as well as guidance for artwork preservation for future generations. This chapter provides an overview of the application of synchrotron radiation to the study of cultural heritage materials.

2.1. Introduction

The potential of using synchrotron radiation for studying cultural heritage materials was recognized as early as 1986 in the now seminal paper by Harbottle et al. (1986). Very high intensities that are orders of magnitude higher than in

Synchrotron Radiation, Cultural Heritage, Biomineralization,
coordinated by Catherine DEJOIE, Pauline MARTINETTO and Nobumichi TAMURA.

laboratory sources and energy tunability are the two main characteristics of synchrotron beams that make them attractive for such applications, as they provide high detection sensitivity, fast data collection and access to spectroscopic techniques such as X-ray absorption Near-Edge Spectroscopy (XANES) and Extended X-ray Absorption Fine Structure (EXAFS), not possible with laboratory sources. High beam collimation and coherence also result in a small probe size and allow for high spatial resolution imaging. Mapping the distribution of trace elements by synchrotron X-ray fluorescence (SXRF) and of minerals by synchrotron X-ray diffraction (SXRD) are the two main techniques that come to mind for cultural heritage materials, with applications for fast analysis of ceramics, paintings and other artifacts for determining their provenance, method of fabrication or age.

The very first published account on the use of synchrotron radiation in archeometry was the measurement of elemental composition in Gaulish coins (Brissaud et al. 1990) using a combination of surface/near surface analysis by Particle-Induced X-ray Emission (PIXE) (ISTN, Saclay), SXRF (LURE, Orsay) and bulk analysis by Neutron Activation Analysis (Orphee reactor, Saclay). The Photon Factory then reported the analysis of Chinese clay dolls and Tenmoku pottery using SXRF for elemental mapping and XANES for determining degree of oxidation (Nakai and Iida 1991; Nakai et al. 1991).

Until the early 2000s, there were very few synchrotron applications in cultural heritage materials, all of them exploratory in nature. These include the use of X-ray fluorescence at the National Synchrotron Light Source (NSLS) on beamline X26 to examine trace element concentrations in dental calculi (precipitates in teeth from bacterial plaques) with potential applications in paleonutrition research (Capasso et al. 1995). X26A has also been used to measure the chemical composition of ancient bronze mirrors from the Han dynasty of China. Specifically, no mercury has been detected, a result consistent with the dealloying hypothesis of the copper from the alpha phase to explain the surface microstructure (Taube et al. 1995). SXRD was furthermore used to get more information about the phase composition in the dealloyed microstructure (Taube et al. 1996).

Before synchrotrons became more widely accessible, the main alternative techniques to map elemental distribution were Particle-Induced X-ray Emission (PIXE), in which a cooled target is hit with MeV particles like protons and alpha particles to generate characteristic X-rays (Johansson 1989) and scanning electron microscopy energy dispersive spectroscopy (SEM-EDS). These techniques had the advantage of lateral resolution and speed, making them more suitable for small samples or archeological materials than laboratory-based XRF (Malmqvist 1986; Carmona et al. 2010; Leon et al. 2012). However, with synchrotron radiation, especially with the advent of third-generation storage rings, these advantages tend to be outweighed by the most glaring disadvantages of these techniques, such as the need for the sample to be in a vacuum, making the use of synchrotron more appealing (Janssens et al. 2000). However, one unique facility worth mentioning is the AGLAE (*Accélérateur Grand Louvre d'analyse élémentaire*) particle accelerator, housed within the Louvre museum in Paris, and managed by the Center for Research and Restoration of Museums of France. It was completed in 1987 and became operational in 1989, thanks to the collaboration with the National Electrostatics Corporation, which powers the accelerator. AGLAE uses two ion sources to produce hydrogen and helium ions, reaching energies of up to 4 MeV and 6 MeV respectively. Its primary purpose is to analyze cultural artifacts, determining their atomic composition and trace elements. Various methods like PIXE, Rutherford backscattering spectrometry (RBS) and gamma-ray emission spectrometry are employed for this purpose (Menu et al. 1990; Calligaro et al. 1998). To further enhance its capabilities and enable safer examination of paintings, an upgrade known as NEW AGLAE was completed in 2017 (Pacheco et al. 2016). While SEM-EDS remains widely used thanks to its high availability, better spatial resolution, low element sensitivity and complementary to synchrotron techniques, the general use of PIXE in archeology and other fields never gained traction, mainly due to the low accessibility and lack of experimental flexibility; its use today remains marginal at most (Lucarelli et al. 2011; Calligaro et al. 2015). The use of synchrotron radiation, on the contrary, is another story.

SXRF measurements clearly dominated the early synchrotron studies of cultural heritage materials. Both NSLS beamline X26A and the Doris III beamline at Hasylab were used to investigate the chemical composition and corrosion of Roman glass from Qumran, Jordan (Janssens et al. 1996; Adams et al. 1997; Aerts et al. 1999, 2000) as well as 15th to 17th century Venetian glass vessels excavated in Antwerp, Belgium (Janssens et al. 1998; De Raedt et al. 2000). The latter study allowed for the categorization of Venetian glass into three groups: one with a high content of calcium oxide dating from the late 15th to early 17th centuries; another with all 16th century sodic glasses and a third with 17th century potassic glass decolored with sulfur dioxide. SXRF at the ELSA storage ring in Bonn was used to analyze the elemental composition of ink on a page of the Gutenberg Bible (1455) and 21 other manuscripts printed between 1470 and 1500. X-ray fluorescence was also used at Spring-8, Japan, to examine the chemical composition of Egyptian vitreous materials, specifically faience and glass, allowing for the identification of two composition clusters distinguishing Egyptian objects from those imported from Mesopotamia (Sato and Nakazawa 2000; Yamahana 2000). The SRI-CAT 2-BM beamline at the Advanced Photon Source (Argonne,

USA) was used to measure elemental compositions in sediments from relic lake Golbasi, Turkey, suggesting environmental changes that affected the early human settlements during the neolithic period (Friedman et al. 2000). The LURE facility (Beamline D1) analyzed the composition of several teeth and bone fragments from Chalcolithic period subjects to determine possible ancient people's living habits. Janssens et al. (2000) published the first review paper on the use of synchrotron X-ray fluorescence in cultural heritage materials, showing the advantages of the technique, some applications and describing the facilities where the data were collected, at Hasylab beamline L at DORIS III (polychromatic X-ray fluorescence), which delivered a focus beam of about 50 microns, and ESRF Beamline u-FID22 (monochromatic X-ray fluorescence) at which a spatial resolution of 3–5 um was achieved.

The first recorded use of synchrotron X-ray microtomography on ancient artifacts was done at NSLS (beamline X-2B) on sherds of a Chinese Song dynasty Jun ware bowl and a modern ash glaze by Jack Troy, showing the distribution of bubbles and precipitates in the glaze, which may be indicative of the firing process (Dunsmuir et al. 1995). Small Angle X-ray Scattering (SAXS) has been used to study the degradation of historic parchment through the analysis of collagen structure (Wess et al. 2001).

The very first workshop bringing together archeologists and synchrotron scientists took place on November 19–20, 1999, at the Daresbury Laboratory, England (Pantos 2000). The workshop demonstrated that archaeometry can benefit greatly from the use of synchrotron radiation. This was followed by another workshop at the Stanford Synchrotron Radiation Laboratory (SSRL), Palo Alto, California on October 18, 2000, and the "First International Symposium on X-ray Archaeometry", at the Waseda University in Tokyo, Japan July 18–20, 2002 (Uda et al. 2005). Since then, a dedicated conference series "Synchrotron and neutron in Art and Archaeology" (SR2A) has been regularly organized every two years since 2006; the first one took place at ESRF, Grenoble, France in 2005 and the second at BESSY, Berlin, Germany in 2006 (Reiche and Sauerborn 2007; Eberhardt et al. 2008). The ninth event in the series occurred in Portsmouth, England, from September 3–7, 2018. However, following this edition, the series had to be put on hold due to the Covid-19 pandemic in 2020. As a substitute, a virtual free event called "SR2A: Hot Topics in Synchrotron Research in Heritage Science" was organized and took place on February 22–24, 2021. The number of publications related to the use of synchrotron radiation to study archeological artifacts and other heritage materials has steadily increased since 2003, and from 2005 onward, 50–100 publications on the topics have been published each year in peer-reviewed academic journals.

This chapter intends to give a brief overview of synchrotron research in the field of cultural heritage by describing some of the artifacts that have been studied. It is not by any means intended to be a comprehensive overview of all research performed in the field, but will focus on a few pertinent examples.

2.2. Synchrotron techniques used in the study of cultural heritage materials

Synchrotron techniques can be roughly classified into spectroscopic, scattering and imaging techniques. Spectroscopic techniques include X-ray fluorescence (XRF), X-ray Absorption Near Edge Spectroscopy (XANES), Extended X-ray Absorption Fine Structure (EXAFS) and Fourier Transform Infrared (FTIR) spectroscopy. Scattering techniques include all the varieties of X-ray diffraction (XRD) techniques, Wide Angle X-ray Scattering (WAXS) and Small Angle X-ray Scattering (SAXS). Imaging techniques include techniques such as X-ray Microscopy, Ptychography, Photoemission Electron Microscopy (PEEM) and X-ray computed tomography (XCT).

X-ray fluorescence (XRF) occurs when the sample is struck by a high-energy X-ray beam, resulting in the ionization of some atoms and the emission of "secondary" X-rays with energies characteristic of the atomic species. Consequently, XRF is commonly used for chemical investigation of materials. If the incident X-ray beam is small, as it is in synchrotrons, the sample can be raster scanned to create fluorescence maps that depict the distribution of the sample's various elements. X-ray absorption spectroscopy (XAS) involves measuring the intensity of the transmitted X-rays through the sample as a function of the incident X-ray photon energy, in the vicinity of the energy necessary to expel an electron from the atom (absorption edge). Small differences in the way X-rays are absorbed at the absorption edge offer information about the material's electronic structure. The technique of XANES refers to XAS performed at energies up to approximately 100 eV from an absorption edge. The energy location of the edge depends on the electronic structure of the atom, and thus, XANES provides qualitative information about the oxidation state of the atom. The intensity of pre-edge characteristics can further reveal the coordination geometry of an atom. The technique of EXAFS refers to XAS done in the 100–1,000 eV region following the edge. EXAFS is used to determine the chemical surroundings of the central atom, through the tiny oscillations in the transmitted signal due to scattering from neighboring atoms. Application of EXAFS includes the identification of poorly crystallized and amorphous phases. More details on spectroscopic techniques are given in Chapter 3.

X-ray diffraction (XRD) occurs when an incident X-ray passes through a periodic arrangement of atoms (crystal). By measuring the position of diffracted X-rays, it is possible to determine how the atoms are ordered in the material. Therefore, X-ray diffraction is widely used to establish the crystal phase makeup of a sample by identifying the mineral species inside it. Powder diffraction is an XRD technique used to analyze crystalline materials by exposing a powdered (or polycrystalline) sample to monochromatic X-rays and measuring the resulting diffraction pattern. The diffraction pattern provides information about the crystal structure, lattice parameters and composition of the material, making it a valuable tool for identifying and characterizing crystalline substances and is therefore widely used for cultural heritage materials. If the X-ray beam is small, X-ray diffraction's raster scanning of the sample reveals the distribution of the mineral species included within it. However, if the crystal size is larger than the beam size, powder diffraction, which relies on large number of randomly oriented crystals to work, becomes impractical. In such cases, the Laue diffraction technique using a polychromatic beam is used instead to map mineral distribution and texture (Chen et al. 2016). X-ray diffraction techniques, such as peak shape analysis, can furthermore be used to measure small deviations from perfect crystallinity. This provides information on defects of the crystal lattice, such as dislocations and mosaicity, which are important indicators of the material's deformation history and the forces it was subjected to. While XRD typically looks at well-ordered materials, other materials that are only partially ordered (e.g. polymers) can be investigated by X-ray scattering techniques. In particular, SAXS examines the X-ray scattering signal emanating from the sample at very low angles and therefore can be used to investigate material ordering at the scale of nanometers. WAXS looks at X-ray scattering signals at higher angles and is used to study the texture of soft and poorly crystalline materials. Pair Distribution Function (PDF) analysis can be applied to such materials by performing a Fourier transform of a powder diffraction pattern. This process allows us to obtain a histogram that represents the distances between atoms within the material of interest.

In XCT, a series of transmission radiographs is taken on a sample placed on a rotation stage. Algorithms are then used to reconstruct a 3D image of the sample under investigation. Contrast in absorption tomography is provided by the difference in attenuation of the materials constituting the sample. If the material is made from materials that do not give much absorption contrast, phase contrast tomography (XPCT) can be used instead, where contrast is provided by the imaginary part of the complex refraction index. Once the 3D image has been reconstructed, the sample can then be virtually dissected on a computer to reveal its internal structure. As for the other techniques, synchrotron provides better spatial resolution, higher data collection speed and better contrast than laboratory based tomographic imaging, which is essential for small samples. Up to now, the investigation of very large samples such as entire paintings were, however, not practical in a synchrotron setting and laboratory tomography techniques with larger X-ray beams were preferable. The recently established BM18 beamline at the ESRF is specifically designed for hierarchical XPCT. This advanced facility offers an alternative approach, enabling researchers to study large objects, even entire human bodies, with exceptional detail, down to submicron levels of resolution (Cotte et al. 2022a).

Pencil beam X-ray tomography, which involves both rotation and raster scanning of the sample, can be combined with various techniques. Instead of the usual outputs (a series of transmission radiographs taken from different angles), corresponding X-ray powder diffraction or X-ray fluorescence maps are used. This fusion of methods allows researchers to obtain a comprehensive 3D distribution of minerals and elements within the sample (Bleuet et al. 2008; Vanmeert et al. 2015; Autran et al. 2020). More details on synchrotron tomographic techniques can be found in Chapter 4 of this book.

PEEM is an imaging technique that uses synchrotron radiation to study the surface structure and chemical composition of materials at the nanoscale. By measuring the energy and distribution of emitted electrons when a sample is illuminated with high-energy photons, PEEM could provide valuable information about the electronic states and chemical properties of the material's surface. Although PEEM has not, to our knowledge, yet been used for the examination of cultural heritage materials, it has found practical applications in the fields of biomineralization and fossil research. For instance, researchers have used PEEM to visualize the orientation of calcite and aragonite crystals in biominerals by manipulating the polarization of the synchrotron X-ray beam (Gilbert et al. 2011).

FTIR spectroscopy (Ugawa et al. 1992) is a more recent addition to the arsenal of synchrotron tools used to study cultural materials, but its use has increased dramatically in recent years. FTIR measures vibrational excitation of molecular groups. Each functional group in a molecule has its own characteristic vibrational energy, and therefore, FTIR spectroscopy can identify them. While synchrotron X-ray diffraction techniques are useful in identifying inorganic compounds in pigments, organic materials generally get damaged by the intense photon flux of hard X-rays. This is where FTIR comes in handy and is often used for identification of organic molecules present in the sample. In FTIR microscopy, the FTIR spectrometer is coupled with a confocal microscope and the intense synchrotron beam allows us to raster scan the sample, collecting FTIR spectra at each point, which can be converted into a 2D map of molecular distribution.

Let us note that more specialized techniques that are variations of the main techniques described above have been occasionally used for cultural heritage materials. For instance, K-edge imaging performed on the biomedical ID17 beamline at ESRF was found to be an interesting technique for the study of paintings. The technique consists of mapping the sample at two different energies around the K-edge of the element of interest. The technique is commonly used in medical imaging but shows promise for cultural heritage materials. The subtraction of one image from the other provides a very high contrast in the element distribution in the material. Measurements on test canvas containing a variety of 17th- to 20th century pigments allow for the easy identification of individual brush strokes, even when they are obscured by others, with potential applications in the reconstruction of hidden or discolored paint layers (Krug et al. 2006).

Another development of interest is synchrotron confocal X-ray microfluorescence. In this technique, the depth from where the X-ray fluorescence originates in the sample is determined by scanning across the sample surface with a polycapillary half-lens in front of an energy dispersive detector. In this configuration, 3D (“depth”) information about elemental distribution in the sample can be obtained, such as the composition and distribution of paint layers in a work of art. Using a setup at the 4W1B beamline at the Beijing Synchrotron Radiation Facility, the technique was employed to study faux bamboo paint in Emperor Qianlong’s Lodge of Retirement in the Forbidden City, China (Wei et al. 2008). Historical records indicate that the Lodge of Retirement was restored twice during the Jiaqing years (CE 1796–1821) and the Guangxu years (CE 1874–1909). The synchrotron analysis shows that the faux bamboo paint has indeed been restored using techniques very similar to the original painting.

2.3. Study of specific pigments in ceramics, sculptures and murals

A large body of synchrotron work is devoted to the analysis of ancient pigments. Ancient artifacts like pottery, paintings, murals and sculptures can lose their colors over time. Research into pigments and how they break down can help bring back the original colors and also help with conservation and preservation issues.

Identification of pigments can also reveal a lot about past techniques used to achieve specific colorations in ceramics or shades in paintings, and this knowledge can help with authentication. One example of study is the investigation of the Chinese Purple ($BaCuSi_2O_6$) pigment from the site of the terracotta warriors from the First Emperor of the Qin dynasty in China (221–210 BCE), by a combination of SEM-EDS, SXRF at beamline 6-2, SSRL and Laue X-ray microdiffraction at beamline 7.3.3, ALS. The crystal structure of the Chinese Purple is similar to the one of Egyptian Blue ($CaCuSi_2O_6$), leading to the hypothesis that the Chinese craftsmen derived the Chinese Purple through a technology transfer from Egypt to China (Berke & Wiedemann 2000). The synchrotron study disproves the hypothesis, and points instead toward an independent invention of the Chinese Purple as a by-product of the fabrication of artificial jades by the Taoists monks (Liu et al. 2007).

In this section, we delve into two examples of pigments used in ancient artifacts: the iron oxides found in Jian ceramics and the Maya Blue used in Mesoamerican murals. Chapter 3 provides comprehensive details on Cadmium and Chromium yellow pigments, while Chapter 4 delves into the use of Lead white. If you seek further examples of synchrotron applications in ancient ceramics, turn to Chapter 5 within this volume.

2.3.1. *Iron oxide pigments in Jian wares*

The first ever published synchrotron study of pigments was by Nakai et al. (1991) on the black glaze of sherds of a Jian (Tenmoku) bowl excavated at a kiln site in southern China. Jian wares were produced during the Song dynasty (960–1279 CE) in giant kilns in the Fujian province. They were renowned for their beauty, originating from their unique black glaze patterns such as "Hare's Fur" and "Oil Spot". Nakai et al. (1991) performed a fluorescence yield K-edge XANES study using BL-4A at the Photon Factory. The XANES obtained in the oil spot pattern is closer to the reference hematite spectrum, whereas the XANES obtained in the surrounding area is closer to olivine. The authors concluded that the Fe in the oil spot is more oxidized than its surroundings and proposed that the oil spots were formed by bubbles of oxidizing gas produced, when the source material was heated up.

A more systematic study on Jian wares was conducted by Dejoie et al. (2014a) using a combination of techniques, including optical microscopy, scanning electron microscopy, transmission electron microscopy, regular X-ray powder diffraction, synchrotron X-ray microdiffraction using both monochromatic and polychromatic radiation, synchrotron X-ray microfluorescence, micro-XANES and micro-Raman

(Figure 2.1). This time, the synchrotron measurements have been performed at BL 2–3 of the Stanford Synchrotron Radiation Lightsource (SSRL) and beamline 12.3.2 at the Advanced Light Source (ALS). The study determined that the pigments responsible for the brownish color of the "Hare's fur" pattern are a mixture of hematite (α-Fe_2O_3) and of the rare metastable phase ε-Fe_2O_3, whereas the "Oil spots" are in majority ε-Fe_2O_3. The study also discovered a significant morphological difference between the ε-Fe_2O_3 in Hare's fur and the oil spot. In the former, they form small, nanometric-sized, randomly oriented crystallites, while in the latter, the phase grows to large, several micron-sized dendritic networks. Due to its high coercivity in the order of 20 kOe, metastable ε-Fe_2O_3 has sparked a lot of interest in magnetic and spintronic applications. However, modern techniques were only able to produce it at the nanoscale and not with high purity. Understanding how the artisans of the Song dynasty were able to stabilize the very large crystals of this phase is therefore of great importance. The secret lies in the high Ca and Mn chemical composition of the glaze and the firing temperature of around 1,250°C. The bubbles bring the iron close to the molten glaze surface, where it begins to precipitate as ε-Fe_2O_3 oxide. The authors conjecture that the glass matrix lowers the free energy of the ε-Fe_2O_3 phase, compared to both γ-Fe_2O_3 and α-Fe_2O_3, significantly slowing its transformation into hematite. The more reducing environment for the "Oil spot" ware prevents this transformation and allows the ε-Fe_2O_3 phase to grow to large crystal sizes. The mineral composition in the Hare's fur Jian ware has been confirmed in a more recent study (Tao et al. 2022) using similar techniques, identifying up to three types of ε-Fe_2O_3 crystals in the glaze surface: leaf-like dendritic structures, flower-like clusters and tightly packed honeycomb structures.

Following this study from 2014, the ε-Fe_2O_3 phase was discovered to be quite common in Chinese ceramics. It was found in an oil spot black glaze bowl from the Xiao Kiln in Anhui province dating from the Jin dynasty (Xu et al. 2017), in the glaze layer of purple-gold glaze porcelain from the Qing dynasty (1636–1912 CE) as stratified crystals of size 1–3 um (Liu et al. 2012), as three-layered microcrystals in sauce glaze porcelain from the Qilizhen kiln, Jiangxi province dating from the Southern Song dynasty (1127–1279 CE) (Wang et al. 2019), in the brown glaze porcelain from the Yaozhou kiln from the Northern Song Dynasty (960–1127 CE) (Wen et al. 2019) and the colored porcelain glaze of the Changsha kiln, dating from the Tang dynasty (618–907 CE) (Shen et al. 2018); and finally in the black glaze from the Qingliangsi kiln, Henan county dating from the Northern Song Dynasty (Liu et al. 2018). In the partially molten surface region of traditionally made unglazed Japanese Bizen stoneware, dendrites of ε-Fe_2O_3 have been discovered growing epitaxially on mullite (Kusano et al. 2008, 2021). In addition, millimeter-sized mullite-ε-Fe_2O_3 clusters blooming in a flower-like fashion on the Hare's fur Jian Ware were identified (Hoo et al. 2020). Interestingly, it was also recently found to be naturally occurring in clinker deposits in the Powder River Basin, Montana (Sprain et al. 2021). An experiment to replicate the ancient Chinese glaze and growth of the ε-Fe_2O_3 phase, using insights gathered from the initial synchrotron studies on the fabrication of Jian ware, has been performed (Holé et al. 2022). Analyses show that ε-Fe_2O_3 crystals begin to grow at high temperature at the beginning of the cooling stage. The growth continues through iron ion pumping from the glaze. The iron depletion of the glaze causes the dissolution of iron rich particles, which were formed during the heating and ion diffusion from deeper in the glaze.

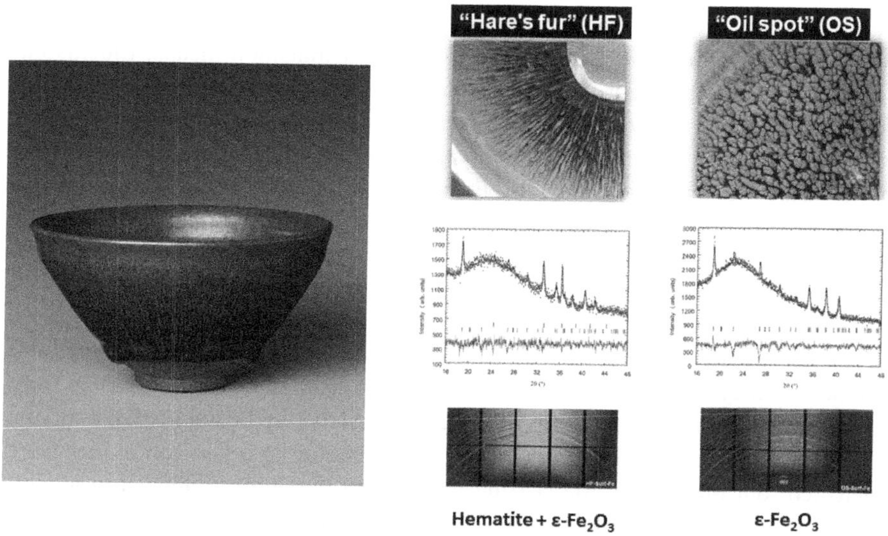

Figure 2.1. *Left: Jian war with "hare's fur" glaze, Southern Song dynasty, 12th century, Metropolitan Museum of Art (public domain). Right: the iron oxides in the Hare's fur pattern are different from the oil spot's pattern as revealed by synchrotron X-ray microdiffraction (modified from Dejoie et al. 2014a). For a color version of this figure, see www.iste.co.uk/dejoie/synchrotron.zip*

2.3.2. *Maya Blue pigment in Mesoamerica*

Maya Blue is another pigment that was extensively studied via synchrotron radiation. This is the pigment used by pre-Columbian Mesoamerican peoples like the Maya to achieve the vivid turquoise blue color in their mural paintings from the 8th century until 1580 during colonization. Maya Blue pigments were first used by the Chupicuaro culture in Western Mexico (600–100 BCE), who also used cinnabar for the first time in Mesoamerica (de Ágredos-Pascual et al. 2019). Maya Blue is of particular interest because it has survived unaltered for over 1,000 years and shows remarkable resistance to harsh conditions, including exposure to high humidity, acids, alkalis and organic solvents, as well as resistance to temperatures up to 250°C. It is made by combining a fibrous phyllosilicate mineral found in clay soil called palygorskite, $Si_8(Mg_2Al_2)O_{20}(OH)_2(OH_2)_4.4H_2O$ (also called attapulgite), and an organic dye, indigo $C_{16}H_{10}N_2O_2$ found in the leaves of the plant *Indigofera suffruticosa* (Van Olphen 1966) (Figure 2.2). The palygosrkite crystal structure is characterized by the presence of micro-tunnels/channels that can be filled with molecules such as zeolitic water (Bradley 1940). Besides palygorskite and indigo, Maya Blue contains impurities such as iron and other transition metals. Elucidating the structure of this pigment would help develop new resistant pigments that are more environmentally friendly than the modern metal-containing ones used in paints. The reasons behind its remarkable durability and color have been elusive, despite more than 50 years of intensive research. Polette et al. (2002) used high-resolution transmission electron microscopy and synchrotron XANES and EXAFS at the Fe K-edge at SSRL beamline SB07-3 on samples from Chitzen Itza and Tulum murals of Mexico and concluded that the Fe oxide phase present in the original Maya Blue was probably Fe_2O_3 in the form of nanocrystals in the channels of palygorskite, which progressively and partially converted to FeO(OH) (goethite) over time and may contribute to the pigment hue. A later study on Maya Blue from samples from Cacaxtla, also using Fe K-edge XANES at beamline ID26 of the ESRF, came to a very different conclusion. Iron, goethite or other iron oxides were not detected, but the iron in Maya Blue was due to the substitutional iron always present in natural palygorskite (partial substitution of Mg and Al in octahedral ions) (Sánchez del Rio et al. 2005).

Palygorskite has two polymorphs, monoclinic and orthorhombic, that are present in different ratios depending on the provenance, which offers a clue as to where the Maya Blue pigment was produced. Separate studies confirm that the source of pre-Columbian palygorskite is most probably mines in the Yucatan peninsula that are still in use in modern times (Arnold et al. 2012). Palygorskite from the mines of the Yucatan peninsula has a unique chemical signature (Sánchez del Río et al. 2009a) that distinguishes it from those elsewhere in the world. Further studies show that the palygorskite used by non-Aztec cultures, such as the Maya, indeed came from the Yucatan peninsula, but conclude that the Aztec culture got its palygorskite elsewhere (Sánchez del Río et al. 2021).

Archaeological samples from six Mexican sites were studied by synchrotron X-ray microfluorescence at the former ID22 beamline (now ID16) at ESRF and PIXE for element composition. In one sample from Cacaxtla, different hues of blue showed different concentrations of Mg, an indication that Maya Blue was mixed with other Ca-based compounds to make it clearer. A sample from the Jiutepec convent shows the only instance where Maya Blue is not the blue pigment but another complex rich in Ba and Zn (Sánchez del Rio et al. 2004). Further samples dating back to the late Aztec period (15th century) in the form of raw pigments prior to their use in artworks also show Maya Blue as a main ingredient; these were investigated by synchrotron high-resolution powder diffraction on BM25A at ESRF (Sánchez del Río et al. 2008). In one case, the organic–inorganic complex is indigo-palygorskite, but in other cases, the clay mineral is sepiolite instead of palygorskite. The indigo-sepiolite pigment is known to be less resistant than the typical indigo-palygorskite complex, so the question remains as to why, in some instances, sepiolite was used instead of palygorskite.

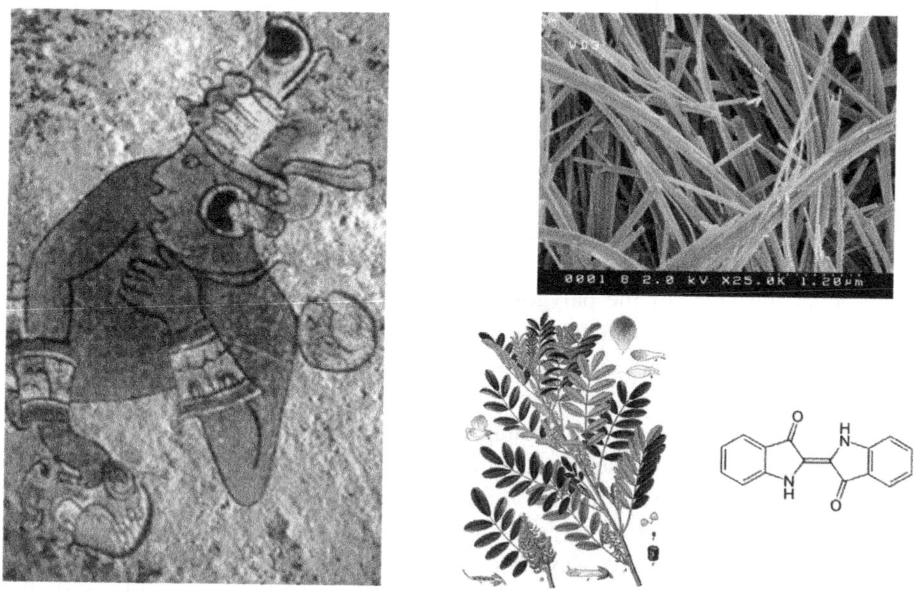

Figure 2.2. *Left: Maya Blue was used in the background on this Mayan painting of a warrior (public domain). Right: Maya Blue is made from a mixture of a clay, palygorskite (top) (Credit: Clay Minerals Society, public domain) and pigment from the American flowering plant Indigofera suffruticosa (bottom) (Credit: Franz Eugen Köhler, Köhler's Medizinal-Pflanzen, public domain). For a color version of this figure, see www.iste.co.uk/dejoie/synchrotron.zip*

The structure of the complex palygorskite-indigo has been studied by Chiari et al. (2003) by molecular modeling and Rietveld refinement on data collected via synchrotron high-resolution powder diffraction at the D8 GILDA beamline of the ESRF. The study shows that the indigo molecules fit into the crystal channels bonded by hydrogen bonds of the palygorskite clay. A neutron diffraction study on Maya Blue pigment prepared with deuterated indigo (Giustetto et al. 2006) confirmed these findings and provided more details on the clay–dye interactions. During heating, indigo molecules lie in the palygorskite microchannels, substituting for the zeolitic water that was previously expelled. Indigo amount is low (<2 wt%), and its occupancy is higher in the orthorhombic polymorph, presumably because of its faster loss of zeolitic water during heating. Synchrotron X-ray powder diffraction at beamline 2-1, SSRL, was one of the tools used (among many others) to obtain how the organic indigo binds to the inorganic palygorskite lattice to form an organic/inorganic complex with the carbonyl oxygen of the indigo bound to a surface Al^{3+} in the Si–O lattice. In addition, indigo may undergo oxidation to dehydro-indigo during preparation. The dehydro-indigo molecule forms a similar but stronger complex with the Al^{3+}. Thus, Maya Blue may vary in color due to the mixed indigo/dehydro-indigo complex (Polette-Niewold et al. 2007).

A later study cast doubt on the model where the zeolitic water is removed first from the palygorskite channels before being replaced by indigo molecules (Chiari et al. 2008). The removal of zeolitic water requires a high temperature of 230–280°C, which is well above the temperature of 100°C at which Maya Blue forms. In that new model, indigo does not enter the channels but gets absorbed in the grooves, which cover the faces of the palygorskite fibers. However, a detailed study using Rietveld refinement and synchrotron powder X-ray diffraction at X7B by NSLS reveals that monoclinic palygorskite loses the majority of its zeolitic water by 150°C (Post and Heaney 2008). In situ measurements using synchrotron powder diffraction and Raman spectroscopy during heating (30–200°C) of a palygorskite–indigo mix show that the formation of the Maya Blue pigment occurs in the 70–130°C interval, when the palygorskite starts to lose zeolitic water, indicated by a reduction of the crystallographic parameter of palygorskite and changes in the C=C and C=O bonds in indigo. The rehydration of Maya Blue is further inhibited by the blocking of the palygorskite tunnel entrances by indigo molecules, which are possibly partially penetrating the tunnels (Sánchez del Río et al. 2009b). Through the integration of new thermogravimetric and synchrotron X-ray diffraction analyses, another research successfully showcased the diffusion of indigo molecules into the channel of palygorskite during the heating process. As a result of this phenomenon, zeolitic water is replaced, leading to the stabilization of the clay's room temperature phases (Dejoie et al. 2011). Further analysis using vibrational spectroscopy up to near-infrared frequencies supports the idea that indigo diffuses inside the palygorskite tunnels in competition with zeolitic water (Tsiantos et al. 2012).

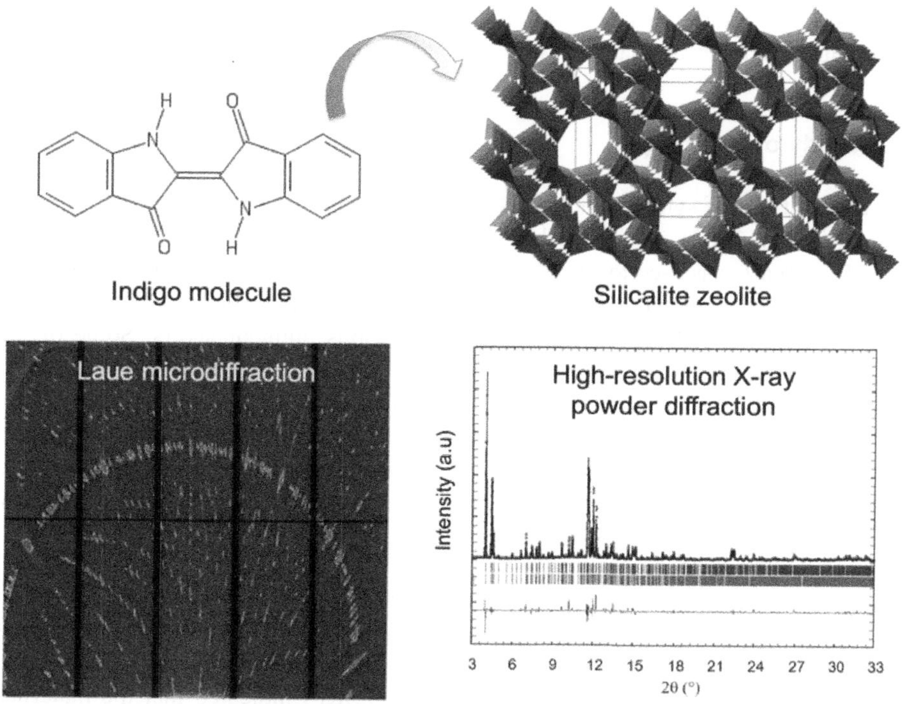

Figure 2.3. *Combining synchrotron high-resolution powder diffraction and Laue microdiffraction, the structure of the artificially made zeolite-indigo pigment which replicates the properties of Maya Blue has been elucidated (modified from Dejoie et al. 2014b). For a color version of this figure, see www.iste.co.uk/dejoie/synchrotron.zip*

According to Rietveld refinement and synchrotron X-ray powder diffraction measurements, the Maya Blue pigment complex indigo-sepiolite is formed similarly to the indigo-palygorskite Maya Blue, with indigo and zeolitic water competing for space inside the channels (Giustetto et al. 2011).

The elucidation of the formation of Maya Blue pigment and the reasons behind its exceptional stability inspired the fabrication of new environmentally resistant pigments. For instance, Dejoie et al. (2014b) have mixed indigo with a silicalite zeolite. A combination of synchrotron Laue microdiffraction performed at beamline 12.3.2, Advanced Light Source and high-resolution powder diffraction performed on beamline BM02 at the ESRF indicates that upon heating, the zeolite undergoes a structural transition from a monoclinic to an orthorhombic phase, with indigo occupying two types of channels within the zeolite crystals (Figure 2.3).

2.4. Study of paintings

The first use of synchrotron radiation in the study of paintings dates from 2002 (Salvadó et al. 2002), when synchrotron X-ray powder diffraction at station 9.6 of SRS, Daresbury, was used to study pigments from three altarpieces by the 14th century Catalan painter Jaume Huguet. The small beam footprint of 100 um allowed for the separation of the diffraction signal from the various pigment layers. Some of the pigments identified were gypsum, weddelite, hydrocerussite, cerussite, tin lead oxide, calumet, paratacamite, atacamite and malachite. The green pigments used by Jaume Huguet were found to be a mixture of green copper compounds synthesized following Theophilus' recipe. He mixed the green pigment with egg tempera and drying oils. The paintings are typically applied in layers of lead white and a yellow pigment known as lead tin oxide.

Other paintings that were studied by synchrotron techniques are the "*Isenheim Altarpiece*" and "*Basel's Crucifixion*" by the famous 16th century German painter Matthias Grünewald. Synchrotron X-ray microfluorescence and micro-XANES at the sulfur K-edge and antimony L-edge (ID21, ESRF) indicated that the materials used to render the metallic aspect of the coat of mail were micrometer pigment grains of stibnite (antimony sulfide), an unusual mineral used in art (Cotte et al. 2007a). Samples from these two paintings were also investigated by synchrotron X-ray microdiffraction (formerly ID22, now ID16, ESRF) and high-resolution powder diffraction (formerly ID31, now ID22, ESRF) and show that the ratio of the two lead carbonates responsible for the white color can be very different from sample to sample and can be used to pinpoint a particular period, geographical origin or know-how (Welcomme et al. 2007).

An extensive study (Gonzalez et al. 2016, 2017) examines the composition and microstructure of lead white pigment in easel paintings from the Middle Ages to the 20th century (including artworks by Leonardo da Vinci, Raphael, Goya, Monet and Van Gogh), using high-resolution synchrotron X-ray diffraction and Rietveld analysis at beamline ID22, ESRF and X-ray microdiffraction mapping at ID13, ESRF. Lead white, a mixture of cerussite and hydrocerussite, was a common pigment in paintings. The analysis revealed variations in the hydrocerussite-to-cerussite ratio, indicating different synthesis processes or post-synthesis treatments. Calcite was found as an extender in some lead white samples. Pigments from the 14th–16th centuries often contained small crystallites, possibly due to postsynthesis treatments. The 19th century pigments had larger crystallites, indicating better corrosion control. Unique compositions were identified, like "blanc de puille" with a high hydrocerussite content.

Rembrandt van Rijn, the famous painter of the Dutch Golden Age, introduced innovative techniques like *impasto*, creating a three-dimensional effect with thick paint protruding from the canvas. To understand his unique recipe, the scientists involved in the lead white pigment study above used the same synchrotron techniques to analyze paint samples from Rembrandt's masterpieces (Gonzalez et al. 2019). Surprisingly, they discovered an uncommon lead compound, plumbonacrite (PN), in the *impasto* layers, which is rare in historical paint. The presence of PN and a high hydrocerussite (HC) to cerussite (C) ratio suggests Rembrandt used an alkaline lead-based medium, possibly treated with litharge (PbO). This finding shed light on Rembrandt's artistic process and offers insights for preserving and conserving his masterpieces.

A microdiffraction and microfluorescence synchrotron study (ID18F beamline, ESRF) of samples from the wall paintings of San Agustin's Church in Cordoba shows how the smaller and more intense synchrotron X-ray beam can detect the full range of iron oxides present in the painting, including goethite, lepidocrocite and hematite, while conventional laboratory X-ray diffraction was only able to see hydrocerussite and cerussite (Herrera et al. 2008).

Vincent van Gogh (1853–1890) is known to paint over previous compositions to save on canvases, with the result that many of his early compositions remain hidden from view. Synchrotron X-ray microfluorescence mapping allows for the imaging of pigment distribution from successive layers, allowing for the reconstruction of earlier compositions that lay beneath the current painting. The technique was successfully applied to the work "*Patch of Grass*" by van Gogh at the beamline L at DORIS III, Hasylab, showing the distribution of Hg (responsible for the vermillion color) and Sb (responsible for the yellow color) with a resolution of about 500 mm and revealing a hidden portrait of a provincial Dutch woman beneath the Parisian style floral painting. The painting was transported directly from the Kröller–Müller Museum, Otterlo, The Netherlands, to the beamline for analysis (Dik et al. 2008, Figure 2.4). The success of the technique to reveal hidden van Gogh compositions was pursued to retrieve other paintings, such as "*Pauline im Weißen Kleid vor Sommerlicher Bauernlandschaf*", an early 19th century painting attributed to Philip-Otto Runge, showing the initial version of the woman's portrait, which was found to be quite different from its final version (Alfeld et al. 2011), and visualize the underpainting (first sketch of a composition) beneath the "*Portrait of an Old Man*" by Rembrandt van Rijn (ca 1630) (Alfeld et al. 2013) or reveal a hidden portrait of Emma Dobigny under the "*Portrait of a Woman*" by Edgar Degas (1834–1927) at the X-ray fluorescence microscopy beamline of the Australian Synchrotron (Thurrowgood et al. 2016). X-ray fluorescence mapping can also be used to reveal high-resolution elemental mapping of a large area of a painting, as demonstrated for a self-portrait of the famous Australian artist Sir Arthur Streeton (1867–1943) at the Australian Synchrotron's X-ray fluorescence microscopy beamline (Howard et al. 2012) and for the "*Blue Room*" painting by Pablo Picasso at G3 Station at the Cornell High Energy Synchrotron Source (CHESS) (Favero et al. 2017). This is only possible at a synchrotron facility, where the high X-ray flux combined with the use of the Maia X-ray detector array (Ryan et al. 2014) allows the recording of maps at a rate of megapixels per hour with high micrometer resolution, thanks to the X-ray's small beam size.

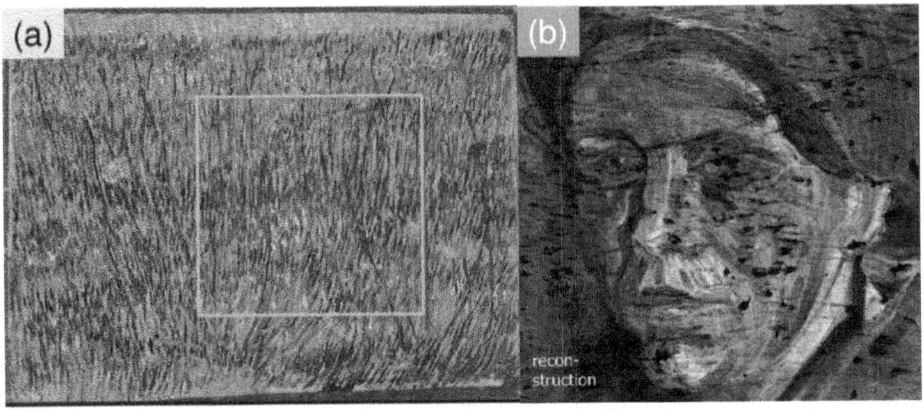

Figure 2.4. *Synchrotron has revealed a hidden portrait of a woman (b) underneath the painting "Patch of Grass" (a) by Vincent van Gogh (courtesy of Joris Dik). For a color version of this figure, see www.iste.co.uk/dejoie/synchrotron.zip*

Synchrotron X-ray diffraction at station 9.5, Daresbury, was used to identify the green pigment in a recently rediscovered 13th century sacred and profane frieze at the palace of Krems, Austria. The green color is due to clinoatacamite

($Cu_2(OH)_3Cl$), mixed with atacamite and a lime matrix (Bidaud et al. 2008). The combination of microdiffraction (BM16, ESRF) and micro-FTIR (station 11.1 at SRS, Daresbury) provides a complete picture of the mineral composition of paint samples, as demonstrated by a study of several samples of 15th century gothic altarpieces from Catalonia (Salvadó et al. 2009) performed on the same beamlines. In particular, it is shown that the evolution of the reactivity of the materials used in the paintings leads to the formation of calcium, lead and copper carboxylates and oxalates in oil and egg yolk tempera and constitutes one of the mechanisms of the degradation process.

The use of synchrotron 3D imaging techniques such as X-ray computed tomography (XCT) has been rather sparse compared to the spectroscopic and diffraction techniques in paintings, but could be useful for some applications. Paint samples from Cuno Amiet's paintings *Winter in Oschwand* (1907) and *Portrait of Max Leu* (1898) were examined by micro-XCT at the MS-X02DA-TOMCAT beamline at the Swiss Light Source, providing information on the degree of porosity and the imaging of microfossils such as foraminifera contained in the chalk. The latter could potentially be useful to determine the source of the chalk used in the paintings (Ferreira et al. 2009; Gervais et al. 2013). A possible use of synchrotron XCT and computed laminography to study ancient altarpieces was explored via a dummy panel that included an inscribed, folded piece of paper behind a wooden inset. Computed laminography is an extension of the tomographic technique to image laterally large objects that has been introduced quite recently at synchrotrons (Helfen et al. 2005). Measurements at beamlines ID17 and ID19 at the ESRF show that the dummy panel can be rendered in 3D in all its details, including the paint layer, the wood grain, the inlay and the folded paper (Krug et al. 2008). The use of computed laminography was further explored to image the different paint layers, pigment morphology, voids, cracks and particles with micron resolution on a mockup painting (Reischig et al. 2013), showing promise as a non-destructive technique to reconstruct the stratigraphy and morphology of paint layers.

2.5. Study of murals and rock art

Synchrotron X-ray fluorescence and diffraction at the ID11 beamline, ESRF, were used to map elemental and mineral distribution in fragments of Roman wall paintings. The size, shape and texture of pigment grains have an effect on the color and luminosity of the painting. Some of the pigments identified are calcite, hematite, goethite and Egyptian blue (Dooryhee et al. 2005). The usefulness of a very small beam has also been demonstrated on the former ID22 beamline (now ID16) at ESRF with the use of microdiffraction with a focal spot of 1-2 mm on a specimen from a mural painting of the Baroque period, across the different paint layers, to identify the different mineral phases composing the pigments (Hochleitner et al. 2004). Multiple techniques were again used in an effort to understand the painting practices of the artist responsible for the 7th century Bamiyan Mural Paintings in Afghanistan (Cotte et al. 2008). Combined with elemental mapping provided by microfluorescence, microdiffraction was used to identify mineral phases, while micro-FTIR was used to identify organic binders and organic–inorganic hybrid materials (ID18F and ID21 at ESRF). Results show that the Bamiyan paintings may be the oldest known use of oil painting in the world. These studies demonstrate the use of multiple synchrotron microprobes performed on several beamlines to obtain key information on the materials and techniques used by ancient painters. Microdiffraction (BM16, ESRF) and mFTIR (station 11.1 at SRS, Daresbury) have also been used to fully identify pigments and support materials, as well as reaction and weathering compounds in Romanesque wall paintings of the church of Saint Eulalia of Unha, Aran Valley, central Pyrenees (Salvadó et al. 2008). The presence of the rare blue pigment aerinite, which has natural compositional variability that could point to its source, indicates a clear connection between the paintings in this church and the Catalan Romanesque paintings from the Pyrenees' southern border.

Images created by prehistoric humans on a hard surface, like a wall, a cliff, or a cave, are known as rock art (or rock paintings). In a study using TEM and XANES (beamline 11-2, SSRL) and micro-XANES (ID21, ESRF), it was discovered that black pigments used in cave painting from the caves of Ekain (Basque Country, Spain), Labastide and Gargas (Hautes-Pyrénées, France) and black "crayon" (i.e. a "pen") from the cave of Combe Saunière (Dordogne, France) are composed of an unusual set of manganese oxides, including manganite, groutite, todorokite and birnessite. This indicates that a few of the manganese ores used to extract these pigments may have traveled quite a distance from the Dordogne and Basque regions and that the pigments from various ores may have been blended together (Chalmin et al. 2006, 2007).

In contrast, hematite is the main component of the black rock art pigment in Jufri Cave in East Kalimantan, Indonesia. A combination of synchrotron X-ray fluorescence, diffraction, XANES, FTIR at Beamline 1.1 W Multiple X-ray Technique (MXT), Synchrotron Light Research Institute, Thailand and SEM-EDS revealed that the hematite in each pigment has different crystallinity and distortion of octahedral symmetry due to differences in the high-temperature heating process, resulting in different color variety (Maryanti et al. 2022).

2.6. Study of cosmetics

Another subject that was the focus of a few synchrotron measurements was the reconstruction of ancient cosmetics production processes, such as the creation of black kohl, which was used by the Egyptians 4,000 years ago. The analysis of the cosmetic traces found in vessels preserved in Egyptian tombs can reveal details about the production methods and the origin of the raw materials. Using a combination of microfluorescence, microdiffraction and micro-XANES (X-ray absorption near edge structure) synchrotron microbeam techniques at the ESRF, trace elements and mineral phases were able to be identified on the same samples of Egyptian cosmetics (Walter et al. 1999; Martinetto et al. 2001; Ungár et al. 2002; Dooryhée et al. 2004). White cerussite ($PbCO_3$) and black galena (PbS), two well-known naturally occurring lead-based compounds, as well as two unexpected synthetic compounds, laurionite ($PbOHCl$) and phosgenite ($Pb_2Cl_2CO_3$), which were produced through wet chemistry through a time-consuming and challenging process, were also discovered (Figure 2.5).

The fact that some medical papyri from the pharaonic period mention the use of makeup as a kind of eye lotion shows that people already desired more from their cosmetics 4,000 years ago than simply highlighting their eyes. These artificial substances were employed as pharmaceuticals to treat eye conditions.

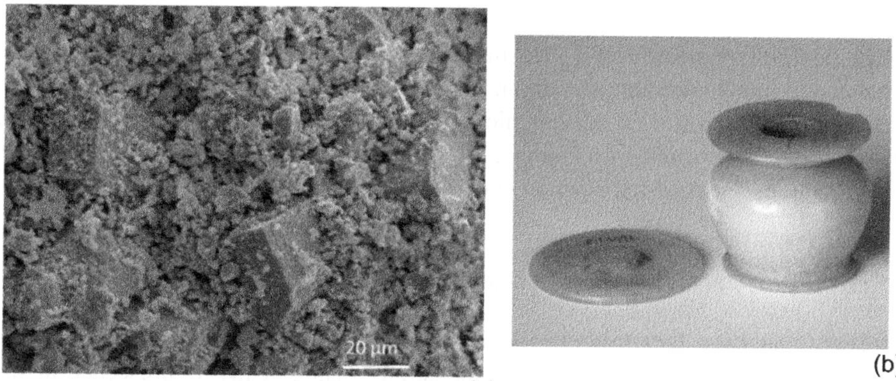

(a) (b)

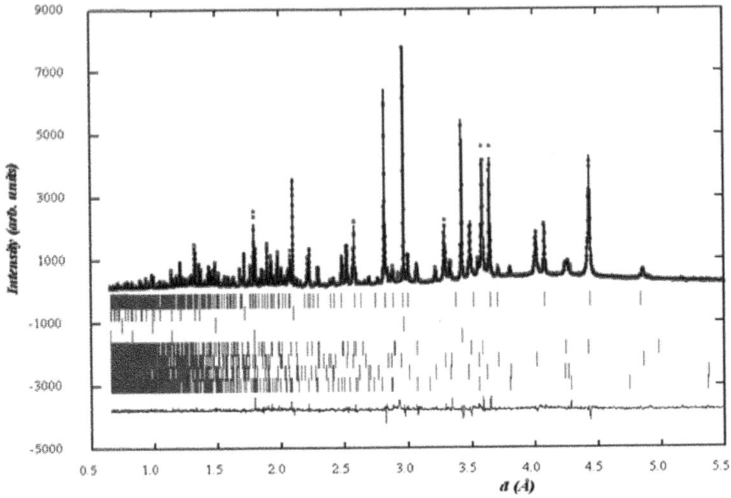

(c)

Figure 2.5. *(a) Scanning electron micrograph of the cosmetic powder taken from the container (Egyptian department of the Louvre Museum) shown in (b) (height = 4.5 cm, diameter = 5.1 cm). Compounds of lead and chloride are aggregated with cubic crystals of galena. (c) X-ray diffraction patterns for the powder: red circles, experimental data points (data BM16, ESRF); black top trace, calculated by Rietveld refinement; blue bottom trace, difference. Green vertical bars indicate Bragg positions of the six identified and quantified phases (galena: 14.5%; cerussite: 22.9%; phosgenite: 32.0%; laurionite: 18.9%; anglesite: 9.3%; gypsum: 2.4%). For a color version of this figure, see www.iste.co.uk/dejoie/synchrotron.zip*

In addition, synchrotron FTIR microscopy was successful in analyzing the organic and mineral components of an Ancient Egyptian cosmetic. Lead soaps, such as lead palmitate, make up the organic phase, which is present in the middle of the cosmetic powder particles. In contrast, phosgenite is mostly found near the surface, indicating that it was added near the end of the preparation (Cotte et al. 2005).

2.7. Study of parchments and manuscripts

Texts that have faded or are hidden can be revealed through the trace of the elements that compose the ink, as demonstrated using X-ray fluorescence at the Kurchatov synchrotron radiation source, Russia, on samples of medieval parchment from the collection of the State Historical Museum. Additional small-angle scattering experiments were performed to diagnose the degradation of the collagen structure in the parchment (Guryeva et al. 2017; Sozontov et al. 2019).

It was demonstrated that handwritten characters on 300-year-old administrative documents can be identified using a variety of X-ray imaging techniques, including X-ray tomography at the TOMCAT beamline of the Swiss Light Source (Albertin et al. 2015). The ancient inks contain heavy elements that produce adequate attenuation contrast for tomographic reconstruction. When the attenuation is insufficient, differential XPCT imaging, where the contrast is obtained through the imaginary part of the complex refractive index (Weitkamp et al. 2005), can be used to reveal the writing.

Combining synchrotron XRF and XANES at the ESRF, carbon black inks on Ancient Egyptiani from various time periods and locales were shown to include copper. No significant differences in the composition of copper-containing carbon inks were detected through different time periods or geographic areas. This makes it likely that the same method for ink manufacture was used throughout Egypt for at least 300 years. It is asserted that the black pigment material (soot) for these inks was derived from technical metallurgy (Christiansen et al. 2017). The same research team performed synchrotron X-ray fluorescence imaging, microdiffraction and micro-FTIR spectroscopy on a large corpus of Roman-era Egyptian papyri inscribed with red and black inks and revealed a previously unknown complex ink composition (Figure 2.6). Iron and ocher give the red color. Lead was often found in red and black inks with phosphate, sulfate, chloride and carboxylate ions. The study indicates that lead was used as a drying agent as opposed to a pigment (Christiansen et al. 2020).

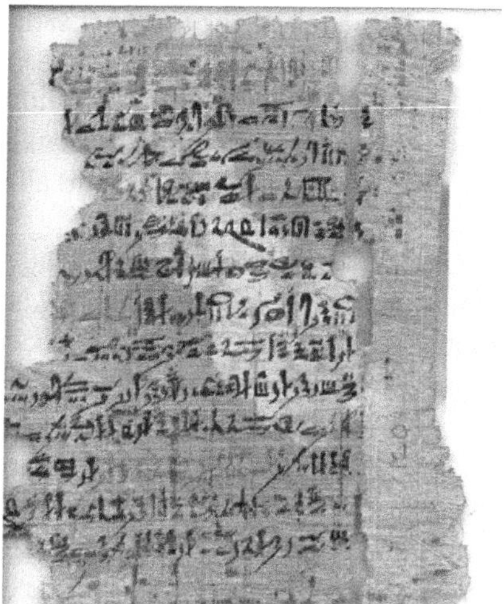

Figure 2.6. *Synchrotron was used to study the red and black ink used in Egyptian writings on papyri. The figure is a New Kingdom text in hieratic script copied from a Middle Kingdom text (public domain). For a color version of this figure, see www.iste.co.uk/dejoie/synchrotron.zip*

A comprehensive study on written and illustrated papyri from the Champollion museum in France was conducted, combining several synchrotron X-ray techniques, including X-ray powder diffraction and fluorescence at ID22, ESRF and X-ray diffraction and fluorescence-computed tomography (XRD-CT and XRF-CT) at ID11, ESRF. The tomographic approach allowed them to distinguish the signal of the raw papyrus from the carbon-based black ink on its surface, enabling them to extract structural contrast based on the black pigment (Autran et al. 2020). This shed new light on the illustration practices of ancient Egyptians who produced "Book of the Dead" manuscripts. The process involved preparatory drawings using red hematite, coloring with traditional Egyptian pigments and final contour lines drawn with carbon-based ink. The study also revealed a delicate balance between standardization and artistic creativity in these illustrations (Autran et al. 2023).

XPCT is being used to decipher texts in charred Herculaneum papyrus scrolls from 79 CE. The complex three-dimensional structure of the scrolls makes traditional unrolling difficult, so virtual unfolding is being explored. XPCT has shown promising results, revealing small portions of the hidden text. A computational platform was proposed for virtual unfolding, and it could lead to interpreting larger portions of the texts. The scrolls' fragile condition makes non-invasive X-ray techniques like micro-XCT and XPCT ideal for analysis (Mocella et al. 2015; Stabile et al. 2021).

2.8. Artwork restoration and preservation effort with synchrotron

Beginning around 2005, the use of synchrotron radiation started to focus on the restoration of works of art. This requires an understanding of the degradation process of cultural heritage materials and finding ways to alleviate or reverse it.

Red Pompeian paintings, made with a pigment called cinnabar (HgS), for example, have been darkening over time and were a major issue for curators because the process was not quite understood. X-ray fluorescence shows a peculiar distribution of chlorine and sulfur. X-ray absorption spectroscopy studies on both the Cl and S K-edge show the presence of two characteristic degradation products: Hg-Cl compounds and gypsum. These findings point to two degradation processes: the reaction of cinnabar with chlorine (such as NaCl, i.e. salt) leading to its decomposition into Hg and S and calcite sulfation. The sulfation process of calcite occurs only at the surface, covering intact cinnabar with a 5 microns thick layer of sulfated black coating (Cotte et al. 2006). A similar study of cinnabar blackening has been performed to restore gothic paintings from a monastery in Pedralbes, Barcelona, Spain (Cotte et al. 2008).

Figure 2.7. *Through synchrotron studies, it was revealed that the yellow color in Vincent van Gogh's painting "View of Arles with Irises" was once brighter than it appears now. This change is attributed to a complex chemical reaction that darkened the pigment containing Chromium (Credit: Wikimedia Commons, public domain). For a color version of this figure, see www.iste.co.uk/dejoie/synchrotron.zip*

Chrome yellow pigments based on lead chromate $PbCrO_4$ can be found in the works of 19th century artists such as Vincent van Gogh, Georges Seurat and John Constable. Chrome yellow darkening was first observed in 1829, when it was discovered that the pigment could change when mixed with oil, lead oxide or Prussian blue. As a result, painters in the 1880s gradually stopped using chrome yellow and switched to more stable pigments, such as cadmium yellow (CdS). The darkening of chrome yellow is commonly attributed to chromium reduction from Cr(IV) to Cr(III) caused by light, temperature or contamination. To investigate artificially aged model samples, a combination of synchrotron radiation-based X-ray techniques such as microfluorescence, microXANES (at Beamline ID21, ESRF) and Electron energy loss spectroscopy (EELS), as well as more traditional spectroscopic methods, was used. Sulfates were discovered to play an important role in promoting the chromium reduction process, which was especially noticeable in samples with higher sulfate concentrations (Monico et al. 2011a). X-ray analysis of two Vincent van Gogh paintings, "*Bank of the Seine*" (1887) and "*View of Arles with Irises*" (1888) (Figure 2.7), indicates that the pigment in them has been similarly altered, which is consistent with the model artificially aged model samples. Reduced Cr was found on the exposed surface of paint layers containing $PbCrO_4$ in both microsamples (Monico et al. 2011b).

A combination of FTIR microscopy and microfluorescence was used to understand the formation of protrusions in a 1610 anonymous oil painting from the Musée du Chatillonais, France. The study found that lead oxide, introduced as a siccative rather than a pigment, is the primary cause of these protrusions, which were determined to be a mixture of unsaponified oil and lead soaps, and that this degradation appears rapidly on paintings, as well as in modern model mixtures and facsimiles (Cotte et al. 2007b).

The restoration of the *Polittico di S'Angelo* (1499), a polyptych composed of six main panels and a *predella*, by the Umbrian painter Niccolò Liberatore, called "*Alunno*", required an understanding of the discoloration (becoming brown) over time of certain pigments such as copper resinate, a transparent green glaze that was obtained by mixing copper pigments with terpenic resins. XANES and EXAFS measurements were performed at the Italian GILDA beamline of the ESRF, complemented with UV and FTIR measurements, on painting samples and pellets of home-made copper resinate. The blackening appears to be due to a strong interaction between the pigment and the organic matrix, leading to the modification of the Cu coordination structure (Cartechini et al. 2008).

A combination of micro-XANES, microfluorescence (ID21, ESRF), microdiffraction (ID18, ESRF) SEM-EDS and Ultraviolet-Induced Visible Fluorescence Photography has helped in understanding the chemical process behind the gradual fading of the bright yellow areas in paintings of James Ensor (1860–1949). The yellow is provided by the pigment cadmium yellow (CdS). The sulfur contained in CdS in its sulfidic form, gets oxidized into its sulfate form. The resulting highly soluble cadmium sulfate ($CdSO_4$) is transported to the surface of the painting and reprecipitated into whitish globules, leading to the fading (Van der Snickt et al. 2009).

Parchments have been used as writing media for over 2,000 years. Museums worldwide house a large collection of parchments from various eras, and it is crucial for their conservation to understand their degradation process. The main component of parchment is collagen, a complex material which has multiple degradation processes that are difficult to study. Small-Angle X-ray Scattering and FTIR spectroscopy performed at the Diamond Light Source, have confirmed that lipids cause the degradation process to accelerate and that iron gall ink has a significant effect on the degradation (Možir et al. 2012).

Ancient wooden artifacts face preservation challenges and are difficult to conserve or restore. One study (Bugani et al. 2009) examines the use of synchrotron radiation micro X-ray computed microtomography (SR-μCT) to investigate the conservation of waterlogged archaeological wood. Although underwater conditions can preserve wooden objects well, anaerobic bacteria can degrade them, necessitating conservation treatments. These treatments involve filling the wood's microcavities with compatible substances. However, assessing the effectiveness of these treatments is complex. SR-μCT offers a non-destructive imaging technique to visualize how impregnating materials are distributed inside the wood. The study shows that all tested impregnating materials effectively penetrated the wood, making SR-μCT a powerful method for assessing the localization of the polymers used in conservation.

2.9. Other cultural heritage studies

FTIR microscopy was used (beamline ID21, ESRF) to identify the organic compounds in individual layers of varnish samples from the late 16th century Venetian lute, offering valuable insights on the little-known varnishing process of ancient musical instruments (Echard et al. 2008). The use of synchrotron X-ray tomography to study musical instruments has been demonstrated at the SYRMEP beamline of the Elettra synchrotron (Sodini et al. 2012). The synchrotron counterpart can offer invaluable complementary information to laboratory computed tomography (Sodoni et al. 2017). While laboratory CT can image (in 3D) an entire instrument with resolution of about 100 microns, synchrotrons can be used to look at finer details such as small damages. However, the limited access to synchrotron and the relatively smaller amount of additional information you can extract from synchrotron sources with respect to the more accessible lab instruments, help explain the limited use of synchrotron radiation to the study of ancient musical instruments. The size of the samples required the use of a large detector, PICASSO (Phase Imaging for Clinical Application with Silicon detector and Synchrotron radiation), which was developed for medical applications (Rigon et al. 2009). The PICASSO detector covers the full width of the synchrotron beam (21 cm) with a 50 microns spatial resolution.

Hairs from Ancient Egyptian mummies have been investigated by SAXS at the D43 beamline of LURE and the ID13 beamline at ESRF, FTIR at MIRAGE-SA5 beamline at LURE and X-ray fluorescence at the D15 beamline at LURE. The structural analyses show a remarkable preservation of keratin supramolecular organization, with almost no discernible modification after two millennia (Bertrand et al. 2003). SXRF at the D15 beamline of LURE was used to investigate hairs belonging to Napoleon Bonaparte and shows high levels of toxic elements, including As, Pb, Hg with considerable variations from hair to hair, which is inconsistent with the hypothesis of arsenic poisoning through ingestion (Chevallier et al. 2006). SXRF, SXRD, XANES and FTIR measurements were performed at the ALS for a forensic study of ancient human hair from Pre-Columbian population who inhabited the Atacama Desert in Chile in 500–1,450 CE and showed a high degree of arsenic contamination consistent with As-polluted water ingestion (Kakoulli et al. 2014). Hair samples from a burial site of Xiongnu nobility in Mongolia (1st century BCE to 1st century CE) studied by SXRF at the Siberian Synchrotron and Terahertz radiation center indicate elevated mercury content from probable exchange with the burial environment (Zvereva et al. 2017). The effect of intense metal work on human health can be seen in the presence of high level of biogenic copper in the hairs of individuals from the third millennium BCE site of Shar-i Sokhta, Iran, through SXRF mapping at beamline ID21, ESRF (Lorentz et al. 2020).

A detailed analysis of fragments from an Egyptian mummy's sarcophagus cartonnage has been conducted using X-ray microfluorescence (μXRF) at the D09B beamline of the Brazilian Synchrotron Light Laboratory (LNLS) to determine the elemental composition of the pigments used in the decorative paintings. The study aimed to verify if the pigments aligned with those used by ancient Egyptian craftsmen during the Roman period. The results indicated that the pigments matched those historically used, such as red ochre, realgar, green earth, Egyptian blue and ivory black (Calza et al. 2007, 2008).

Synchrotron white-beam (Laue) X-ray microdiffraction (μXRD) offers advantages for small heterogeneous samples. It provides detailed grain maps for different mineral phases, aiding in the identification of complex mineralogies. This approach was applied at the Advanced Light Source to analyze ancient Egyptian pigments from a sarcophagus dating from the 5th Dynasty to early 26th Dynasty (747–600 BCE), successfully identifying cuprorivaite, quartz and calcite in the blue pigment, and malachite and goethite in the green–yellow pigment (Lynch et al. 2007) (Figure 2.8).

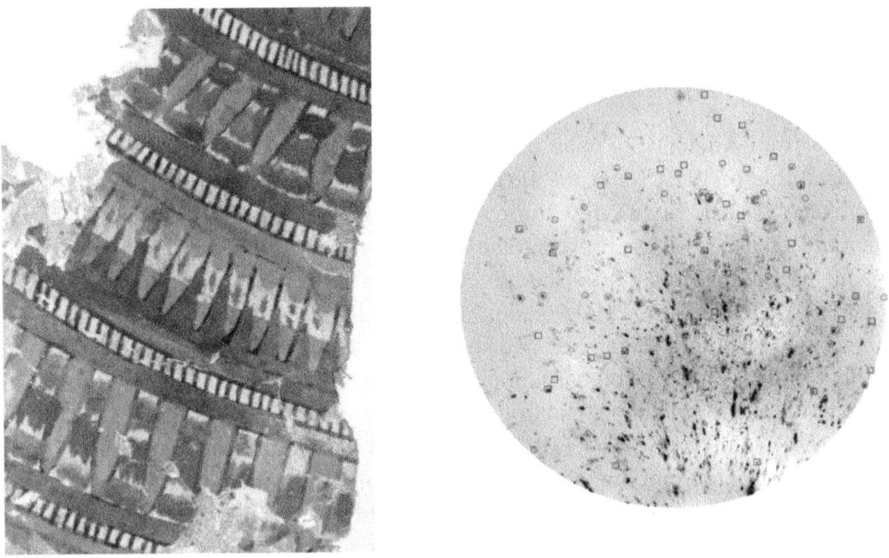

Figure 2.8. *Using synchrotron Laue microdiffraction, researchers examined the decoration on an Egyptian coffin lid and discovered reflections from calcite (depicted as blue squares), quartz (shown as red circles) and cuprorivaite (represented by green triangles) (modified from Lynch et al. 2007). For a color version of this figure, see www.iste.co.uk/dejoie/synchrotron.zip*

Regarding Egyptian sarcophagi and mummies, there has been a fascinating study involving the Hawara Portrait Mummy 4, an ancient Egyptian portrait mummy from the Roman era. This mummy was thoroughly examined using non-invasive techniques, namely computed tomography (CT) and CT-guided synchrotron X-ray diffraction mapping, the latter at the APS beamline 1-ID. It marked the first time X-ray diffraction was applied to objects within a mummy without the need for invasive measures. The CT data provided insights into the remains of a 5-year-old child, consistent with the portrayed female on the portrait. In addition, the diffraction analysis revealed the presence of modern dual-phase steel wires used for stabilization and a calcite inclusion serving as protection for the evisceration site. This groundbreaking research showcased the potential of combining CT and X-ray diffraction for non-invasive mummy analysis, which may open up exciting possibilities for further studies on mummies, their contents and even uncovering metallic or ceramic artifacts concealed within their wrappings (Stock et al. 2020).

An ancient Egyptian crocodile mummy, dating to the Roman period, was analyzed using synchrotron multiscale microtomography at the BM05 beamline of the ESRF (Porcier et al. 2019). This advanced technology allowed for a virtual autopsy of the animal without disturbing the bones, flesh, balms and linen bandages. Through this technique, researchers were able to examine the bones and tissue of the specimen, leading to the determination of the cause of death and the last meal(s) of the crocodile. Surprisingly, the findings suggest that this particular crocodile was hunted while it was living in the wild, marking the first evidence of this method being used to obtain animals for mummification. Additional study by the same group revealed more specifics on the mummification process (Berruyer et al. 2020). Apart from the head and limb extremities, all internal organs, muscles and most of the skeleton were removed to prevent further decay. This unexpected finding challenges previous assumptions about mummification techniques and demonstrates the diversity of practices used by the Ancient Egyptians.

There are countless more applications of synchrotron radiation to the study of cultural materials than the ones mentioned in this chapter. For instance, synchrotron X-ray microdiffraction has been instrumental for deciphering how the Romans made their concrete and understanding why they last so long (Jackson et al. 2014; Seymour et al. 2022; Chapter 7 of this book). The use of synchrotron for the study of ancient metallic objects is described in Chapter 6.

2.10. Study in paleontology

Paleontology, the study of fossils and past life, is one of these fields that has been revolutionized by the development of modern 3D visualization techniques.

Tomography and the development of powerful 3D visualization software make it possible to examine the internal structure of fossils in detail, without needing to destroy them (Cunningham et al. 2014). Skeletal elements are usually found disassembled, and with modern technology, it is now possible to reconstruct the skeleton virtually on a computer and account for diagenic distortions, rather than having to painfully extract each element from the rocks. Computed tomography has been used to study the evolution of the middle ear and neocortex in mammals by 3D imaging the chain of bones attached to the mandible and cranium in various mammalian ancestors (Rowe 1996). A virtual skull of the sauropod *Diplodocus* obtained through computed tomography of a real specimen was used for finite element analysis and biomechanical modeling of the feeding behavior of this dinosaur (Young et al. 2012). The use of synchrotron X-ray tomography in paleontology came more recently. The high flux, coherence and collimation of synchrotron radiation sources provide high voxel resolution, faster data collection and access to new capabilities such as phase contrast microtomography than what lab sources can provide, allowing us to look at smaller samples with intimate details. One example of the use of synchrotron 3D imaging is the reconstruction of the multielement apparatus of conodonts, an extinct group of vertebrate relatives that are used as index fossils thanks to their widespread distribution and rapid evolution during the Paleozoic era (Martínez-Pérez et al. 2016; Agematsu et al. 2017; Huang et al. 2019). Another is the study of the internal microstructures of fossil embryos from fossils dating from the Cambrian period (Donoghue et al. 2006; Chen and Dong 2008). Synchrotron micro-XPCT proved particularly useful in fossil bone histology, revealing the bone architecture and ontogenic record (Curtin et al. 2012; Sanchez et al. 2012) and allowing the 3D imaging of fossil inclusions contained in opaque and semi-opaque amber (Lak et al. 2008). The 3D imaging of coprolites (fossil feces) using synchrotron micro-XPCT can reveal their fossil composition, giving insights on paleoecosystems (Qvarnström et al. 2017). Micro-XPCT was also employed to examine the wood structure of *Armoricaphyton chateaupannense*, the earliest known woody plant from the Devonian period in France. By analyzing the CT scans, researchers determined the hydraulic properties of this ancient wood. The findings suggest that early woody plants developed axis rigidity through the formation of low-density woods (Strullu-Derrien et al. 2014). Through the utilization of synchrotron X-ray tomography, intricate details of the pharyngeal skeleton in the Devonian placoderm, *Paraplesiobatis heinrichi*, have been unveiled, allowing us to establish the presence of a minimum of four branchial arches in this ancient fish. This research not only sheds light on the origins of jaws but also carries the potential to yield crucial insights into the phylogenetics of enigmatic taxa (Brazeau et al. 2017).

Tullimonstrum gregarium, commonly known as the Tully monster (Figure 2.9) and originating from the Carboniferous period in Mazon Creek, Illinois, represents the quintessential enigmatic fossil. Despite being documented since 1955 through thousands of fossils, which include mostly intact soft body impressions, the true nature of this creature continues to elude the scientific community. Notably, two significant studies, involving meticulous examination of numerous fossils, concluded that *Tullimonstrum* was a vertebrate (Clements et al. 2016; McCoy et al. 2016). The latter of these studies pointed out that the presence of spheroidal and cylindrical melanosomes within *Tullimonstrum*, forming vestiges of the retinal pigment epithelium, presents compelling evidence for classifying it as a vertebrate. However, a more recent investigation (Rogers et al. 2019) delved into the arrangement of melanosomes across various species, challenging the notion that melanosome layering is an exclusive trait of vertebrates. It was revealed that such layering can also be observed in cephalopods. The outcome of this investigation, facilitated by synchrotron X-ray fluorescence and XANES techniques, emphasized that the chemical composition of melanosomes found in the eyes of *Tullimonstrum* bears a stronger resemblance to contemporary cephalopods, rather than vertebrates.

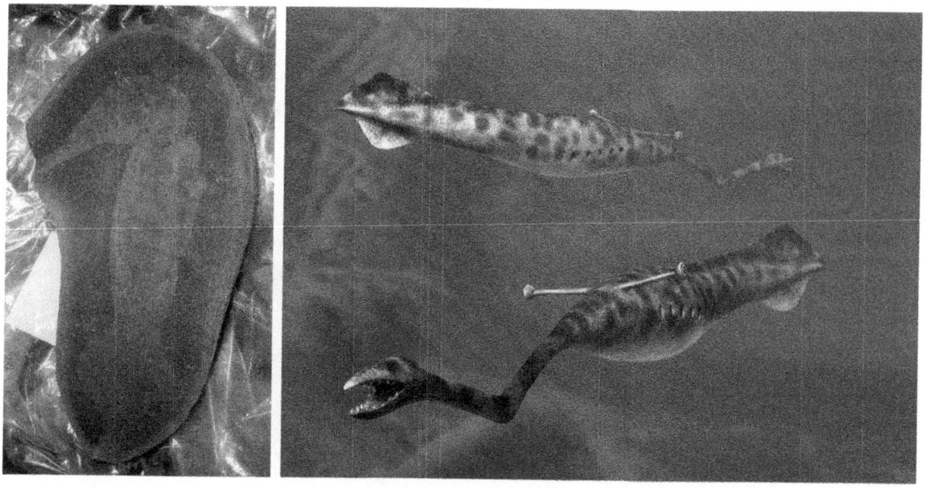

Figure 2.9. *The enigmatic fossil, Tullimonstrum gregarium, at the Museo di Storia Naturale, Milano (left, credit Ghedoghedo, Wikimedia Commons, Creative Commons license CC BY-SA 3.0), and its reconstruction (right) (Credit: N. Tamura). For a color version of this figure, see www.iste.co.uk/dejoie/synchrotron.zip*

Synchrotron X-ray fluorescence imaging and XANES spectroscopy are also particularly useful in mapping the distribution of trace elements in fossils. One spectacular application was the mapping of trace elements and organic sulfur, which revealed the preservation of color melanin pigments in the Berlin fossil feather specimen of the ancient bird *Archaeopteryx*, inferring a black and white plumage (Manning et al. 2013) (Figure 2.10). These techniques can also be used effectively to test hypotheses about the nature of some enigmatic fossils. Spicule-like structures from the Ediacaran Doushantuo Formation have been interpreted as belonging to the earliest demosponges, a class of sponges that includes most of the modern forms. However, a combination of synchrotron X-ray fluorescence mapping and XANES spectroscopy indicates that the structures are carbonaceous in composition and rectangular in transverse section, a morphology and mineralogy that are inconsistent with the cylindrical siliceous spicules of demosponges (Muscente et al. 2015).

When combined with X-ray diffraction, which provides mineralogical information, X-ray fluorescence and XANES provide key information in the taphonomic and mineralization processes, leading to the fossil, offering insights on the preservation process, the environment where the animal or plant lived and on their soft tissues (Gueriau et al. 2020a, 2020b; Hall et al. 2020). Adding synchrotron FTIR to the mix, a comprehensive analysis of preserved blood vessel structures in a *Tyrannosaurus Rex* bone reveals their endogenic nature and adds credence to the possibility of preservation of microvascular tissue through deep time (Boatman et al. 2019).

Figure 2.10. *The analysis of the Berlin fossil feather (left, Credit: H. Raab, licensed under Creative Commons Attribution-Share Alike 3.0 Unported) allowed the reconstruction of the plumage color of the ancient Jurassic bird Archaeopteryx (reconstruction by N. Tamura). For a color version of this figure, see www.iste.co.uk/dejoie/synchrotron.zip*

Biomineralization, the process by which organisms form minerals, plays a crucial role in understanding the fossil record. It is essential for interpreting fossils in geochemistry, understanding evolution and analyzing how geological events shaped the fossil record. Traditionally, geologists mainly used fossils for climate reconstruction, overlooking the significance of biomineral formation. However, recent research has brought attention to the importance of biomineralization in paleontology and geosciences. Advanced techniques like atomic force microscopy and synchrotron imaging have shed new light on both modern and fossil biominerals (Pérez-Huerta et al. 2018). One studied biomineral is nacre found in pearls and mollusk shells. The technique of PEEM at the ALS beamline 11.0.1, has been used to study fossil and modern nacre from Pinnidae seashells (Gilbert et al. 2017). Nacre has periodic organic sheets alternating with aragonite tablet layers. The thickness of nacre tablets (TT) has a significant impact on its properties, and this study investigates the correlation between TT and temperature (T) in modern and fossil bivalve shells. The findings suggest that TT can serve as a valuable paleothermometer for estimating ancient shallow-water paleoenvironments. To gain further insights into the application of synchrotron radiation for understanding the biomineralization process, refer to Chapter 8 of this book.

2.11. Conclusion

Since the early 2000s, the use of synchrotron techniques to examine cultural artifacts has grown in popularity. Using micrometer- to tens-nanometer-sized synchrotron X-ray beams, the fine structures of these typically heterogeneous materials were analyzed in exquisite detail. X-ray microfluorescence, which displays the elemental composition, XANES, which reveals oxidation states, and X-ray microdiffraction, which identifies mineral phases, are the most often used techniques for this purpose. On occasion, however, other techniques such as X-ray computed tomography, which can image the internal 3D structures of an object by absorption contrast or phase contrast, Small Angle X-ray Scattering, which can assess the crystallinity and orientation of poorly crystallized materials such as collagen, and Fourier-Transform Infrared Spectroscopy, which can investigate the structure of organic materials, have also been used. The penetrating capacity of hard X-rays has also been used to non-destructively reveal hidden features and objects within works of art, such as compositions that have been painted over, handwriting on parchment scrolls that have become too fragile to unwrap and microfossils embedded within rocks. Typically, non-synchrotron techniques such as scanning electron microscopy, transmission electron microscopy, Raman spectroscopy and neutron diffraction are used in conjunction with a number of these techniques for a full analysis of the material of interest.

Due to its capacity to investigate minute amounts of materials to determine their composition and structure, synchrotron techniques were particularly useful for detecting pigments in ceramics, statues, paintings, rock art, inks, cosmetics and fossilized soft tissues. Pigment identification is necessary for determining production method, material provenance, determining authenticity and replicating the original appearance of an artwork. Since the late 2000s, the emphasis has shifted to the conservation and preservation of historical objects through the study of the deterioration process of these pigments and of the medium where the ink or paint has been deposited, by comprehending their underlying degradation processes and inventing strategies to mitigate them. The analysis of cultural materials, fossils and ancient artifacts is projected to continue to rely heavily on synchrotron techniques. Today, almost every synchrotron facility in the world has measurements touching cultural heritage and paleontology materials, and a few beamlines are almost exclusively used for the study of cultural heritage materials (such as ID21 at ESRF; Cotte et al. 2017, PUMA (Photons Utilisés pour les Matériaux Anciens) at SOLEIL; Tack et al. 2019). Access to beamlines for cultural heritage study has been facilitated through a specific allocation mode such as the recently implemented "historical materials block allocation group" mode at the ESRF with access to two beamlines ID13 and ID22 (Cotte et al. 2022b). More potent synchrotron sources, such as those generated by free-electron lasers (XFEL) and diffraction-limited storage rings (DLSR), have made it possible to examine these chemicals in even more details.

2.12. Acknowledgments

The Advanced Light Source at the Lawrence Berkeley National Laboratory is supported by the Office of Science, Office of Basic Energy Sciences, Scientific User Facilities Division of the US Department of Energy under contract No. DE-AC02-05CH11231. The project Streamline has received funding from the European Union's Horizon 2020 research and innovation programme under the INFRADEV grant agreement no. 870313.

2.13. References

Adams, F., Adriaens, A., Aerts, A., De Raedt, I., Janssens, K., Schalm, O. (1997). Micro and surface analysis in art and archaeology. *Journal of Analytical Atomic Spectrometry*, 12(3), 257–265.

Aerts, A., Janssens, K., Adams, F., Wouters, H. (1999). Trace-level microanalysis of Roman glass from Khirbet Qumran, Israel. *Journal of Archaeological Science*, 26(8), 883–891.

Aerts, A., Janssens, K., Velde, B.D., Adams, F., Wouters, H. (2000). Analysis of the composition of glass objects from Qumrân, Israel, and comparison with other Roman glass from western Europe. *MOM Éditions*, 33(1), 113–121.

Agematsu, S., Uesugi, K., Sano, H., Sashida, K. (2017). Reconstruction of the multielement apparatus of the earliest Triassic conodont, *Hindeodus parvus*, using synchrotron radiation X-ray micro-tomography. *Journal of Paleontology*, 91(6), 1220–1227.

de Ágredos-Pascual, M.L.V., Roldán-García, C., Murcia-Mascarós, S., Barber, D.J., Sánchez, M.G.J., Faugere, B., Darras, V. (2019). Multianalytical characterization of pigments from funerary artefacts belongs to the Chupicuaro Culture (Western Mexico): Oldest Maya blue and cinnabar identified in Pre-Columbian Mesoamerica. *Microchemical Journal*, 150, 104101.

Albertin, F., Astolfo, A., Stampanoni, M., Peccenini, E., Hwu, Y., Kaplan, F., Margaritondo, G. (2015). Ancient administrative handwritten documents: X-ray analysis and imaging. *Journal of Synchrotron Radiation*, 22(2), 446–451.

Alfeld, M., Janssens, K., Appel, K., Thijsse, B., Blaas, J., Dik, J. (2011). A portrait by Philipp Otto Runge? Visualizing modifications to the painting using synchrotron-based X-ray fluorescence elemental scanning. *Zeitschrift für Kunsttechnologie und Konservierung*, 25(1), 157–163.

Alfeld, M., Siddons, D.P., Janssens, K., Dik, J., Woll, A., Kirkham, R., van de Wetering, E. (2013). Visualizing the 17th century underpainting in Portrait of an Old Man by Rembrandt van Rijn using synchrotron-based scanning macro-XRF. *Applied Physics A*, 111(1), 157–164.

Arnold, D.E., Bohor, B.F., Neff, H., Feinman, G.M., Williams, P.R., Dussubieux, L., Bishop, R. (2012). The first direct evidence of pre-columbian sources of palygorskite for Maya Blue. *Journal of Archaeological Science*, 39(7), 2252–2260.

Autran, P.O., Dejoie, C., Bordet, P., Hodeau, J.L., Dugand, C., Gervason, M., Anne, M., Martinetto, P. (2020). Revealing the nature of black pigments used on ancient Egyptian papyri from Champollion collection. *Analytical Chemistry*, 93(2), 1135–1142.

Autran, P.O., Dejoie, C., Dugand, C., Gervason, M., Bordet, P., Hodeau, J.L., Anne, M., Martinetto, P. (2023). Illustrating papyrus in ancient Egypt. *Scientific Reports*, 13(1), 524.

Berke, H. and Wiedemann, H.G. (2000). The chemistry and fabrication of the anthropogenic pigments Chinese blue and purple in ancient China. *East Asian Science, Technology, and Medicine*, (17), 94–120.

Berruyer, C., Porcier, S.M., Tafforeau, P. (2020). Synchrotron "virtual archaeozoology" reveals how Ancient Egyptians prepared a decaying crocodile cadaver for mummification. *PLoS One*, 15(2), e0229140.

Bertrand, L., Doucet, J., Dumas, P., Simionovici, A., Tsoucaris, G., Walter, P. (2003). Microbeam synchrotron imaging of hairs from ancient Egyptian mummies. *Journal of Synchrotron Radiation*, 10(5), 387–392.

Bidaud, E., Halwax, E., Pantos, E., Sipek, B. (2008). Analyses of a green copper pigment used in a thirteenth-century wall painting. *Studies in Conservation*, 53(2), 81–92.

Bleuet, P., Welcomme, E., Dooryhée, E., Susini, J., Hodeau, J.L., Walter, P. (2008). Probing the structure of heterogeneous diluted materials by diffraction tomography. *Nature Materials*, 7(6), 468–472.

Boatman, E.M., Goodwin, M.B., Holman, H.Y.N., Fakra, S., Zheng, W., Gronsky, R., Schweitzer, M.H. (2019). Mechanisms of soft tissue and protein preservation in *Tyrannosaurus rex*. *Scientific Reports*, 9(1), 1–12.

Bradley, W. (1940). The structural scheme of attapulgite. *American Mineralogist: Journal of Earth and Planetary Materials*, 25(6), 405–410.

Brazeau, M.D., Friedman, M., Jerve, A., Atwood, R.C. (2017). A three-dimensional placoderm (stem-group gnathostome) pharyngeal skeleton and its implications for primitive gnathostome pharyngeal architecture. *Journal of Morphology*, 278(9), 1220–1228.

Brissaud, I., Chevallier, P., Dardenne, C., Deschamps, N., Frontier, J.P., Gruel, K., Taccoen, A., Tarrats, A., Wang, J.X. (1990). Analysis of Gaulish coins by proton induced X-ray emission, synchrotron radiation X-ray fluorescence and neutron activation analysis. *Nuclear Instruments and Methods in Physics Research Section B: Beam Interactions with Materials and Atoms*, 49(1–4), 305–308.

Bugani, S., Modugno, F., Łucejko, J.J., Giachi, G., Cagno, S., Cloetens, P., Janssens, K., Morselli, L. (2009). Study on the impregnation of archaeological waterlogged wood with consolidation treatments using synchrotron radiation microtomography. *Analytical and Bioanalytical Chemistry*, 395, 1977–1985.

Calligaro, T., Dran, J.C., Hamon, H., Moignard, B., Salomon, J. (1998). An external milli-beam for archaeometric applications on the AGLAE IBA facility of the Louvre museum. *Nuclear Instruments and Methods in Physics Research Section B: Beam Interactions with Materials and Atoms*, 136, 339–343.

Calligaro, T., Gonzalez, V., Pichon, L. (2015). PIXE analysis of historical paintings: Is the gain worth the risk? *Nuclear Instruments and Methods in Physics Research Section B: Beam Interactions with Materials and Atoms*, 363, 135–143.

Calza, C., Anjos, M.J., de Souza, S.M., Brancaglion Jr, A., Lopes, R.T. (2007). X-ray microfluorescence analysis of pigments in decorative paintings from the sarcophagus cartonnage of an Egyptian mummy. *Nuclear Instruments and Methods in Physics Research Section B: Beam Interactions with Materials and Atoms*, 263(1), 249–252.

Calza, C., Anjos, M.J., Mendonça de Souza, S.M.F., Brancaglion Jr, A., Lopes, R.T. (2008). X-ray microfluorescence with synchrotron radiation applied in the analysis of pigments from ancient Egypt. *Applied Physics A*, 90, 75–79.

Capasso, L.D.J.T., Di Tota, G., Jones, K.W., Tuniz, C. (1995). Synchrotron radiation microprobe analysis of human dental calculi from an archaeological site: A new possible perspective in palaeonutrition studies. *International Journal of Osteoarchaeology*, 5(3), 282–288.

Carmona, N., Ortega-Feliu, I., Gomez-Tubio, B., Villegas, M.A. (2010). Advantages and disadvantages of PIXE/PIGE, XRF and EDX spectrometries applied to archaeometric characterisation of glasses. *Materials Characterization*, 61(2), 257–267.

Cartechini, L., Miliani, C., Brunetti, B.G., Sgamellotti, A., Altavilla, C., Ciliberto, E., D'Acapito, F. (2008). X-ray absorption investigations of copper resinate blackening in a XV century Italian painting. *Applied Physics A*, 92(1), 243–250.

Chalmin, E., Vignaud, C., Salomon, H., Farges, F., Susini, J., Menu, M. (2006). Minerals discovered in paleolithic black pigments by transmission electron microscopy and micro-X-ray absorption near-edge structure. *Applied Physics A*, 83(2), 213–218.

Chalmin, E., Farges, F., Vignaud, C., Susini, J., Menu, M., Brown Jr, G.E. (2007). Discovery of unusual minerals in Paleolithic black pigments from Lascaux (France) and Ekain (Spain). *AIP Conference Proceedings*, 882(1), 220–222.

Chen, F. and Dong, X. (2008). The internal structure of Early Cambrian fossil embryo *Olivooides* revealed in the light of synchrotron X-ray tomographic microscopy. *Chinese Science Bulletin*, 53(24), 3860–3865.

Chen, X., Dejoie, C., Jiang, T., Ku, C.S., Tamura, N. (2016). Quantitative microstructural imaging by scanning Laue X-ray micro- and nanodiffraction. *MRS Bulletin*, 41(6), 445–453.

Chevallier, P., Ricordel, I., Meyer, G. (2006). Trace element determination in hair by synchrotron X-ray fluorescence analysis: Application to the hair of Napoleon I. *X-Ray Spectrometry: An International Journal*, 35(2), 125–130.

Chiari, G., Giustetto, R., Ricchiardi, G. (2003). Crystal structure refinements of palygorskite and Maya Blue from molecular modelling and powder synchrotron diffraction. *European Journal of Mineralogy*, 15(1), 21–33.

Chiari, G., Giustetto, R., Druzik, J., Doehne, E., Ricchiardi, G. (2008). Pre-columbian nanotechnology: Reconciling the mysteries of the maya blue pigment. *Applied Physics A*, 90(1), 3–7.

Christiansen, T., Cotte, M., Loredo-Portales, R., Lindelof, P.E., Mortensen, K., Ryholt, K., Larsen, S. (2017). The nature of ancient Egyptian copper-containing carbon inks is revealed by synchrotron radiation based X-ray microscopy. *Scientific Reports*, 7(1), 1–8.

Christiansen, T., Cotte, M., de Nolf, W., Mouro, E., Reyes-Herrera, J., De Meyer, S., Vanmeert, F., Salvadó, N., Gonzalez, V., Lindelof, P.E. et al. (2020). Insights into the composition of ancient Egyptian red and black inks on papyri achieved by synchrotron-based microanalyses. *Proceedings of the National Academy of Sciences*, 117(45), 27825–27835.

Clements, T., Dolocan, A., Martin, P., Purnell, M.A., Vinther, J., Gabbott, S.E. (2016). The eyes of *Tullimonstrum* reveal a vertebrate affinity. *Nature*, 532(7600), 500–503.

Cotte, M., Dumas, P., Richard, G.L., Breniaux, R., Walter, P. (2005). New insight on ancient cosmetic preparation by synchrotron-based infrared microscopy. *Analytica Chimica Acta*, 553(1–2), 105–110.

Cotte, M., Susini, J., Metrich, N., Moscato, A., Gratziu, C., Bertagnini, A., Pagano, M. (2006). Blackening of Pompeian cinnabar paintings: X-ray microspectroscopy analysis. *Analytical Chemistry*, 78(21), 7484–7492.

Cotte, M., Welcomme, E., Solé, V.A., Salomé, M., Menu, M., Walter, P., Susini, J. (2007a). Synchrotron-based X-ray spectromicroscopy used for the study of an atypical micrometric pigment in 16th century paintings. *Analytical Chemistry*, 79(18), 6988–6994.

Cotte, M., Checroun, E., Susini, J., Walter, P. (2007b). Micro-analytical study of interactions between oil and lead compounds in paintings. *Applied Physics A*, 89(4), 841–848.

Cotte, M., Susini, J., Solé, V.A., Taniguchi, Y., Chillida, J., Checroun, E., Walter, P. (2008). Applications of synchrotron-based micro-imaging techniques to the chemical analysis of ancient paintings. *Journal of Analytical Atomic Spectrometry*, 23(6), 820–828.

Cotte, M., Pouyet, E., Salomé, M., Rivard, C., De Nolf, W., Castillo-Michel, H., Fabris, T., Monico, L., Janssens, K., Wang, T. et al. (2017). The ID21 X-ray and infrared microscopy beamline at the ESRF status and recent applications to artistic materials. *Journal of Analytical Atomic Spectrometry*, 32(3), 477–493.

Cotte, M., Dollman, K., Fernandez, V., Gonzalez, V., Vanmeert, F., Monico, L., Dejoie, C., Burghammer, M., Huder, L., Fisher, S. et al. (2022a). New opportunities offered by the ESRF to the cultural and natural heritage communities. *Synchrotron Radiation News*, 35(5), 3–9.

Cotte, M., Gonzalez, V., Vanmeert, F., Monico, L., Dejoie, C., Burghammer, M., Huder, L., de Nolf, W., Fisher, S., Fazlic, I. et al. (2022b). The "historical materials BAG": A new facilitated access to synchrotron X-ray diffraction analyses for cultural heritage materials at the European synchrotron radiation facility. *Molecules*, 27(6), 1997.

Cunningham, J.A., Rahman, I.A., Lautenschlager, S., Rayfield, E.J., Donoghue, P.C. (2014). A virtual world of paleontology. *Trends in Ecology & Evolution*, 29(6), 347–357.

Curtin, A.J., MacDowell, A.A., Schaible, E.G., Roth, V.L. (2012). Noninvasive histological comparison of bone growth patterns among fossil and extant neonatal elephantids using synchrotron radiation X-ray microtomography. *Journal of Vertebrate Paleontology*, 32(4), 939–955.

De Raedt, I., Vekemans, B., Janssens, K., Adams, F. (2000). Synchrotron light through ancient glass. *Europhysics News*, 31(6), 15–15.

Dejoie, C., Martinetto, P., Dooryhee, E., Brown, R., Blanc, S., Bordat, P., Strobel, P., Odier, P., Porcher, F., Sanchez del Rio, M. et al. (2011). Diffusion of indigo molecules inside the palygorskite clay channels. *MRS Online Proceedings Library (OPL),* 1319, mrsf10-1319.

Dejoie, C., Sciau, P., Li, W., Noé, L., Mehta, A., Chen, K., Luo, H., Kunz, M., Tamura, N., Liu, Z. (2014a). Learning from the past: Rare ε-Fe2O3 in the ancient black-glazed Jian (Tenmoku) wares. *Scientific Reports*, 4(1), 1–9.

Dejoie, C., Martinetto, P., Tamura, N., Kunz, M., Porcher, F., Bordat, P., Brown, R., Dooryhée, E., Anne, M., McCusker, L.B. (2014b). Crystal structure of an indigo@silicalite hybrid related to the ancient Maya blue pigment. *The Journal of Physical Chemistry C*, 118(48), 28032–28042.

Dik, J., Janssens, K., Van Der Snickt, G., Van Der Loeff, L., Rickers, K., Cotte, M. (2008). Visualization of a lost painting by Vincent van Gogh using synchrotron radiation based X-ray fluorescence elemental mapping. *Analytical Chemistry*, 80(16), 6436–6442.

Donoghue, P.C., Bengtson, S., Dong, X.P., Gostling, N.J., Huldtgren, T., Cunningham, J.A., Yin, C., Yue, Z., Peng, F., Stampanoni, M. (2006). Synchrotron X-ray tomographic microscopy of fossil embryos. *Nature*, 442(7103), 680–683.

Dooryhée, E., Martinetto, P., Walter, P., Anne, M. (2004). Synchrotron X-ray analyses in art and archaeology. *Radiation Physics and Chemistry*, 71(3–4), 863–868.

Dooryhée, E., Anne, M., Bardies, I., Hodeau, J.L., Martinetto, P., Rondot, S., Salomon, J., Vaughan, G.B.M., Walter, P. (2005). Non-destructive synchrotron X-ray diffraction mapping of a Roman painting. *Applied Physics A*, 81(4), 663–667.

Dooryhée, E., Menu, M., Susini, J. (2006). Synchrotron radiation in art and archaeology. Special Issue. *Applied Physics A*, 83(2), 163–265.

Dunsmuir, H., Vandiver, P.B., Chlanelll, R.R., Deckman, H.W., Hardenbergh, J.H. (1995). X-ray microtomography of ceramic artifacts. *MRS Online Proceedings Library (OPL)*, 352.

Eberhardt, W., Kanngießer, B., Reiche, I. (2008). Synchrotron radiation in art and archaeology 2006 – a truly interdisciplinary event. *Applied Physics A*, 90, 1.

Echard, J.P., Cotte, M., Dooryhee, E., Bertrand, L. (2008). Insights into the varnishes of historical musical instruments using synchrotron micro-analytical methods. *Applied Physics A*, 92, 77–81.

Favero, P.A., Mass, J., Delaney, J.K., Woll, A.R., Hull, A.M., Dooley, K.A., Finnefrock, A.C. (2017). Reflectance imaging spectroscopy and synchrotron radiation X-ray fluorescence mapping used in a technical study of The Blue Room by Pablo Picasso. *Heritage Science*, 5(1), 1–18.

Ferreira, E.S., Boon, J.J., Van der Horst, J., Scherrer, N.C., Marone, F., Stampanoni, M. (2009). 3D synchrotron X-ray microtomography of paint samples. *O3A: Optics for Arts, Architecture, and Archaeology II* (Vol. 7391, pp. 192–199). SPIE.

Friedman, E.S., Sato, Y., Alatas, A., Johnson, C.E., Wilkinson, T.J., Yener, K.A., Lai, B., Jennings, G., Mini, S.M., Alp, E.E. (2000). An X-ray fluorescence study of lake sediments from ancient Turkey using synchrotron radiation. *Advances in X-Ray Analysis*, 42, 151–160.

Gervais, C., Boon, J.J., Marone, F., Ferreira, E.S. (2013). Characterization of porosity in a 19th century painting ground by synchrotron radiation X-ray tomography. *Applied Physics A*, 111(1), 31–38.

Gilbert, P.U.P.A., Young, A., Coppersmith, S.N. (2011). Measurement of c-axis angular orientation in calcite (CaCO3) nanocrystals using X-ray absorption spectroscopy. *Proceedings of the National Academy of Sciences*, 108(28), 11350–11355.

Gilbert, P.U.P.A., Bergmann, K.D., Myers, C.E., Marcus, M.A., DeVol, R.T., Sun, C.Y., Blonsky, A.Z., Tamre, E., Zhao, J., Karan, E.A. et al. (2017). Nacre tablet thickness records formation temperature in modern and fossil shells. *Earth and Planetary Science Letters*, 460, 281–292.

Giustetto, R., Levy, D., Chiari, G. (2006). Crystal structure refinement of Maya Blue pigment prepared with deuterated indigo, using neutron powder diffraction. *European Journal of Mineralogy*, 18(5), 629–640.

Giustetto, R., Levy, D., Wahyudi, O., Ricchiardi, G., Vitillo, J.G. (2011). Crystal structure refinement of a sepiolite/indigo Maya Blue pigment using molecular modelling and synchrotron diffraction. *European Journal of Mineralogy*, 23(3), 449–466.

Gonzalez, V., Calligaro, T., Wallez, G., Eveno, M., Toussaint, K., Menu, M. (2016). Composition and microstructure of the lead white pigment in Masters paintings using HR Synchrotron XRD. *Microchemical Journal*, 125, 43–49.

Gonzalez, V., Wallez, G., Calligaro, T., Cotte, M., De Nolf, W., Eveno, M., Ravaud, E., Menu, M. (2017). Synchrotron-based high angle resolution and high lateral resolution X-ray diffraction: Revealing lead white pigment qualities in old masters paintings. *Analytical Chemistry*, 89(24), 13203–13211.

Gonzalez, V., Cotte, M., Wallez, G., van Loon, A., De Nolf, W., Eveno, M., Keune, K., Noble, P., Dik, J. (2019). Unraveling the composition of Rembrandt's impasto through the identification of unusual plumbonacrite by multimodal X-ray diffraction analysis. *Angewandte Chemie*, 131(17), 5675–5678.

Gueriau, P., Réguer, S., Leclercq, N., Cupello, C., Brito, P.M., Jauvion, C., Morel, S., Charbonnier, S., Thiaudière, D., Mocuta, C. (2020a). Visualizing mineralization processes and fossil anatomy using synchronous synchrotron X-ray fluorescence and X-ray diffraction mapping. *Journal of the Royal Society Interface*, 17(169), 20200216.

Gueriau, P., Bernard, S., Farges, F., Mocuta, C., Dutheil, D.B., Adatte, T., Bomou, B., Godet, M., Thiaudière, D., Charbonnier, S. (2020b). Oxidative conditions can lead to exceptional preservation through phosphatization. *Geology*, 48(12), 1164–1168.

Guryeva, P.V., Demkiv, A.A., Sozontov, E.A. (2017). Mapping a text on ancient parchment by X-ray fluorescence analysis using a synchrotron source. *Journal of Surface Investigation: X-ray, Synchrotron and Neutron Techniques*, 11, 167–168.

Hall, J.G., Smith, E.F., Tamura, N., Fakra, S.C., Bosak, T. (2020). Preservation of erniettomorph fossils in clay-rich siliciclastic deposits from the Ediacaran Wood Canyon Formation, Nevada. *Interface Focus*, 10(4), 20200012.

Harbottle, G., Gordon, B.E., Jones, K.W. (1986). Use of synchrotron radiation in archaeometry. *Nuclear Instruments and Methods in Physics Research Section B: Beam Interactions with Materials and Atoms*, 14(1), 116–122.

Helfen, L., Baumbach, T., Mikulik, P., Kiel, D., Pernot, P., Cloetens, P., Baruchel, J. (2005). High-resolution three-dimensional imaging of flat objects by synchrotron-radiation computed laminography. *Applied Physics Letters*, 86(7), 071915.

Herrera, L.K., Cotte, M., De Haro, M.J., Duran, A., Justo, A., Perez-Rodriguez, J.L. (2008). Characterization of iron oxide-based pigments by synchrotron-based micro X-ray diffraction. *Applied Clay Science*, 42(1–2), 57–62.

Hochleitner, B., Schreiner, M., Drakopoulos, M., Snigireva, I., Snigirev, A. (2004). Analysis of paint layers by light microscopy, scanning electron microscopy and synchrotron induced X-ray micro-diffraction. In *Cultural Heritage Conservation and Environmental Impact Assessment by Non-Destructive Testing and Micro-Analysis*. CRC Press, Boca Raton, 179–190.

Holé, C., Ren, Z., Wang, F., Zhu, J., Wang, T., Sciau, P. (2022). Study of the growth mechanism of ε-Fe2O3 crystals in Chinese sauce glaze replications. *Materials Today Communications*, 33, 104329.

Hoo, Q., Liang, Y., Yan, X., Wang, X., Cao, T., Cao, X. (2020). Millimeter-sized flower-like clusters composed of mullite and ε-Fe2O3 on the Hare's Fur Jian Ware. *Journal of the European Ceramic Society*, 40(12), 4340–4347.

Howard, D.L., de Jonge, M.D., Lau, D., Hay, D., Varcoe-Cocks, M., Ryan, C.G., Kirkham, R., Moorhead, G., Paterson, D., Thurrowgood, D. (2012). High-definition X-ray fluorescence elemental mapping of paintings. *Analytical Chemistry*, 84(7), 3278–3286.

Huang, J.Y., Martinez-Perez, C., Hu, S.X., Donoghue, P.C., Zhang, Q.Y., Zhou, C.Y., Wen, W., Benton, M.J., Luo, M., Yao, H.-Z. et al. (2019). Middle Triassic conodont apparatus architecture revealed by synchrotron X-ray microtomography. *Palaeoworld*, 28(4), 429–440.

Jackson, M.D., Landis, E.N., Brune, P.F., Vitti, M., Chen, H., Li, Q., Kunz, M., Wenk, H.-R., Monteiro, P.J.M., Ingraffea, A.R. (2014). Mechanical resilience and cementitious processes in Imperial Roman architectural mortar. *Proceedings of the National Academy of Sciences*, 111(52), 18484–18489.

Janssens, K., Aerts, A., Vincze, L., Adams, F., Yang, C., Utui, R., Malmqvist, K., Jones, K.W., Radtke, M., Garbe, S. et al. (1996). Corrosion phenomena in electron, proton and synchrotron X-ray microprobe analysis of Roman glass from Qumran, Jordan. *Nuclear Instruments and Methods in Physics Research Section B: Beam Interactions with Materials and Atoms*, 109, 690–695.

Janssens, K., Deraedt, I., Schalm, O., Veeckman, J. (1998). Composition of 15–17th century archaeological glass vessels excavated in Antwerp, Belgium. *Modern Developments and Applications in Microbeam Analysis*, Love, G., Nicholson, W.A.P., Armigliato, A. (eds). Springer, Vienna, 253–267.

Janssens, K., Vittiglio, G., Deraedt, I., Aerts, A., Vekemans, B., Vincze, L., Wei, F., De Ryck, I., Schalm, O., Adams, F. et al. (2000). Use of microscopic XRF for non-destructive analysis in art and archaeometry. *X-Ray Spectrometry: An International Journal*, 29(1), 73–91.

Johansson, S.A. (1989). PIXE: A novel technique for elemental analysis. *Endeavour*, 13(2), 48–53.

Kakoulli, I., Prikhodko, S.V., Fischer, C., Cilluffo, M., Uribe, M., Bechtel, H.A., Fakra, S.C., Marcus, M.A. (2014). Distribution and chemical speciation of arsenic in ancient human hair using synchrotron radiation. *Analytical Chemistry*, 86(1), 521–526.

Krug, K., Dik, J., Den Leeuw, M., Whitson, A.D., Tortora, J., Coan, P., Nemoz, C., Bravin, A. (2006). Visualization of pigment distributions in paintings using synchrotron K-edge imaging. *Applied Physics A*, 83(2), 247–251.

Krug, K., Porra, L., Coan, P., Wallert, A., Dik, J., Coerdt, A., Bravin, A., Elyyan, M., Reischig, P., Helfen, L. et al. (2008). Relics in medieval altarpieces? Combining X-ray tomographic, laminographic and phase-contrast imaging to visualize thin organic objects in paintings. *Journal of Synchrotron Radiation*, 15(1), 55–61.

Kusano, Y., Fujii, T., Takada, J., Fukuhara, M., Doi, A., Ikeda, Y., Takano, M. (2008). Epitaxial growth of ϵ-Fe2O3 on mullite found through studies on a traditional Japanese stoneware. *Chemistry of Materials*, 20(1), 151–156.

Kusano, Y., Nakata, H., Peng, Z., Maki, R.S., Ogawa, T., Fukuhara, M. (2021). Studying and utilizing traditional technologies: Microstructure and formation mechanism of ε-Fe2O3 on traditional Japanese Bizen stoneware. *ACS Applied Materials & Interfaces*, 13(32), 38491–38498.

Lak, M., Néraudeau, D., Nel, A., Cloetens, P., Perrichot, V., Tafforeau, P. (2008). Phase contrast X-ray synchrotron imaging: Opening access to fossil inclusions in opaque amber. *Microscopy and Microanalysis*, 14(3), 251–259.

Leon, Y., Sciau, P., Bouquillon, A., Pichon, L., de Parseval, P. (2012). PIXE (particle induced X-ray emission): A non-destructive analysis method adapted to the thin decorative coatings of antique ceramics. *Nuclear Instruments and Methods in Physics Research Section B: Beam Interactions with Materials and Atoms*, 291, 45–52.

Liu, Z., Mehta, A., Tamura, N., Pickard, D., Rong, B., Zhou, T., Pianetta, P. (2007). Influence of Taoism on the invention of the purple pigment used on the Qin terracotta warriors. *Journal of Archaeological Science*, 34(11), 1878–1883.

Liu, Z.W., Zhang, Z.G., Li, H., Wu, W., Hua, J.J., Cheng, G.F., Lin, C.C. (2012). Microstructure characteristics of the black-glazed shreds excavated from the Qingliangsi Kiln. *Key Engineering Materials*, 492, 112–117.

Liu, Z., Jia, C., Li, L., Li, X., Ji, L., Wang, L., Lei, Y., Wei, X. (2018). The morphology and structure of crystals in Qing Dynasty purple-gold glaze excavated from the Forbidden City. *Journal of the American Ceramic Society*, 101(11), 5229–5240.

Lorentz, K.O., de Nolf, W., Cotte, M., Ioannou, G., Foruzanfar, F., Zaruri, M.R., Sajjadi, S.M.S. (2020). Synchrotron radiation micro X-ray fluorescence (SR-μXRF) elemental mapping of ancient hair: Metals and health at 3rd millennium BCE Shahr-i Sokhta, Iran. *Journal of Archaeological Science*, 120, 105193.

Lucarelli, F., Nava, S., Calzolai, G., Chiari, M., Udisti, R., Marino, F. (2011). Is PIXE still a useful technique for the analysis of atmospheric aerosols? The LABEC experience. *X-Ray Spectrometry*, 40(3), 162–167.

Lynch, P.A., Tamura, N., Lau, D., Madsen, I., Liang, D., Strohschnieder, M., Stevenson, A.W. (2007). Application of white-beam X-ray microdiffraction for the study of mineralogical phase identification in ancient Egyptian pigments. *Journal of Applied Crystallography*, 40(6), 1089–1096.

Malmqvist, K.G. (1986). Comparison between PIXE and XRF for applications in art and archaeology. *Nuclear Instruments and Methods in Physics Research Section B: Beam Interactions with Materials and Atoms*, 14(1), 86–92.

Manning, P.L., Edwards, N.P., Wogelius, R.A., Bergmann, U., Barden, H.E., Larson, P.L., Schwartz-Wings, D., Egerton, V.M., Sokaras, D., Mori, R.A. et al. (2013). Synchrotron-based chemical imaging reveals plumage patterns in a 150 million year old early bird. *Journal of Analytical Atomic Spectrometry*, 28(7), 1024–1030.

Martinetto, P., Anne, M., Dooryhée, E., Drakopoulos, M., Dubus, M., Salomon, J., Simionovici, A., Walter, P. (2001). Synchrotron X-ray micro-beam studies of ancient Egyptian make-up. *Nuclear Instruments and Methods in Physics Research Section B: Beam Interactions with Materials and Atoms*, 181(1–4), 744–748.

Martínez-Pérez, C., Rayfield, E.J., Botella, H., Donoghue, P.C. (2016). Translating taxonomy into the evolution of conodont feeding ecology. *Geology*, 44(4), 247–250.

Maryanti, E., Ilmi, M., Nurdini, N., Setiawan, P., Syah, Y.M., Saiyasombat, C., Kadja, G. (2022). Hematite as unprecedented black rock art pigment in Jufri Cave, East Kalimantan, Indonesia: The microscopy, spectroscopy, and synchrotron X-ray-based investigation. *Archaeological and Anthropological Sciences*, 14(7), 1–19.

McCoy, V.E., Saupe, E.E., Lamsdell, J.C., Tarhan, L.G., McMahon, S., Lidgard, S., Mayer, P., Whalen, C.D., Soriano, C., Finney, L. et al. (2016). The 'Tully monster' is a vertebrate. *Nature*, 532(7600), 496–499.

Menu, M., Calligaro, T., Salomon, J., Amsel, G., Moulin, J. (1990). The dedicated accelerator-based IBA facility AGLAE at the Louvre. *Nuclear Instruments and Methods in Physics Research Section B: Beam Interactions with Materials and Atoms*, 45(1–4), 610–614.

Mocella, V., Brun, E., Ferrero, C., Delattre, D. (2015). Revealing letters in rolled Herculaneum papyri by X-ray phase-contrast imaging. *Nature Communications*, 6(1), 5895.

Mommsen, H., Beier, T., Dittmann, H., Heimermann, D., Hein, A., Rosenberg, A., Boghardt, M., Hanebutt-Benz, E.-M., Halbey, H. (1996). X-ray fluorescence analysis with synchrotron radiation on the inks and papers of incunabula. *Archaeometry*, 38(2), 347–357.

Monico, L., Van der Snickt, G., Janssens, K., De Nolf, W., Miliani, C., Verbeeck, J., Tian, H., Tan, H., Dik, J., Radepont, M. et al. (2011a). Degradation process of lead chromate in paintings by Vincent van Gogh studied by means of synchrotron X-ray spectromicroscopy and related methods. 1. Artificially aged model samples. *Analytical Chemistry*, 83(4), 1214–1223.

Monico, L., Van der Snickt, G., Janssens, K., De Nolf, W., Miliani, C., Dik, J., Radepont, M., Hendriks, E., Geldof, M., Cotte, M. (2011b). Degradation process of lead chromate in paintings by Vincent van Gogh studied by means of synchrotron X-ray spectromicroscopy and related methods. 2. Original paint layer samples. *Analytical Chemistry*, 83(4), 1224–1231.

Možir, A., Gonzalez, L., Kralj Cigić, I., Wess, T.J., Rabin, I., Hahn, O., Strlič, M. (2012). A study of degradation of historic parchment using small-angle X-ray scattering, synchrotron-IR and multivariate data analysis. *Analytical and Bioanalytical Chemistry*, 402, 1559–1566.

Muscente, A.D., Michel, F.M., Dale, J.G., Xiao, S. (2015). Assessing the veracity of Precambrian 'sponge' fossils using in situ nanoscale analytical techniques. *Precambrian Research*, 263, 142–156.

Nakai, I. and Iida, A. (1991). Application of SR-XRF imaging and micro-XANES to meteorites, archaeological objects and animal tissues. *Advances in X-ray Analysis*, 35(B), 1307–1315.

Nakai, I., Taguchi, I., Yamasaki, K. (1991). Chemical speciation of archaeological objects by XRFXANES analysis using synchrotron radiation. *Analytical Sciences*, 7(Supplement), 365–368.

Pacheco, C., Lemasson, Q., Moignard, B., Pichon, L., Radepont, M., Gourier, D. (2016). D'aglae à new aglae. La science au service de l'histoire de l'art et de la préservation des biens culturels, *Technè.* 43, 63–69.

Pantos, M. (2000). Synchrotron radiation in archaeometry. *Synchrotron Radiation News*, 13(3), 6–10.

Pérez-Huerta, A., Coronado, I., Hegna, T.A. (2018). Understanding biomineralization in the fossil record. *Earth-Science Reviews*, 179, 95–122.

Polette, L.A., Meitzner, G., Yacaman, M.J., Chianelli, R.R. (2002). Maya blue: Application of XAS and HRTEM to materials science in art and archaeology. *Microchemical Journal*, 71(2–3), 167–174.

Polette-Niewold, L.A., Manciu, F.S., Torres, B., Alvarado Jr, M., Chianelli, R.R. (2007). Organic/inorganic complex pigments: Ancient colors Maya Blue. *Journal of Inorganic Biochemistry*, 101(11–12), 1958–1973.

Porcier, S.M., Berruyer, C., Pasquali, S., Ikram, S., Berthet, D., Tafforeau, P. (2019). Wild crocodiles hunted to make mummies in Roman Egypt: Evidence from synchrotron imaging. *Journal of Archaeological Science*, 110, 105009.

Post, J.E. and Heaney, P.J. (2008). Synchrotron powder X-ray diffraction study of the structure and dehydration behavior of palygorskite. *American Mineralogist*, 93(4), 667–675.

Qvarnström, M., Niedźwiedzki, G., Tafforeau, P., Žigaitė, Ž., Ahlberg, P.E. (2017). Synchrotron phase-contrast microtomography of coprolites generates novel palaeobiological data. *Scientific Reports*, 7(1), 1–6.

Reiche, I. and Sauerborn, M. (2007). Meeting report: Synchrotron radiation in art and archeology 2006. *Synchrotron Radiation News*, 20(1), 44–45.

Reischig, P., Helfen, L., Wallert, A., Baumbach, T., Dik, J. (2013). High-resolution non-invasive 3D imaging of paint microstructure by synchrotron-based X-ray laminography. *Applied Physics A*, 111(4), 983–995.

Rigon, L., Arfelli, F., Astolfo, A., Bergamaschi, A., Dreossi, D., Longo, R., Menk, R.-H., Schmidt, B., Vallazza, E., Castelli, E. (2009). A single-photon counting "edge-on" silicon detector for synchrotron radiation mammography. *Nuclear Instruments and Methods in Physics Research Section A: Accelerators, Spectrometers, Detectors and Associated Equipment*, 608(1), S62–S65.

Rogers, C.S., Astrop, T.I., Webb, S.M., Ito, S., Wakamatsu, K., McNamara, M.E. (2019). Synchrotron X-ray absorption spectroscopy of melanosomes in vertebrates and cephalopods: Implications for the affinity of Tullimonstrum. *Proceedings of the Royal Society B*, 286(1913), 20191649.

Rowe, T. (1996). Coevolution of the mammalian middle ear and neocortex. *Science*, 273(5275), 651–654.

Ryan, C.G., Siddons, D.P., Kirkham, R., Li, Z.Y., De Jonge, M.D., Paterson, D.J., Kuczewski, A., Howard, D.L., Dunn, P.A., Borg, S. (2014). Maia X-ray fluorescence imaging: Capturing detail in complex natural samples. *Journal of Physics: Conference Series*, 499(1), 012002.

Salvadó, N., Pradell, T., Pantos, E., Papiz, M.Z., Molera, J., Seco, M.E., Vendrell-Saz, M. (2002). Identification of copper-based green pigments in Jaume Huguet's Gothic altarpieces by Fourier transform infrared microspectroscopy and synchrotron radiation X-ray diffraction. *Journal of Synchrotron Radiation*, 9(4), 215–222.

Salvadó, N., Butí, S., Pantos, E., Bahrami, F., Labrador, A., Pradell, T. (2008). The use of combined synchrotron radiation micro FT-IR and XRD for the characterization of Romanesque wall paintings. *Applied Physics A*, 90(1), 67–73.

Salvadó, N., Butí, S., Nicholson, J., Emerich, H., Labrador, A., Pradell, T. (2009). Identification of reaction compounds in micrometric layers from gothic paintings using combined SR-XRD and SR-FTIR. *Talanta*, 79(2), 419–428.

Sanchez, S., Ahlberg, P.E., Trinajstic, K.M., Mirone, A., Tafforeau, P. (2012). Three-dimensional synchrotron virtual paleohistology: A new insight into the world of fossil bone microstructures. *Microscopy and Microanalysis*, 18(5), 1095–1105.

Sánchez del Río, M.S., Martinetto, P., Somogyi, A., Reyes-Valerio, C., Dooryhée, E., Peltier, N., Alianelli, L., Moignard, B., Pichon, L., Calligaro, T. et al. (2004). Microanalysis study of archaeological mural samples containing Maya blue pigment. *Spectrochimica Acta Part B: Atomic Spectroscopy*, 59(10–11), 1619–1625.

Sánchez del Río, M.S., Sodo, A., Eeckhout, S.G., Neisius, T., Martinetto, P., Dooryhee, E., Reyes-Valerio, C. (2005). Fe K-edge XANES of Maya blue pigment. *Nuclear Instruments and Methods in Physics Research Section B: Beam Interactions with Materials and Atoms*, 238(1–4), 50–54.

Sánchez del Río, M., Gutierrez-Leon, A., Castro, G.R., Rubio-Zuazo, J., Solís, C., Sanchez-Hernandez, R., Robles-Camacho, J., Rojas-Gaytan, J. (2008). Synchrotron powder diffraction on Aztec blue pigments. *Applied Physics A*, 90(1), 55–60.

Sánchez del Río, M., Suárez, M., Garciá-Romero, E. (2009a). The occurrence of palygorskite in the Yucatán peninsula: Ethno-historic and archaeological contexts. *Archaeometry*, 51(2), 214–230.

Sánchez del Río, M., Boccaleri, E., Milanesio, M., Croce, G., van Beek, W., Tsiantos, C., Chryssikos, G.D., Gionis, V., Kacandes, G.H., Suárez, M. et al. (2009b). A combined synchrotron powder diffraction and vibrational study of the thermal treatment of palygorskite–indigo to produce Maya blue. *Journal of Materials Science*, 44(20), 5524–5536.

Sánchez del Río, M., García-Rivas, J., Suárez, M., García-Romero, E. (2021). Crystal–chemical and diffraction analyses of Maya blue suggesting a different provenance of the palygorskite found in Aztec pigments. *Archaeometry*, 63(4), 738–752.

Sato, Y. and Nakazawa, M. (2000). The development of non-destructive analysis for archaeological samples using high energy synchrotron radiation and laser electron photon. *Radioisotopes (Tokyo)*, 49(3), 131–135.

Seymour, L.M., Tamura, N., Jackson, M.D., Masic, A. (2022). Reactive binder and aggregate interfacial zones in the mortar of Tomb of Caecilia Metella concrete, 1C BCE, Rome. *Journal of the American Ceramic Society*, 105(2), 1503–1518.

Shen, B., Sciau, P., Wang, T., Brunet, M., Li, J., Lu, W., Zhu, T. (2018). Microstructural study of colored porcelains of Changsha kiln using imaging and spectroscopic techniques. *Ceramics International*, 44(15), 18528–18534.

Sodini, N., Dreossi, D., Chen, R., Fioravanti, M., Giordano, A., Herrestal, P., Rigon, L., Zanini, F. (2012). Non-invasive microstructural analysis of bowed stringed instruments with synchrotron radiation X-ray microtomography. *Journal of Cultural Heritage*, 13(3), S44–S49.

Sodini, N., Dreossi, D., Giordano, A., Kaiser, J., Zanini, F., Zikmund, T. (2017). Comparison of different experimental approaches in the tomographic analysis of ancient violins. *Journal of Cultural Heritage*, 27, S88–S92.

Sozontov, E.A., Demkiv, A.A., Guryeva, P.V., Peters, G.S., Kolobylina, N.N., Oukhanova, E.V., Yatsishina, E.B. (2019). Ancient parchments: Structural diagnostics and visualization of textual fragments of manuscripts–A natural-science approach. *Journal of Surface Investigation: X-ray, Synchrotron and Neutron Techniques*, 13, 366–370.

Sprain, C.J., Feinberg, J.M., Lamers, R., Bono, R.K. (2021). Characterization of magnetic mineral assemblages in clinkers: Potential tools for full vector paleomagnetic studies. *Geochemistry, Geophysics, Geosystems*, 22(9), e2021GC009795.

Stabile, S., Palermo, F., Bukreeva, I., Mele, D., Formoso, V., Bartolino, R., Cedola, A. (2021). A computational platform for the virtual unfolding of Herculaneum Papyri. *Scientific Reports*, 11(1), 1695.

Stock, S.R., Stock, M.K., Almer, J.D. (2020). Combined computed tomography and position-resolved X-ray diffraction of an intact Roman-era Egyptian portrait mummy. *Journal of the Royal Society Interface*, 17(172), 20200686.

Strullu-Derrien, C., Kenrick, P., Tafforeau, P., Cochard, H., Bonnemain, J.L., Le Hérissé, A., Lardeux, H., Badel, E. (2014). The earliest wood and its hydraulic properties documented in c. 407-million-year-old fossils using synchrotron microtomography. *Botanical Journal of the Linnean Society*, 175(3), 423–437.

Tack, P., Bazi, B., Vekemans, B., Okbinoglu, T., Van Maldeghem, F., Goderis, S., Schöder, S., Vincze, L. (2019). Investigation of (micro-) meteoritic materials at the new hard X-ray imaging PUMA beamline for heritage sciences. *Journal of Synchrotron Radiation*, 26(6), 2033–2039.

Tao, S., Liu, S., Yuan, Y., Dong, J., Li, Q. (2022). A microstructural and compositional study of ε-Fe2O3 crystals in the Hare's Fur Jian Ware. *Crystals*, 12(3), 367.

Taube, M., Chase, W.T., Davenport, A.J., Jardine, A.P. (1995). Investigation and replication of the surface microstructure of early Chinese Black Bronze mirrors. *MRS Online Proceedings Library (OPL)*, 352.

Taube, M., Davenport, A.J., King, A.H., Chase, W.T. (1996). Selective dissolution in copper-tin alloys: Formation of corrosion-resistant patina on ancient Chinese bronze mirrors. *MRS Online Proceedings Library (OPL)*, 432

Thurrowgood, D., Paterson, D., De Jonge, M.D., Kirkham, R., Thurrowgood, S., Howard, D.L. (2016). A hidden portrait by Edgar Degas. *Scientific Reports*, 6(1), 1–10.

Tsiantos, C., Tsampodimou, M., Kacandes, G.H., Sánchez del Río, M., Gionis, V., Chryssikos, G.D. (2012). Vibrational investigation of indigo–palygorskite association(s) in synthetic Maya blue. *Journal of Materials Science*, 47(7), 3415–3428.

Uda, M., Demortier, G., Nakai, I. (eds) (2005). *X-Rays for Archaeology*. Springer, Dordrecht.

Ugawa, A., Ishii, H., Yakushi, K., Okamoto, H., Mitani, T., Watanabe, M., Sakai, K., Suzui, K., Kato, S. (1992). Design of an instrument for far-infrared microspectroscopy using a synchrotron radiation source. *Review of Scientific Instruments*, 63(1), 1551–1554.

Ungár, T., Martinetto, P., Ribárik, G., Dooryhée, E., Walter, P., Anne, M. (2002). Revealing the powdering methods of black makeup in Ancient Egypt by fitting microstructure based Fourier coefficients to the whole X-ray diffraction profiles of galena. *Journal of Applied Physics*, 91(4), 2455–2465.

Van der Snickt, G., Dik, J., Cotte, M., Janssens, K., Jaroszewicz, J., De Nolf, W., Groenewegen, J., Van der Loeff, L. (2009). Characterization of a degraded cadmium yellow (CdS) pigment in an oil painting by means of synchrotron radiation based X-ray techniques. *Analytical Chemistry*, 81(7), 2600–2610.

Van Olphen, H. (1966). Maya Blue: A clay-organic pigment? *Science*, 154(3749), 645–646.

Vanmeert, F., Van der Snickt, G., Janssens, K. (2015). Plumbonacrite identified by X-ray powder diffraction tomography as a missing link during degradation of red lead in a Van Gogh painting. *Angewandte Chemie International Edition*, 54(12), 3607–3610.

Walter, P., Martinetto, P., Tsoucaris, G., Bréniaux, R., Lefebvre, M.A., Richard, G., Talabot, J., Dooryhée, E. (1999). Making make-up in ancient Egypt. *Nature*, 397, 483–484.

Wang, L., Wang, Y., Zhang, M., Li, Q., Wu, J., Liu, Z., Li, L., Wei, X. (2019). Three-dimensional microstructure of ε-Fe2O3 crystals in ancient Chinese sauce glaze porcelain revealed by focused ion beam scanning electron microscopy. *Analytical Chemistry*, 91(20), 13054–13061.

Wei, X., Lei, Y., Sun, T., Lin, X., Xu, Q., Chen, D., Zou, Y., Jiang, Z., Huang, Y., Yu, X. et al. (2008). Elemental depth profile of faux bamboo paint in forbidden city studied by synchrotron radiation confocal μ-XRF. *X-Ray Spectrometry: An International Journal*, 37(6), 595–598.

Weitkamp, T., Diaz, A., David, C., Pfeiffer, F., Stampanoni, M., Cloetens, P., Ziegler, E. (2005). X-ray phase imaging with a grating interferometer. *Optics Express*, 13(16), 6296–6304.

Welcomme, E., Walter, P., Bleuet, P., Hodeau, J.L., Dooryhee, E., Martinetto, P., Menu, M. (2007). Classification of lead white pigments using synchrotron radiation micro X-ray diffraction. *Applied Physics A*, 89(4), 825–832.

Wen, R., Wang, D., Wang, L., Dang, Y. (2019). The colouring mechanism of the Brown glaze porcelain of the Yaozhou Kiln in the Northern Song Dynasty. *Ceramics International*, 45(8), 10589–10595.

Wess, T.J., Drakopoulos, M., Snigirev, A., Wouters, J., Paris, O., Fratzl, P., Collins M., Hillier, J., Nielsen, K. (2001). The use of small-angle X-ray diffraction studies for the analysis of structural features in archaeological samples. *Archaeometry*, 43(1), 117–129.

Xu, Y., Qin, Y., Ding, F. (2017). Characterization of the rare oil spot glazed bowl excavated from the Xiao kiln site of north China. *Ceramics International*, 43(12), 8636–8642.

Yamahana, K. (2000). Synchrotron radiation analysis on ancient Egyptian vitreous materials. *Proceedings of the 25th Linear Accelerator Meeting in Japan*, Hameji, 13, 1.

Young, M.T., Rayfield, E.J., Holliday, C.M., Witmer, L.M., Button, D.J., Upchurch, P., Barrett, P.M. (2012). Cranial biomechanics of *Diplodocus* (Dinosauria, Sauropoda): Testing hypotheses of feeding behaviour in an extinct megaherbivore. *Naturwissenschaften*, 99(8), 637–643.

Zvereva, V.V., Trunova, V.A., Sorokoletov, D.S., Polosmak, N.V. (2017). Mercury in archeological hair samples from Xiongnu burials (Noin-Ula, Mongolia): SR XRF and CXRM analysis. *X-Ray Spectrometry*, 46(6), 563–568.

3

Application of Full-field X-ray Absorption Spectroscopy Imaging in Transmission Mode to Study Cultural Heritage Samples

Emeline POUYET[1], Letizia MONICO[2], Philippe SCIAU[3] and Marine COTTE[1,4]

[1] *Laboratoire d'archéologie moléculaire structurale, Sorbonne University, Paris, France*

[2] *CNR-SCITEC, Perugia, Italy*

[3] *CEMES, Toulouse, France*

[4] *European Synchrotron Radiation Facility, Grenoble, France*

Synchrotron radiation-based microanalytical techniques have become common tools for the study of cultural heritage objects. In particular, micro X-ray Diffraction (μXRD), micro X-ray Fluorescence (μXRF), micro X-ray Absorption Near-Edge Structure (μXANES) spectroscopy and micro Fourier Transform infrared (μFTIR) spectroscopies are employed alone or in combination, to determine with (sub)micrometer lateral resolution the heterogeneous chemical nature and distribution of components from artwork fragments. μXRD, μXRF and μFTIR are routinely achieved over 2D or even 3D areas. However, 2D-μXANES is typically limited to the acquisition of (i) full XANES spectra on a few points of interest (limited spatial coverage) or (ii) several full 2D μXRF map at few fixed energies (limited spectral coverage). Full-field X-ray near edge structure (FF-XANES) instruments in transmission mode offer an alternative to the approaches mentioned, as it allows for the acquisition of hyperspectral imaging data sets in transmission mode and thus combines spatial and spectral information with micrometric resolution over millimetric 2D areas. In this chapter, the opportunities offered by the FF-XANES approach in the Cultural Heritage field are illustrated by a number of case studies (i.e. painting,

Synchrotron Radiation, Cultural Heritage, Biomineralization,
coordinated by Catherine DEJOIE, Pauline MARTINETTO and Nobumichi TAMURA.

ink and ceramic) with questions regarding both artistic techniques and transformation/degradation processes of the constitutive materials.

3.1. Introduction

Considering the heterogeneity of artistic materials, chemical selectivity and imaging capability are of key importance when analyzing artworks (at both the object and the fragment length scales). Therefore, complementary chemical and spatially resolved techniques, which combine spectroscopy and microscopy, are very well suited to attain a full 2D or even 3D description of the composition of artistic materials. In this context, the extremely bright and collimated synchrotron radiation (SR) source offers unique lateral resolution (a few microns down to some tens of nanometers) combined with energy source tunability (Bertrand et al. 2012). Among SR-based techniques, micro Fourier Transform InfraRed (μFTIR) spectroscopy, micro X-ray Fluorescence (μXRF), micro X-ray Absorption Near Edge structure (μXANES) spectroscopy and micro X-ray Diffraction (μXRD) have provided combined information on the identification and the location of compounds of interest at the sub-micrometer scale (Cotte et al. 2017). However, while SR μXRF, μFTIR and μXRD are routinely performed in mapping modes, 2D-μXANES is less commonly used.

With respect to the analysis of artworks, the XANES technique is generally employed to understand processes taking place during the creation and possible subsequent transformation/degradation of specific constitutive materials of artifacts. All such processes are indeed usually based on changes in the oxidation state and/or modifications of the local chemical environment (e.g. coordination number, site symmetry, oxidation state, etc.) that can be accessed and mapped by 2D-XANES analyses (Cotte et al. 2010, 2017).

Currently, different methods are used to obtain chemical mapping (combination of spectroscopy and microscopy) with applications in the Cultural Heritage (CH) field. The first approach – also the most common – is based on a multi-spectral mapping approach in fluorescence mode, also named Selectively Induced X-ray Emission Spectroscopy (SIXES) (Sakurai et al. 1988). It consists of the acquisition of n μXRF maps (generally only three to four maps), at n specific energies E_i providing the maximum contrast between the different expected speciation of the probed element. Consequently, the efficiency of this approach directly depends on an a priori knowledge of the few species present in the sample, to determine the energies E_i and calculate the corresponding species distribution (Brinza et al. 2014). For example, SIXES mapping at the calcium K-edge was recently used for probing the distribution of different calcium-based compounds (i.e. amorphous calcium carbonate, vaterite, calcite, dolomite and others) in consolidated limestones (Monico et al. 2020b). This approach is also commonly used at the sulfur K-edge to provide direct information on sulfur oxidation state. Since the position of the K-edge white line of sulfur species varies over 11 eV, from 2.472 keV for sulfides up to 2.483 keV for sulfates, an easy distinction and mapping of the different oxidation states of sulfur is achieved (Cotte et al. 2006; Van der Snickt et al. 2009, 2012; Radepont et al. 2011). However, this situation is not systemically encountered. Furthermore, the field of view (FOV) and number of maps are limited by dwell times, which are usually in the range of a hundred of milliseconds per point with standard energy-dispersive detectors and electronics. Accordingly, this approach is rather qualitative and some compounds may be completely over-looked. The development of ultra-fast detectors with large detection solid-angle, such as the Maia, allowed for considerably decreasing pixel acquisition time (down to less than 1 ms) and increasing the number of μXRF maps (n >100), thus open the possibility to perform full-spectral (FS)-XANES imaging in fluorescence mode (Ryan et al. 2010, 2018; Monico et al. 2015a). For example, such approach was used for studying the darkening of chrome yellow pigments ($PbCr_{1-x}S_xO_4$, $0 \leq x \leq 0.8$). Cr K-edge FS-XANES imaging investigations of a series of microsamples obtained from paintings by Vincent van Gogh and artificially aged paint mock-ups confirmed that photo-reduction process of the original Cr^{VI} to Cr^{III} compounds is responsible for lead chromates darkening and permitted to obtain 2D Cr-phase maps with micrometric lateral resolution. It is expected that state-of-the-art XRF detectors and electronics will allow for the generalization of FS-XANES imaging in 2D and possibly 3D XRF mode for studying further CH degradation phenomena.

This approach, however, can present several drawbacks for the analyses of thick cross-sections for which the element of interest is present in a high concentration. The results can ultimately be affected by self-absorption effects (Pouyet et al. 2015b), further limiting XANES data interpretation in complex systems. Moreover, the heterogeneity of the samples can prevent semi-quantitative information to be gained on the elements of interest. For highly concentrated and heterogeneous samples, measuring XANES in transmission mode instead of XRF mode appears as a valuable alternative, providing that samples can be prepared accordingly (see section 3.2).

In this context, full-field X-ray near edge structure (FF-XANES) imaging and Transmission X-ray Microscopy (TXM) are well-established approaches in the soft X-ray domain with application in the field of CH and paleontology, for example, for the study at the C K-edge of fossil spores (Bertrand et al. 2012), the Fe L-edges XANES study of a 450-year-old corroded archaeological iron nail (Michelin et al. 2013) or the C-K and N-K Edges XANES study of gelatine distribution at the scale of a paper fiber (Gimat et al. 2021).

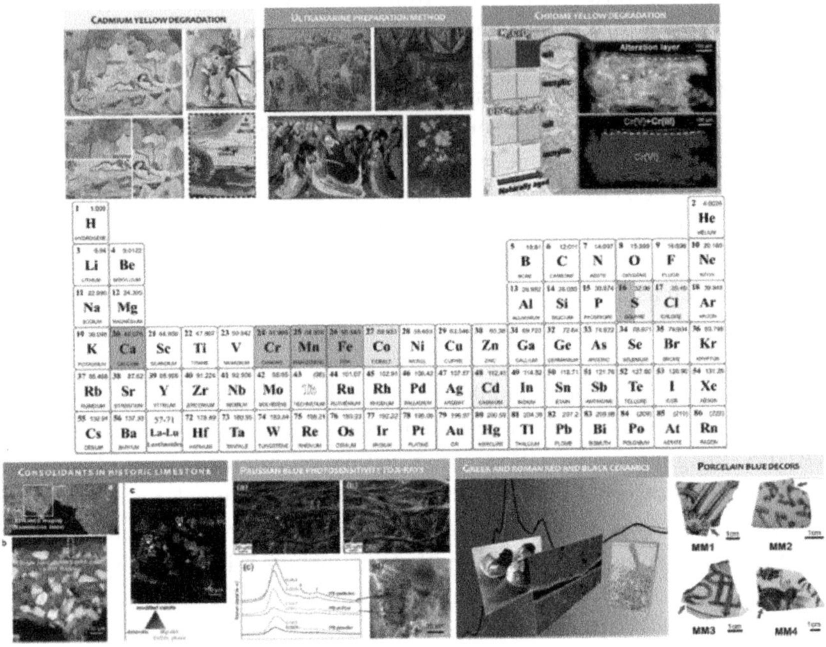

Figure 3.1. *FF-XANES application in the tender and hard X-ray domain for CH studies of: cadmium yellows degradation (illustration from Pouyet et al. (2015)), ultramarine preparation method (illustration from Gambardella et al. (2020)), chrome yellows degradation (illustration from Monico et al. (2019)), consolidants in limestones (illustration from Monico et al. (2020)), Greek and Roman red and black ceramics (illustration from Meirer et al. (2013)), porcelain blue decors (illustration from Wang et al. (2016)) and Prussian blue photosensitivity to X-rays (illustration from Gervais et al. (2013a)). For a color version of this figure, see www.iste.co.uk/dejoie/synchrotron.zip*

Extending 2D-XANES in the tender or hard X-ray ranges have been explored over the last decade. Two instrumental approaches are available: one based on X-ray microscopy and another based on X-ray lens-less imaging. In both cases, a 2D camera is used to provide imaging information (sample X-ray radiography) on a static sample, while scanning the energy of the incoming X-ray beam across the absorption edge of the element of interest. While the FF is a lens-less approach, in a TXM, a condenser focuses the beam on the sample, and an objective projects the image of the sample onto a detector. The 3D (2D image combined with 1D energy) XANES stacks consist of a series of normalized images that characterize the sample transmittance over the absorption edge of the element of interest. The energy resolution of the XANES spectrum depends on the number of frames in the stack and is in practice only limited by the energy resolution of the monochromator. Thanks to the simultaneous acquisition of millions of pixels, over millimetric areas with (sub-)micron resolution, this technique is very well suited for the study of heterogeneous samples, in which the element of interest is concentrated, a case study frequently encountered for CH materials.

Several FF-XANES applications for CH studies have been published over the last decade (Pouyet et al. 2015; Wang et al. 2016; Monico et al. 2019; Sciau and Wang 2019; Gambardella et al. 2020; Monico et al. 2020). As shown in Figure 3.1, most of these experiments were carried out on paints and pigments (see example in section 3.4.1), but the set-up was also successfully used to analyze ceramics (see example in section 3.4.2) and, more recently, consolidants in limestones (Monico et al. 2020). A selection of representative applications is further discussed in the following chapter to provide with an overview of the potential offered by FF-XANES for studying CH materials.

3.2. The worldwide context of tender and hard X-ray domain FF-XANES instruments

For tender or hard X-ray measurements, several devices have been developed worldwide, notably at Stanford Synchrotron Radiation Lightsource (SSRL) (Andrews et al. 2010; Meirer et al. 2011), Angströmquelle Karlsruhe (ANKA) (Mangold et al. 2013), the National Synchrotron Light Source (NSLS) (Karen Chen-Wiegart et al. 2012), the Advanced Photon Source (APS) (De Andrade et al. 2016) and the European Synchrotron Radiation Facility (ESRF)(De Andrade et al. 2011; Fayard et al. 2013).

The beamline 6-2c, at SSRL, accommodates a commercial TXM from X-Radia optimized for photon energies from 5 to 14 keV with spatial resolution down to 30 nm and an energy resolution $\Delta E/E$ of 10^{-4} (Andrews et al. 2010; Meirer et al. 2011). The TXM uses a capillary condenser to focus the beam to a spot with size down to a few tens of microns. A Fresnel zone plate placed between the sample and the scintillator crystal achieves a magnification of around 50x, increased 10 to 20 times by the optical objective lens of a CCD camera (coupled with a scintillator crystal to collect the projection images). It achieves a single flat FOV of about 15 x 15 μm^2 or 30 x 30 μm^2, depending on the type of X-ray optics.

At NSLS (beamline 18ID-FXI) and APS (sector 32-I), the development of in-house FF hard X-ray TXM with spatial resolution below 30 nm and 2D and 3D imaging capabilities have opened up the possibility of conducting dynamic chemical studies at the nanoscale (Karen Chen-Wiegart et al. 2012; De Andrade et al. 2016).

	Energy range (keV)	FOV	Pixel size	Typical total scan time for a single FOV @ Fe K-edge	Pixel binning
SSRL 6-2c	5–14	15 × 15 μm^2 30 × 30 μm^2	15 × 15 nm^2 30 × 30 nm^2	76 min	2 x 2
APS 32 ID	8–15	50 × 50 μm^2 10 × 10 μm^2	40 × 40 nm^2 16 × 16 nm^2		
NSLS 18ID-FIX	5–11	40 × 40 μm^2	30 nm		
ANKA NANO	3–30	20 × 20 μm^2 40 × 40 μm^2	30 × 30 nm^2 50 × 50 nm^2		4 × 4
ESRF ID21	2–9	648 × 768 μm^2 1.3 × 1.5 mm^2	0.3 × 0.3 μm^2 0.6 × 0.6 μm^2	55 min	No binning
ANKA XAS/SUL	3–27	13 × 11 μm^2	5 × 5 μm^2		2 x 2 μm^2

Table 3.1. *Experimental parameters used for TXM-XANES imaging systems. Values refer to typical acquisition of Fe K-edge XANES stacks from Andrews et al. (2010); De Andrade et al. (2011, 2016); Meirer et al. (2011); Karen Chen-Wiegart et al. (2012); Fayard et al. (2013); Mangold et al. (2013)*

At the ANKA XAS and SUL-X beamlines, the full uniform and unfocused beam (9 x 1 mm^2 and 0.5 x 0.3 mm^2) is used to perform sample radiographs. Using a single crystal scintillator coupled with a microscope objective (with 3.6x magnification and a 2 × 2 binning) onto a PCO camera, transmission images of the sample are acquired with an effective pixel size of 5 × 5 μm^2 (Mangold et al. 2013).

FF acquisitions are also performed at another ANKA beamline, the NANO beamline, with a wiggler source and double crystal monochromator Si(111) ($\Delta E / E = 10^{-4}$) coupled with Carl-Zeiss FF TXM microscope (Carl-Zeiss-TXM) (Bauer et al. 2015). This latter uses a condenser as focusing optics with an FOV of 20 × 20 μm^2 and 40 × 40 μm^2 and a resolution of 30 nm and 50 nm at magnification factors of 74 and 36 respectively.

Similar to ANKA XAS/SUL instrument, the ESRF ID21 FF-XANES instrument was designed on a vacuum chamber and lens-less approach, giving access to the full ID21 energy range down to 2.1keV (Fayard et al. 2013; Salomé et al. 2013). The beam size was defined by slits (typically 1.5 × 1.5 mm^2). Sample radiographies were acquired via a detecting ensemble made of a scintillator (Lu_2SiO_5:Tb), a magnifying long working distance optical objective (Mitutoyo) and a CMOS camera (CMOS PCO edge camera). It accommodated pixel size ranging from 0.3 × 0.3 μm^2 to 0.6 × 0.6 μm^2 and FOV from 648 × 768 μm^2 up to 1.3 × 1.5 mm^2.

3.3. Typical acquisition, processing and sample preparation strategies for CH studies: the example of the FF-XANES set-up at beamline ID21 (ESRF)

3.3.1. *Sample preparation*

FF-XANES is based on the measurement of the sample absorption in transmission acquisition mode. Consequently, it requires the preparation of the section with optimal thickness to ensure reliable results. If the sample is too thin, very few photons are absorbed by the sample resulting in a poor Signal-to-Noise Ratio (SNR) signal. Conversely, if the sample is too thick, very few photons make it through the sample, resulting in high noise value in the transmitted signal, and ultimate distortions of the XANES spectra. Two main rules are thus followed to estimate the optimal thickness of the section to be analyzed (Calvin 2013). First, a maximum absorbance of the sample of around 2–2.5 should not be exceeded, i.e. the sample transmittance should be higher than 8% over the energy range of interest. Second, a variation of the absorbance below and above the edge (i.e. *edge jump* or *step*) of about 1 is preferred (Pouyet et al. 2015). These values guide the selection of the sample thickness depending on the element and the absorption edge of interest, the concentration of this element and the overall chemical composition of the matrix.

For the examples presented in this chapter, optimum thicknesses range from 2 to 60 μm and require the preparation of thin sections. As reaching such thickness ranges can be challenging for leanly bound materials (Pouyet et al. 2015), adapted sample preparation strategies are carried out.

Concerning paintings, the cutting approach is favored. The usual method consists of embedding samples into polymeric resin before using a microtome to cut thin sections at the desired thickness. When samples are large enough (millimetric) and have enough intrinsic holding, they are directly cut from their polycarbonate support using the microtome, similarly to the SES protocol described elsewhere (Pouyet et al. 2015).

To obtain thin sections of harder material, such as ceramics and mortars, polishing is usually preferred. For the study of ceramics, a preparation method derived from transmission electron microscopy is implemented. The main steps of the sample preparation are developed and illustrated in Meirer et al. (2011). In summary, after several cutting steps, two material strips are glued together (surface against surface), using *Gatan G-1* epoxy. The sandwich is cut into ~500 µm slice. A slice is temporarily mounted on a movable support in brass with a dental wax. After polishing the first side, the sample is flipped and glued with wax. The slice is polished down to have the same thickness as standard copper grid, 40–60 µm. Finally, a copper washer is glued to the final thin section for safeguarding the samples and facilitating their handling.

Finally, when the sample does not present a layered stratigraphy along the beam path, and its intrinsic thickness respects the above guidelines, no further sample preparation is performed. This approach was chosen for the study on Prussian blue (hereafter denoted "PB"). For this work, a Whatman paper 180 µm thick immersed into PB colloidal suspension, close to the composition of soluble PB, $KFe^{III}[Fe^{II}CN_6]$, resulted into a sample presenting an unknown total density, avoiding theoretical calculation for thickness determination. The experimental approach was consequently preferred to assess if the sample preparation was mandatory. With a minimum transmittance value of 30% and an edge jump of 0.7, the original thickness of the material of interest was suitable for direct measurements.

The sections and samples analyzed in the application section below (section 3.3.3 and 3.3.4) are systematically placed between one or two Ultralene polymer foils (4 µm thick, from Spex Sample Prep), and, in a few cases, fixed on sulfur-free adhesive tape to provide mechanical stability during mounting and analyses.

3.3.2. *Acquisition and data pre-processing*

The data presented in this chapter are recorded following the acquisition and processing strategies developed at ID21. Regarding the data acquisition chain, a preliminary image, named *dark image*, is first acquired without X-ray beam to correct for the offset created by the dark current in the camera (Figure 3.2a). The monochromator is set at the first value of the energy scan and a first image of the beam is acquired, without sample in the FOV, namely a *reference image* (Figure 3.2b). The sample is then moved into the FOV thanks to an in-out vertical sample translation and a radiography of the sample is acquired, namely a *data image* (Figure 3.2c). To correct the spatial heterogeneity of the beam intensity, a so-called flat field correction is performed, and the sample transmission *data image* is normalized by the corresponding *reference image* (Figure 3.2d). A corrected image acquired at specific energy E_i can be expressed according to equation [3.1]:

$$\text{Transmittance}\ (E_i) = \frac{\text{I data}\ (E_i) - \text{I dark}}{\text{I reference}\ (E_i) - \text{I dark}} \qquad [3.1]$$

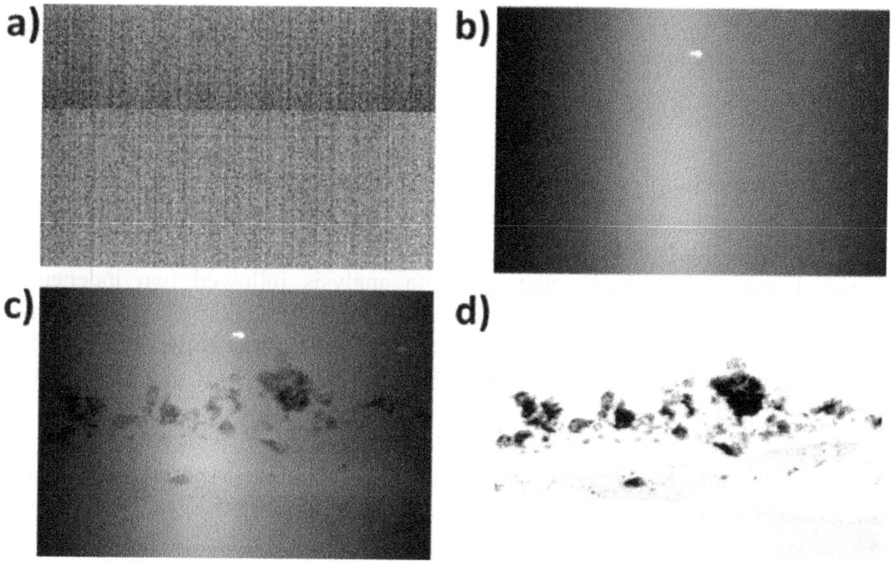

Figure 3.2. *Example of a) a dark image; b) a flat-field image (reference); (c) a sample image (data) to calculate (d) the transmission image following equation [3.1]*

The acquisition is then repeated for all energy steps defined over the full energy range covered by XANES acquisition. This approach allows us to easily define different energy regions with different energy steps and consequently energy resolution (usually down to 0.1 eV in the edge region and up to 1–2 eV in the post-edge regions) and optimized acquisition time. Usually, about 300 normalized radiographies are acquired over a few hundreds of eV, with variable energy steps.

To cope with sample repositioning after in-out motion, an accurate procedure of realignment of the normalized images based on key-points extraction has been applied. The method called *feature-based registration* is based on the establishment of correspondence between distinctive points in the image stack. With key points associated with the scale of the features and their orientation, this method achieves alignment precision superior to 0.1 pixel (Larson et al. 2013). Based on a widely used panoramic image stitching algorithm (Freestone 1987; Brill and Cahill 1988), the Scale Invariant Feature Transform (SIFT) has been adapted to FF-XANES (Tajima et al. 2014) and used for the different results shown below. The set of normalized and aligned images are then analyzed and post-processed using different software packages, such as PyMca (Sole et al. 2007) and TXM-Wizard (Liu et al. 2012).

3.3.3. *Data analysis workflow*

Prior to the analysis of precious and complex historical samples, the ID21 FF-XANES set-up and related data processing and analyses chain are first tested and improved on a mock-up sample modeling a two-layer painting sample. The mock-up is made of two manganese-based paint layers (Figure 3.3a), containing MnO_2 and $MnSO_4$, mixed with linseed oil (Fayard et al. 2013). 2D FF-XANES analyses are performed across the Mn K-edge and data analysis followed two independent approaches: (i) principle component analysis (PCA) combined to subsequent k-means clustering (Figure 3.3c); and (ii) least squares linear combination (LSLC) fitting with MnO_2 and $MnSO_4$ standards and subsequent R-factor correlation analysis (Figure 3.3d).

Both methods provide the distribution of the different Mn-based compounds with a micrometric resolution and an FOV of ~1 × 1 mm^2. For each single pixel XANES spectrum extracted from the two Mn layers (Figure 3.3b), LSLC fitting with low R-factor values (between 0 and 0.1) is obtained. The latter procedure allows a highly sensitive and quantitative 2D mapping of the two-manganese species and confirms its potential for the investigation of historical paint microsamples.

3.3.4. *A low fluence chemical imaging strategy for dedicated case studies*

SR analyses are usually presented as nondestructive. However, in some cases, artistic materials may be damaged during X-ray beam exposure (Bertrand et al. 2015). This effect can be increased when using micro-and nano-probes, which, by focusing the high photon flux into a micro (nano) metric site, release high photon doses to the sample. Among the different SR-based techniques, μXANES is particularly demanding because in addition to the microprobe, spectral information requires acquisition over hundreds of energy points. This in turn may modify the existing species and/or generate new ones, seriously compromising data interpretation (Ganio et al. 2018; Monico et al. 2020c).

This effect may be monitored by following the data evolution over successive acquisitions (XANES, XRD, etc.). If noticed, a few strategies exist to minimize the damages, all aiming to reduce the total photon dose absorbed by the sample: (i) decrease the exposure time (e.g. using more efficient detectors); (ii) decrease the beam intensity (e.g. with attenuators), (iii) by using cryo-preservation; or (iv) by regularly changing the area of the sample exposed to the beam (this is limited to samples presenting an homogeneous composition over the points analyzed).

Such radiation damage issues are reported for several organo iron-compounds, including iron gall ink (Kanngiesser et al. 2004; Wilke et al. 2009), hemoglobin (Mazel et al. 2007) and Prussian blue pigments (Gervais et al. 2013; Lu 2014). More particularly, Prussian blue (PB, iron(III) hexacyanoferrate(II)), used in Europe from the 18th–19th centuries as an artist's pigment and dye for a wide range of application (Wang et al. 2016), poses several problems of conservation because of its pronounced photosensitivity which results in a fading of the original intense blue color (Geng 1993; Samain et al. 2011) due to the reduction of Fe (III) into Fe (II). However, the investigation of the natural degradation process by XANES is limited by the fact that the exposure to X-rays during analyses leads to similar spectral changes in the Fe K-edge XANES spectra. The "live" fading of PB studied in paper-based model artworks by time-resolved energy-dispersive XANES, at the synchrotron SOLEIL, highlighted the need for low fluence chemical imaging strategy to decorrelate radiation damages induced by the analytical X-ray probe, from the natural aging of the material (Heinzer et al. 2006; Gervais et al. 2013, 2015).

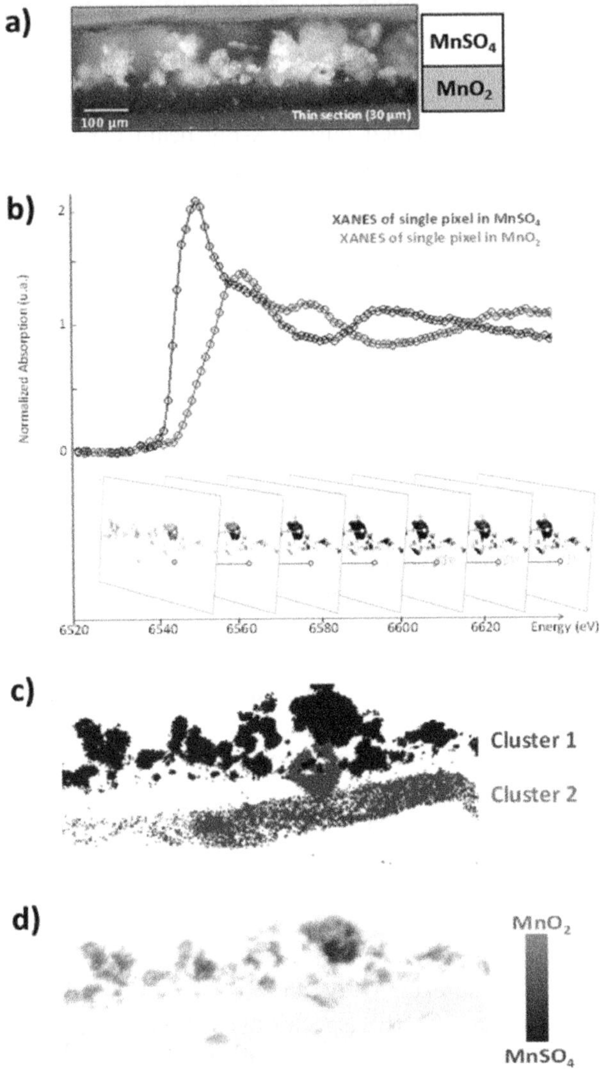

Figure 3.3. *a) Scheme and optical image of a 30 µm thick section of the binary mock-up sample, composed of MnO_2 and $MnSO_4$ layers; b) XANES spectra of a single pixel from the MnO_2 and $MnSO_4$ layers (from 6.54 to 6.64 keV in 100 energy steps of 1eV, with 0.6 x 0.6 μm^2 pixel size, in 45 min); c) PCA and subsequent clustering results showing the localization of the two Mn-based species; d) LSLC results using MnO_2 and $MnSO_4$ reference standards for localization and quantification of Mn-based species into the sample stratigraphy. For a color version of this figure, see www.iste.co.uk/dejoie/synchrotron.zip*

	Pinhole	Zone plate	Full-field
Acquisition time	18 min	10 min	4 h
Scan time	1 min	1 min	1 h
Beamsize	200 µm (diameter)	0.12 µm (v.) 0.97 µm (h.)	950 µm (v.) 400 µm (h.)
Fluence rate (ph/(s.µm^2))	7.6 10^5	4.3 10^{10}	3.6 10^5
Fluence (ph/ µm^2)	2.8 10^9	5.8 10^{13}	5.3 10^9

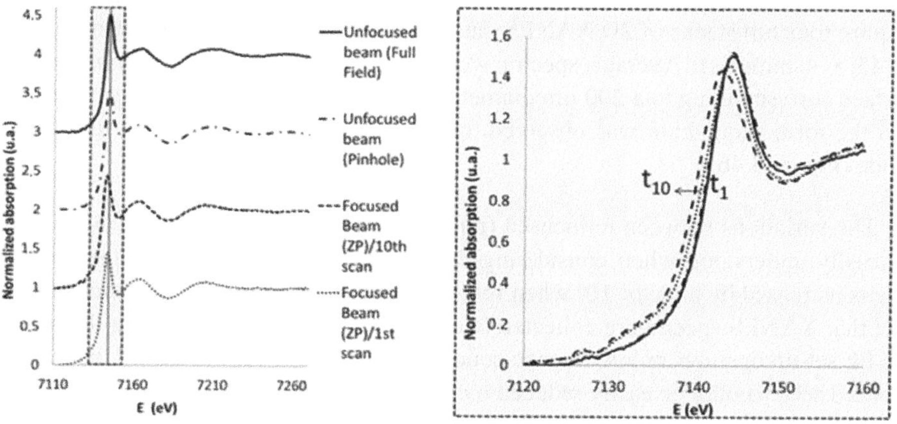

Figure 3.4. *Top: comparison of the acquisition conditions in focused mode (zone plate); unfocused mode (pinhole); and full-field mode for the recording of a complete XANES spectrum. Bottom: zoom of the Fe K-edge XANES pre-edge region highlighting photo-reduction of a Prussian blue pigment at high fluence rate focused mode, in comparison with two low fluence rate unfocused modes (pinhole, full-field). Acquisitions were carried out from 7,100 to 7,280 eV in 500 steps, with energy resolution of 0.3 eV.*

In this context, a 100% cellulosic Whatman paper impregnated with PB pigment was analyzed using the ID21 FF-XANES set-up and results were thus compared to those obtained using two other standard acquisition set-ups (Figure 3.4a) in nominal conditions of flux at Fe K-edge: unfocused beam (detection in transmission mode with a photodiode) and focused beam in the SXM (detection in transmission mode with a photodiode), summarized in Figure 3.4a. An unfocused beam defined using a pinhole (fluence: 2.8 10^9 ph/µm^2) was used to acquire 20 consecutive scans over

18 min. The comparison of the 20 XANES spectra allowed concluding that no visible spectral changes specific of a photo-reduction of the Fe-based species were observed in the present configuration (Figure 3.4b). A microfocused beam obtained with a zone-plate (fluence: 5.8 10^{13} ph/µm^2) was used to acquire a series of 10 consecutive scans over 9 minutes. In these conditions, important temporal spectral changes could be observed: the edge position shifted to lower energies and the spectral pattern above and below the edge was modified. By performing a comparison with the spectrum acquired in the unfocused mode, it can be noted that even after the first scan, the X-rays induced a photo-reduction process of iron (Figure 3.4b).

In FF-XANES mode, the beam size (fluence: 5.3 10^9 ph/µm^2) was used to acquire four full stacks of 2D XANES data (corresponding to a total acquisition time of 45 × 4 minutes). Average spectra were calculated for each acquisition over a surface corresponding to a 200 µm diameter disk, and no difference between the first and the forth acquisition was observed, nor with the spectra acquired in unfocused mode (Figure 3.4b).

The variations between unfocused (pinhole or FF) and focused beam modes can be easily understood when considering that, in the present conditions, the fluence rate is increased by a factor 10^5 when focusing the beam (Figure 3.4a). Thanks to the fact that XANES spectra are collected over a large area, without focusing the beam, the FF set-up operates in low fluence conditions. Obviously, the radiation damage in focused set-up could be easily reduced by reducing the beam intensity.

3.4. Applications of FF-XANES to study CH samples

3.4.1. *Painting composition and degradation: historical cases and paint mock-ups*

3.4.1.1. *Cadmium yellow pigments*

Cadmium sulfide-based compounds ($CdS/Cd_{1-x}Zn_xS$, with $0 < x \leq 0.35$) are a class of synthetic yellow pigments frequently used by artists of the late 19th to early 20th century. Despite having a good reputation regarding permanency, physical manifestations of oxidative degradation of the CdS-based pigment have been observed over the past decade in numerous impressionist and early modernist paintings (Leone et al. 2005; Van der Snickt et al. 2009, 2012; Plahter and Topalova-Casadiego 2011; Mass et al. 2013b; Pouyet et al. 2015; Miliani et al. 2018; Monico et al. 2020). This degradation appears in many different ways: paint chalking (i.e. drying and crumbling), flaking, spalling, fading and, in its most advanced cases, as a thick

(20–50 μm) ivory-tan alteration crust covering the original yellow paint. In this context, Cd L_3-edge FF-XANES imaging was combined with 2D-μFTIR, -μXRF and -μXRD measurements in order to identify the nature of Cd-based compounds associated with the degradation process of the cadmium yellow paint and to obtain their stratigraphic distribution. In particular, an original study on paintings by Matisse addresses one of the long-standing questions about cadmium yellow alteration process: the origin and roles of cadmium carbonates and cadmium sulfates found in the altered paint layers from Matisse's *The Joy of Life* and *Flower Piece* (1906, The Barnes Foundation) (Pouyet et al. 2015). These compounds have often been identified as photo-oxidation products of the original CdS-based pigment, but could also be residual starting reagents from a wet process synthesis of the pigment itself. FF-XANES imaging appeared as an ideal way to extend previous studies of similar samples where single-point μXANES of a few selected spots was carried out and so to eliminate some of the ambiguities that are inherently present when only a point-by-point analysis is performed. SR 2D X-ray and infrared analyses allowed us to identify and map the location of cadmium carbonates, cadmium chlorides, cadmium oxalates, cadmium sulfates and cadmium sulfides through thin sections of altered cadmium yellow paints. The nature and corresponding distribution of the identified cadmium compounds suggest that cadmium carbonates ($CdCO_3$), oxalates (CdC_2O_4) and sulfates ($CdSO_4 \cdot nH_2O$) are degradation products of the original cadmium yellow paints in *The Joy of Life*. In a different way, in *Flower Piece*, cadmium carbonates ($CdCO_3$) appear to have been a ((partially) unreacted) starting reagent for the yellow paint.

The same methodological approach has been recently extended to study *The Scream* (circa 1910) by Edvard Munch (Munch Museum, Oslo) and of a set of artificially aged mock-ups by means of combined molecular spectroscopic techniques and SR X-ray microspectroscopic methods (Monico et al. 2018, 2020). Such investigations allowed more specific information on the factors triggering the degradation process of cadmium yellow paints to be determined. Among the analyzed oil paints mock-ups, a set was prepared using an early 20th century cadmium yellow pigment powder (named as "$7914_{Cd(OH)Cl}$"), which was specifically selected for having chemical properties similar to those observed in selected degraded cadmium yellow paints of *The Scream* (circa 1910) (Monico et al. 2020).

To assess the effect of different environmental parameters on the oxidation process of the CdS-based pigments, the mock-ups were exposed to different accelerated aging conditions: (i) UVA-visible light at environmental relative humidity (RH) (~45%) and (ii) at RH ≥ 95% (without exposure to light). While the exposure to light did not induce any significant changes, high moisture conditions in combination with the presence of cadmium chloride-compounds (i.e. Cd(OH)Cl, a leftover starting reagent from a wet process synthesis of CdS) resulted in a highly relevant phenomenon: the formation of S^{VI}-rich aggregates ($CdSO_4$/$CdSO_4{\cdot}nH_2O$) within the S^{-II} (CdS)-based matrix, as shown by the S speciation distributions (Figures 3.5a and 3.6b). Such newly formed Cd-compounds are better visualized in the SR μXRD maps and FF-XANES Cd-phase images (Figures 3.5c and 3.5d), in which, within the hexagonal-CdS/$CdCO_3$-based paint matrix, minor amounts of $CdC_2O_4{\cdot}3H_2O$, $CdSO_4{\cdot}nH_2O$ and $CdCl_2{\cdot}nH_2O$ are observed and localized in small aggregates (ca. 10 μm in diameter). Notably, the quantitative FF-XANES analysis (Figures 3.5e and 3.5f) shows that the relative amount of Cd(OH)Cl, $CdCl_2{\cdot}nH_2O$ and $CdCO_3$ achieves values up to ~80%, ~45% and ~70% respectively, whereas that of hexagonal-CdS, $CdC_2O_4{\cdot}3H_2O$ and $CdSO_4{\cdot}nH_2O$ is not above ~30%–35%. Note that $CdC_2O_4{\cdot}3H_2O$, $CdSO_4{\cdot}nH_2O$ and $CdCl_2{\cdot}nH_2O$ compounds were not detected by μXRD, possibly due to their amorphous nature.

Overall, the results prove that the original CdS-based pigment converts into cadmium sulfates in the presence of chloride-containing compounds in high-moisture conditions. This takes place even in the absence of light and does not occur on chlorine-free oil paint mock-ups aged under similar conditions (Monico et al. 2018). The results also show that exposure to moisture causes cadmium chloride-compounds to migrate through the paint along with the formation of cadmium sulfates. The photodegradation of CdS, on the contrary, does not appear to be influenced by the absence or presence of cadmium chlorine-compounds.

3.4.1.2. *Chrome yellow pigments*

Cr K-edge FF-XANES imaging has been also recently used in the context of the study of degradation mechanism of chrome yellows ($PbCr_{1\text{-}x}S_xO_4$, $0 \leq x \leq 0.8$), a class of pigments often encountered in paintings by Vincent van Gogh (Monico et al. 2013b, 2015, 2019; Vanmeert et al. 2018).

Earlier SR-based Cr K-edge μXANES investigations (in XRF mode) in combination with electron paramagnetic spectroscopy (EPR) analysis of artificially aged chrome yellow paint mock-ups, allowed us to demonstrate that the darkening of chrome yellows is due to a $Cr^{VI} \rightarrow Cr^{III}$ photoreduction with paramagnetic Cr^V-species that arise from the interaction between the pigment and the oil binder (Monico et al. 2015, 2016). The results also show that monoclinic and/or orthorhombic $PbCr_{1-x}S_xO_4$ ($0 < x \leq 0.8$) varieties discolor more readily than monoclinic S-free $PbCrO_4$, possibly as the result of the highest solubility of $PbCr_{1\text{-}x}S_xO_4$ solid solutions in the binding medium (Monico et al. 2013).

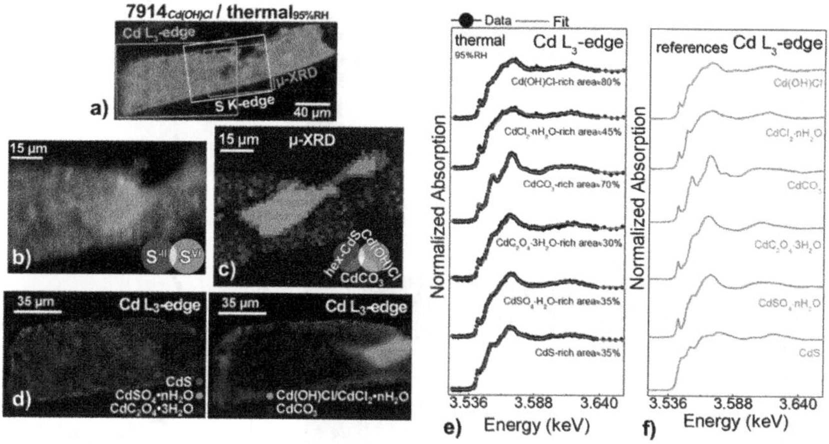

Figure 3.5. *a) Photomicrograph of $7914_{Cd(OH)Cl}$ thin section after thermal aging (RH ≥ 95%, T = 40°C, 90 days). RGB composite b) SR µXRF maps of S^{-II}/S^{VI} (step size: 1 × 1 µm^2, exp. time: 100 ms/pixel) and c) SR µXRD images of hex-CdS/Cd(OH)Cl/$CdCO_3$ (step size: 1.5 × 1.5 µm^2, exp. time: 1 s/pixel, energy: 21 keV) acquired from the regions shown in a). d) Cd-phase maps obtained from the LSLC fit of the FF-XANES stack using the profiles of CdS (red), $CdSO_4{\cdot}nH_2O$/(Cd,Cl)-chlorides (green) and $CdC_2O_4{\cdot}3H_2O/CdCO_3$ (blue). e) Selection of the Cd L3-edge FF-XANES spectra (black) extracted from selected areas of the phase maps shown in d) (number of averaged pixels: 10–20; pixel, sizes: 0.7 × 0.7 µm^2) and corresponding LSLC results (cyan) obtained using the profiles of Cd-references shown in f). Modified from Monico et al. (2020). For a color version of this figure, see www.iste.co.uk/dejoie/synchrotron.zip*

To assess the effects of different organic binding media on the degradation pathways of chrome yellows and any possible correlation between the solubility of the pigment in the binding medium and its (photo)chemical reactivity in the organic binder, recent studies have been performed on a series of paint mock-ups, prepared by employing chrome yellow types of different solubility in water and using either linseed oil and acrylic as binders.

Additional paints made up of the highly water-soluble orthorhombic K_2CrO_4 were also studied and used as benchmarks to provide a more in-depth understanding of the effects of the compound's solubility on the chromate reduction pathways (Monico et al. 2019).

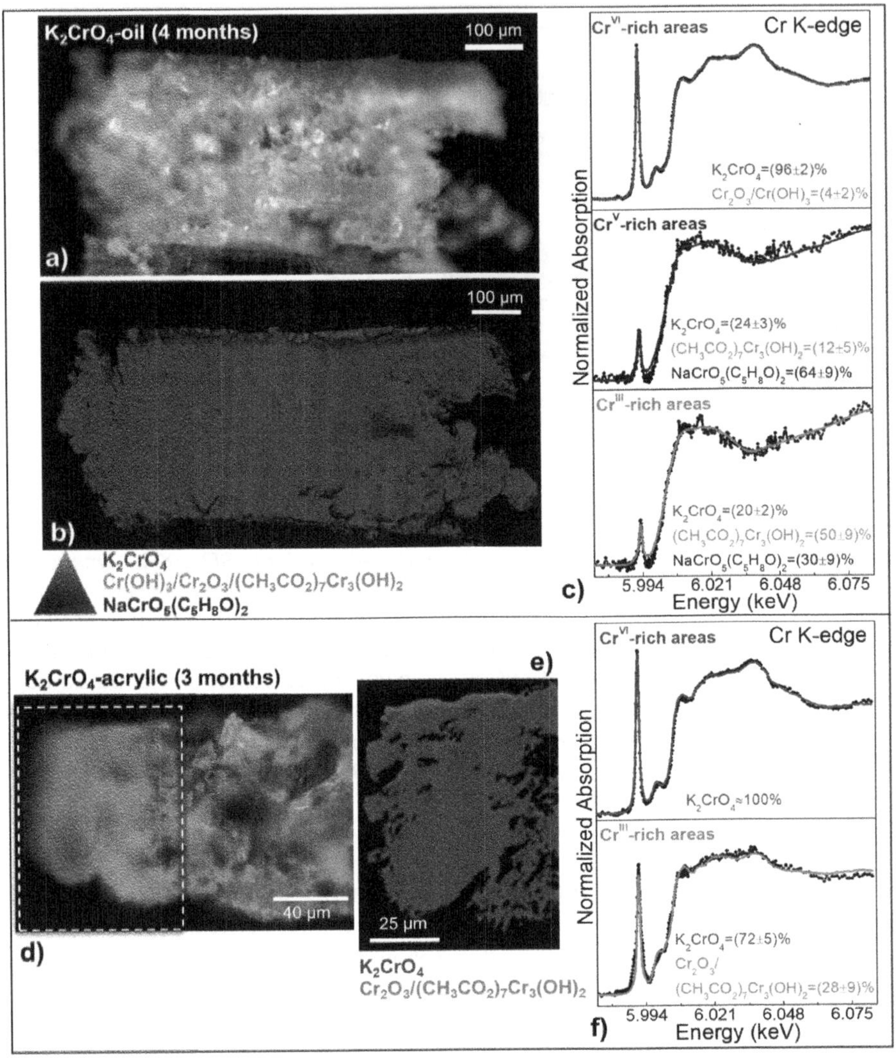

Figure 3.6. *Photomicrographs of thin sections obtained from a naturally aged a) K_2CrO_4-oil paint (four months) and d) K_2CrO_4-acrylic paint (three months). b),e) Cr-phase maps obtained from the LSLC fit of the FF-XANES stack using the spectra of (red) K_2CrO_4, (green) $Cr(OH)_3/Cr_2O_3/(CH_3CO_2)_7Cr_3(OH)_2$ and (blue) $NaCrO_5(C_5H_8O)_2$. c),f) Selection of the Cr K-edge FF-XANES spectra (black) extracted from selected areas of the phase maps shown in b),e) (number of averaged pixels: 7–15; pixel sizes: 0.7 × 0.7 μm^2) and corresponding LSLC results (red, green, blue). Modified from Monico et al. (2019). For a color version of this figure, see www.iste.co.uk/dejoie/synchrotron.zip*

In the context of such research, Cr K-edge FF-XANES imaging experiments in transmission mode were not feasible on chrome yellow samples due to some intrinsic properties of the paint matrix itself, such as the presence of lead and the variable content of Cr (e.g. too low in the case of the $PbCr_{0.2}S_{0.8}O_4$ compound). A series of measurements performed on thin sections of variable thickness (down to around 2 µm; results not reported) did not enable us to obtain FF-XANES datasets with an acceptable SNR of the XANES spectra, especially in the case of the Cr-poorest $PbCr_{0.2}S_{0.8}O_4$ pigment. Therefore, XANES analyses of chrome yellows were feasible only in XRF mode.

However, transmission mode Cr K-edge FF-XANES imaging experiments were successfully carried out on thin sections (around 5 µm thick) of a three months naturally aged K_2CrO_4-oil paint and a three months naturally aged K_2CrO_4-acrylic paint (Monico et al. 2019). As illustrated in the photomicrographs of Figures 3.6a and 3.6b, a brownish layer of around 20 µm in thickness is apparent only at the surface of the K_2CrO_4-oil paint. The corresponding quantitative Cr-phase maps and XANES spectra (Figures 3.6b and 3.6c) show that K_2CrO_4 (red) is homogeneously present as the main component of the yellow paint, where it reaches relative abundance up to around 95%; different Cr^{III}-compounds ($Cr(OH)_3/Cr_2O_3/(CH_3CO_2)_7Cr_3(OH)_2$) (green) and a Cr^{V}-compound ($NaCrO_5(C_5H_8O)_2$) (blue) are instead localized in the uppermost brownish layer, where they achieve relative amounts up to about 50% and 65% respectively. A similar Cr^{VI}- and Cr^{III}-compounds distribution is observable for K_2CrO_4-acrylic paint (Figures 3.6d–3.6f). Nevertheless, the superficial alteration layer is characterized by a thickness (less than 5 µm) and a relative abundance of reduced chromium (between 20 and 30%) which is much smaller than those detected in the K_2CrO_4-oil paint (Figures 3.6a and 3.6b).

To summarize, the results reveal that the chromium reduction is more favored when the yellow powder is mixed with the oil than the acrylic binder and that is more significant at the paint surface. To note is that naturally aged K_2CrO_4 paints show a similar behavior as the equivalent ones prepared using lead chromate-compounds. Such experimental evidences lead to the conclusion that in the same binding medium, the extent of chromate reduction may be positively correlated with the solubility of the pigment. From the methodological point of view, the results indicate that, even though it is not relevant to study lead-based chrome yellow pigments, transmission mode Cr K-edge FF-XANES imaging may find application for further studies on the darkening process of other chromate-based yellow pigments, for which the main constituents are lighter elements than Pb (e.g. zinc yellows, $K_2O \cdot 4ZnCrO_4 \cdot 3H_2O$).

3.4.2. *Combined instrumentation for a representative multi-scale characterization: the example of estimating firing conditions in ancient Roman ceramics*

Ceramics and more generally artifacts obtained by firing techniques have been extensively studied by XAS techniques. Indeed, most of the optical effects in glass and ceramics (color, opacity, luster effect, etc.) are obtained by mastering the redox state of key elements (most often transition metals). The determination of the oxidation state of these elements and their distribution can be exploited to understand and model the optical effects and to provide information regarding their manufacturing conditions (choice of ingredients, successive temperatures and atmosphere, etc.) (Rapantzikos and Balas 2005). The ability of XANES measurements to yield information on the local structure of an element of interest even when the material has completely vitrified into a glassy matrix provides a very good complement to other investigations, such as XRD and Raman spectroscopy. Many XAS experiments are carried out with millimeter, or sub-millimeter X-ray probes, giving averaged bulk information. However, firing conditions are often not suitable for the formation of large crystals, being more favorable to give rise to the production of micrometer and nanometer crystallites (Sciau and Goudeau 2015). Micrometric beams have been used, as, for example, for the spatial discrimination of crystals and glass in Egyptian and Roman glasses opacified with calcium antimonite crystals (Lahlil et al. 2010, 2011) or for the study of Mn speciation over brown-black Mn-rich corrosion bodies, before and after conservation treatment (Cagno et al. 2011).

Fe K-edge μXANES and μXRF techniques have been applied to study the local mineralogy and to map the Fe oxidation state of remarkable roman vessels in order to understand the firing technology used for relative production (Sciau et al. 2011). In this context, the energies at 7,122 and 7,134 eV were selected to give the highest contrast between Fe^{2+} and Fe^{3+} respectively. However, this mapping strategy does not allow for differentiating different Fe^{3+} phases (e.g. hematite and smectite) and only single-point μXANES investigations permitted to obtain specific information on the nature of such compounds. Consequently, FF-XANES rapidly appeared as a natural progression to further improve the spectro–imaging combination for the characterization of these materials at the micrometric scale (Sciau and Wang 2019).

Thus, two complementary FF-XANES instruments were used to get new insights into possible differences in the manufacturing of two types of Roman black gloss ceramics: a B-type Campanian ceramic (with black-gloss coating and a non-vitrified body, 1st century BCE) (Figure 2.7a) and a pre-sigillata ceramic (with a black gloss slip on a non-vitrified red/brown paste, from 10 BCE to 10 CE). By probing the different oxidation state of iron species, it was expected to obtain information on temperature and oxygen fugacity fluctuations and variations during the firing processes (Meirer et al. 2013). As these changes are often heterogeneously distributed and frequently confined to the near surface regions and to the small interfacial region between the slip and the body of the ceramics, their characterization necessitates a probe combining sub-micrometer resolution together with millimeter FOV. The zone plate-based SSRL TXM instrument (with an FOV of 30 μm^2 and a lateral resolution of 30 nm) allowed us to obtain iron distribution and speciation at the interface region with nanometer resolution, along with porosity and nanostructure mapping. The lens-less FF microscope at ID21 (with an FOV from 648 × 768 μm^2 μm and a pixel size of 0.3 × 0.3 μm^2) allowed us to record a dataset with increased FOV and SNR of Fe K-edge XANES spectra.

Two sherds, one from each production, were sampled. Sample sections were prepared by following the protocol proposed in (Meirer et al. 2013). It is important to note that the first step of the preparation consists of gluing two pieces of the same ceramics face to face. Data are therefore acquired simultaneously over the two pieces of ceramics.

The Campanian sherd was first analyzed at SSRL. The results presented in Figure 3.7b show a distinct phase composition with mainly hercynite ($FeAl_2O_4$, black phase, Fe^{2+}) in the slip and maghemite (γ-Fe_2O_3, red-brown phase, Fe^{3+}) in the body. The same sample was further investigated by performing X-ray tomography in order to map microcracks and porosity distribution in the slip and in the body. The superposition of spectral and morphological results suggests that the oxidation from Fe^{2+} to Fe^{3+} from the body of the ceramic to the slip has occurred. The combined information on Fe-speciation and ceramic porosity shows possible pathways for fast transport of oxygen through microcracks and open micropores to oxidize iron in the ceramic next to these channels, quickly converting Fe minerals from hercynite to maghemite, and then slowly diffusing through the denser slip to push the re-oxidation front deeper into the slip. With lateral resolution 10 times lower than SSRL instrument, the lens-less ID21 instrument allows acquiring complementary information thanks to a different geometry and a larger field of view (at least 20 times higher) increasing SNR XANES values and statistic relevance.

Despite the lower spatial resolution, both the high concentration of hercynite in the slip and the predominant presence of maghemite in the body were confirmed (Figure 3.7c). The increased FOV with a larger pixel size and improved mineralogical specificity allow us to detect the very heterogeneous and shallow oxidation of the surface of the slip (presence of a thin superficial layer of maghemite). Furthermore, it was possible to note the presence of small islands of hematite in the body (Figure 3.7c).

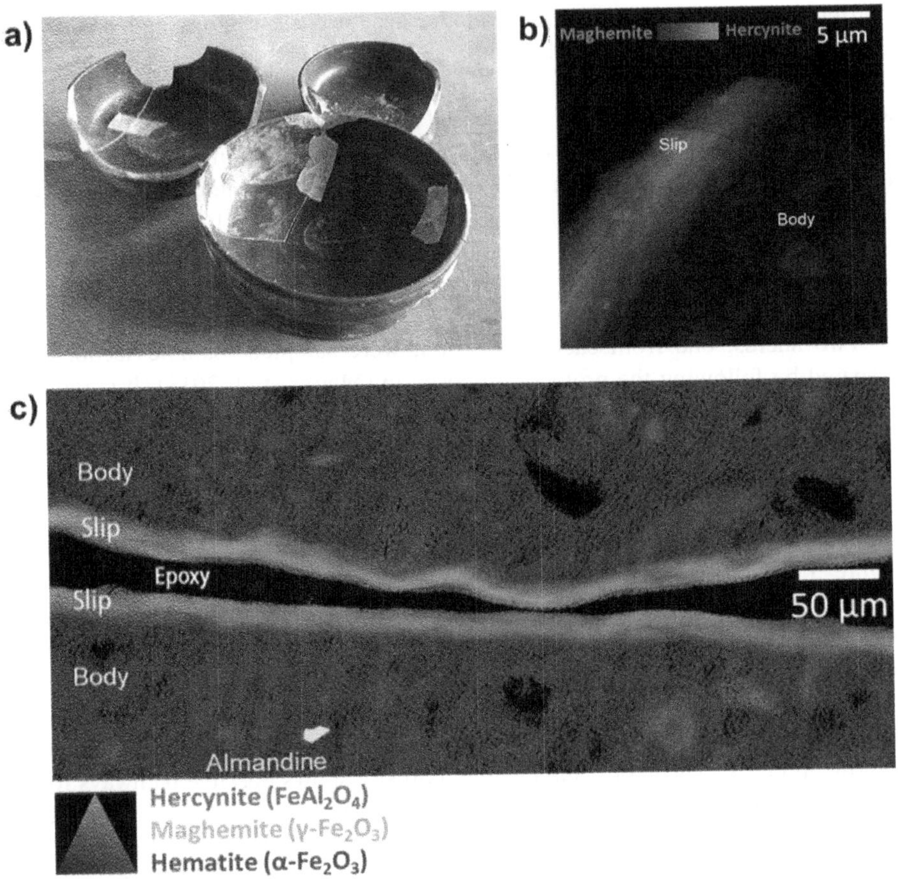

Figure 3.7. *FF-XANES of the thin section of a fragment from the Campanian sample. a) Campanian ceramic; b) phase map obtained from processing of data acquired at BL6.2 (SSRL) Hercynite (green), maghemite (blue); c) phase map obtained from processing of data acquired at ID21 (ESRF) Hercynite (green), maghemite (blue) and hematite (red), almandine (yellow) (Meirer et al. 2013). For a color version of this figure, see www.iste.co.uk/dejoie/synchrotron.zip*

Moreover, thanks to the extremely large number of acquired XANES, the LCLS fit results reveal the presence of an additional iron phase in the body. Indeed, an area of the sample presented unsatisfactory fitting results in comparison to the experimental spectrum, when considering only hercynite, maghemite and hematite. Consequently, the set of these three references did not represent all of the iron phases present in the cluster. By adding the spectrum of almandine ($Fe_3Al_2(SiO_4)_3$, Fe^{2+}) in the previous dataset, fit results showed an improved match with the sum spectrum from the cluster. Consequently, a new iron phase was identified which helps in determining firing temperature and duration conditions for this specific sherd. In conclusion, the distribution of hercynite (black, Fe^{2+}), hematite and maghemite (both red-brown, Fe^{3+}) achieved with a large FOV explained the visible colors of the ceramic and presented a very pictorial evidence that this vessel was once reduced and then partially re-oxidized. Moreover, chemical differences between the two ceramics allow us to establish assumption on different firing conditions. Indeed, the thinner surface layer of maghemite on the slip, its larger amount and the presence of patches of un-oxidized Fe^{2+} minerals in the body of the Campanian sherd suggest that the final re-oxidation step was performed under either lower oxygen fugacity or even more likely at significantly lower temperature and/or short duration in comparison with the pre-sigillata sherd.

More broadly, this example demonstrates the benefits of combining nano and micro FF-XANES, i.e. large FOV (i.e. increased sample heterogeneity probed) with sub-micrometric pixel size, to correlate the macroscopic properties of an object to its intrinsic structure.

3.5. Conclusion

In this work, SR-based FF-XANES in transmission mode allows for 2D imaging at the micrometric scale combined with full XANES spectral information. A full dataset is typically recorded on millimetric areas with (sub)micrometric resolution within a few tenth of minutes (depending on the energy resolution needed and the composition of the sample studied). Through the investigations of large areas of the sample and the extraction of XANES profiles from all locations in the examined region, XANES imaging contributes to deepening our knowledge on the nature and distribution of compounds of interest and extends the possibility of the point-by-point approach.

As illustrated by the examples described in this chapter, this method was successfully used for semi-quantitative speciation analysis of element of interest present in a wide range of artworks (i.e. painting, ink and ceramic) with micrometer lateral resolution. In combination with other conventional spectroscopy techniques, this approach has proven essential in providing supplementary spatially resolved chemical information that is crucial to answering questions about artistic processes and degradation mechanisms.

Moreover, as illustrated by the roman ceramic case study, the large FOV can also be exploited to analyze several samples simultaneously, providing that samples are mounted closely. This method is consequently fully applicable to a large corpus of heterogeneous samples, a condition often encountered when dealing with archeological problematic. The combination of high-resolution/small FOV TXM and low-resolution/large-FOV full-field or dispersive instruments makes it possible to perform multi-scale analyses by combining millimetric field of view with nanometer resolution and ends into relevant information at the object scale.

From a technical point of view, the main limitation of this approach is that, as all other transmission approaches, this technique is limited to samples presenting element of interest in sufficiently high concentration with respect to a matrix with relatively low absorption in the energy domain of interest and may necessitate thin section sample preparation. If those conditions cannot be fulfilled, performing XRF mode FS-XANES mapping should be considered. With the development of fourth generation synchrotron, of larger active area fluorescence detector and fast electronics, it is expected that this acquisition approach will soon represent a standard strategy for CH samples, usually prepared and mounted as thick cross-sections. To go further into the details, X-ray spectro-ptychography (ptychographic-X-ray absorption fine structure, XAFS) recently opened up nanoscale chemical imaging and structural analysis for heterogeneous materials that can benefit the CH community (Pattammattel et al. 2020).

3.6. Acknowledgments

The European Synchrotron Radiation Facility is acknowledged for providing beamtimes at the ID21 beamline (HG-64, HG-95 and in-house beamtimes). Cadmium yellow degradation work on Matisse paintings is supported by: the Andrew W. Mellon Foundation, the Barnes Foundation, the Lenfest Foundation and the National Science Foundation DMR 0415838. The research on cadmium yellow and chromate yellow mock-ups was financially supported by the European research project IPERION-CH (funded by the European Commission, H2020-INFRAIA-2014–2015; grant agreement no. 654028), the project AMIS (program Dipartimenti di Eccellenza 2018–2022, funded by MIUR and University of Perugia) and the program "Ricerca di Base 2017" (funded by University of Perugia). We would also like to express our thanks to DESY-P06 beamline, a member of the Helmholtz Association HGF (experiment nos. I-20130221 EC and I-20160126 EC), and the project CALIPSOplus (Grant Agreement no. 730872, EU Framework Programme for Research and Innovation HORIZON 2020) for the grants received for performing μXRD investigations. Barbara Buckley of the Barnes Foundation and Unn Plahter of the University of Oslo are thanked for their many helpful discussions. Corinne Sanchez of "Archéologie des Sociétés Méditerranéennes" (Lattes, France) is also thanked for the ceramic samples.

3.7. References

Andrews, J.C., Almeida, E., van der Meulen, M.C.H., Alwood, J.S., Lee, C., Liu, Y., Chen, J., Meirer, F., Feser, M., Gelb, J. et al. (2010). Nanoscale X-ray microscopic imaging of mammalian mineralized tissue. *Microscopy and Microanalysis*, 16, 327–336.

Bauer, S., de Biasi, L., Glatthaar, S., Toukam, L., Geßwein, H., Baumbach, T. (2015). In operando study of the high voltage spinel cathode material $LiNi_{0.5}Mn_{1.5}O_4$ using two dimensional full-field spectroscopic imaging of Ni and Mn. *Physical Chemistry Chemical Physics*, 17(25), 16388–16397.

Bertrand, L., Cotte, M., Stampanoni, M., Thoury, M., Marone, F., Schöder, S. (2012). Development and trends in synchrotron studies of ancient and historical materials. *Physics Reports*, 519(2), 51–96.

Bertrand, L., Schöeder, S., Anglos, D., Breese, M.B., Janssens, K., Moini, M., Simon, A. (2015). Mitigation strategies for radiation damage in the analysis of ancient materials. *TrAC Trends in Analytical Chemistry*, 66, 128–145.

Brill, R.H. and Cahill, N.D. (1988). A red opaque glass from Sardis and some thoughts on red opaques in general. *Journal of Glass Studies*, 30, 16–27.

Brinza, L., Schofield, P.F., Hodson, M.E., Weller, S., Ignatyev, K., Geraki, K., Quinn, P.D., Mosselmans, J.F.W. (2014). Combining μXANES and μXRD mapping to analyse the heterogeneity in calcium carbonate granules excreted by the earthworm *Lumbricus terrestris*. *Journal of Synchrotron Radiation*, 21(1), 235–241.

Cagno, S., Nuyts, G., Bugani, S., De Vis, K., Schalm, O., Caen, J., Helfen, L., Cotte, M., Reischig, P., Janssens, K. (2011). Evaluation of manganese-bodies removal in historical stained glass windows via SR-μ-XANES/XRF and SR-μ-CT. *Journal of Analytical Atomic Spectrometry*, 26, 2442–2451.

Calvin, S. (2013). *XAFS for Everyone*. CRC Press, Boca Raton.

Cotte, M., Susini, J., Metrich, N., Moscato, A., Gratziu, C., Bertagnini, A., Pagano, M. (2006). Blackening of Pompeian cinnabar paintings: X-ray microspectroscopy analysis. *Analytical Chemistry*, 78(21), 7484–7492.

Cotte, M., Susini, J., Dik, J., Janssens, K. (2010). Synchrotron-based X-ray absorption spectroscopy for art conservation: Looking back and looking forwards. *Accounts of Chemical Research*, 43(6), 705–714.

Cotte, M., Pouyet, E., Salomé, M., Rivard, C., De Nolf, W., Castillo-Michel, H., Fabris, T., Monico, L., Janssens, K., Wang, T. (2017). The ID21 X-ray and infrared microscopy beamline at the ESRF: Status and recent applications to artistic materials. *Journal of Analytical Atomic Spectrometry*, 32(3), 477–493.

De Andrade, V., Susini, J., Salomé, M., Beraldin, O., Rigault, C., Heymes, T., Lewin, E., Vidal, O. (2011). Submicrometer hyperspectral X-ray imaging of heterogeneous rocks and geomaterials: Applications at the Fe K-edge. *Analytical Chemistry*, 83(11), 4220–4227.

De Andrade, V., Deriy, A., Wojcik, M.J., Gürsoy, D., Shu, D., Fezzaa, K., De Carlo, F. (2016). Nanoscale 3D imaging at the advanced photon source. *SPIE Newsroom*, 10(2.1201604), 006461.

Fayard, B., Pouyet, E., Berruyer, G., Bugnazet, D., Cornu, C., Cotte, M., Andrade, V.D., Chiaro, F.D., Hignette, O., Kieffer, J. et al. (2013). The new ID21 XANES full-field end-station at ESRF. *Journal of Physics: Conference Series*, 425, 192001.

Freestone, I.C. (1987). Composition and microstructure of early opaque red glass. Early vitreous materials. *British Museum*, 173–191 [Online]. Available at: https://www.researchgate.net/publication/285870112_Composition_and_microstructure_of_early_opaque_red_glass.

Gambardella, A.A., Cotte, M., de Nolf, W., Schnetz, K., Erdmann, R., van Elsas, R., Gonzalez, V., Wallert, A., Iedema, P.D., Eveno, M. (2020). Sulfur K-edge micro-and full-field XANES identify marker for preparation method of ultramarine pigment from lapis lazuli in historical paints. *Science Advances*, 6(18), eaay8782.

Ganio, M., Pouyet, E.S., Webb, S.M., Schmidt Patterson, C.M., Walton, M.S. (2018). From lapis lazuli to ultramarine blue: Investigating Cennino Cennini's recipe using sulfur K-edge XANES. *Pure and Applied Chemistry*, 90(3), 463–475.

Geng, B. (1993). *Identification of Porcelain of Ming and Qing Dynasties*. Zi-Jin-Cheng Publishing House and Liang-Mu Publishing House, Beijing.

Gervais, C., Languille, M.-A., Reguer, S., Gillet, M., Pelletier, S., Garnier, C., Vicenzi, E.P., Bertrand, L. (2013a). Why does Prussian blue fade? Understanding the role(s) of the substrate. *Journal of Analytical Atomic Spectrometry*, 28(10), 1600–1609.

Gervais, C., Languille, M.A., Reguer, S., Gillet, M., Vicenzi, E.P., Chagnot, S., Baudelet, F., Bertrand, L. (2013b). "Live" Prussian blue fading by time-resolved X-ray absorption spectroscopy. *Applied Physics A*, 111(1), 15–22.

Gervais, C., Languille, M.A., Moretti, G., Reguer, S. (2015). X-ray photochemistry of Prussian blue cellulosic materials: Evidence for a substrate-mediated redox process. *Langmuir*, 31(29), 8168–8175.

Gimat, A., Michelin, A., Belhadj, O., Pellizzi, E., Massiani, P., Rouchon, V. (2021). Paper sizing with gelatine: From the macro-to the nano-scale. *Cellulose*, 28, 2419–2432.

Heinzer, S., Krucker, T., Stampanoni, M., Abela, R., Meyer, E.P., Schuler, A., Schneider, P., Müller, R. (2006). Hierarchical microimaging for multiscale analysis of large vascular networks. *NeuroImage*, 32(2), 626–636.

Kanngiesser, B., Hahn, O., Wilke, M., Nekat, B., Malzer, W., Erko, A. (2004). Investigation of oxidation and migration processes of inorganic compounds in ink-corroded manuscripts. *Spectrochimica Acta B*, 59, 1511–1516.

Karen Chen-Wiegart, Y.-C., Harris, W.M., Lombardo, J.J., Chiu, W.K., Wang, J. (2012). Oxidation states study of nickel in solid oxide fuel cell anode using X-ray full-field spectroscopic nano-tomography. *Applied Physics Letters*, 101(25), 253901.

Lahlil, S., Biron, I., Cotte, M., Susini, J., Menguy, N. (2010). Synthesis of calcium antimonate nano-crystals by the 18th dynasty Egyptian glassmakers. *Applied Physics A*, 98(1), 1–8.

Lahlil, S., Cotte, M., Biron, I., Szlachetko, J., Menguy, N., Susini, J. (2011). Synthesizing lead antimonate in ancient and modern opaque glass. *Journal of Analytical Atomic Spectrometry*, 26(5), 1040–1050.

Larson, D.J., Prosa, T.J., Ulfig, R.M., Geiser, B.P., Kelly, T.F. (2013). *Local Electrode Atom Probe Tomography*. Springer, New York.

Leone, B., Burnstock, A., Jones, C., Hallebeek, P., Boon, J., Keune, K. (2005). The deterioration of cadmium sulphide yellow artists' pigments. In *ICOM Committee for Conservation 14th Triennial Meeting*, The Hague, 12–16 September. James & James, London, and Earthscan, Oxford.

Liu, Y., Meirer, F., Williams, P.A., Wang, J., Andrews, J.C., Pianetta, P. (2012). TXM-Wizard: A program for advanced data collection and evaluation in full-field transmission X-ray microscopy. *Journal of Synchrotron Radiation*, 19(2), 281–287.

Lu, Y. (2014). *A History of Chinese Science and Technology*. Springer, Berlin, Heidelberg.

Mangold, S., Steininger, R., dos Santos Rolo, T., Göttlicher, J. (2013). Full field spectroscopic imaging at the ANKA-XAS-and-SUL-X-Beamlines. *Journal of Physics: Conference Series*, 430(1), 012130.

Mass, J., Sedlmair, J., Patterson, C.S., Carson, D., Buckley, B., Hirschmugl, C. (2013a). SR-FTIR imaging of the altered cadmium sulfide yellow paints in Henri Matisse's Le Bonheur de vivre (1905–6) – Examination of visually distinct degradation regions. *Analyst*, 138(20), 6032–6043.

Mass, J.L., Opila, R., Buckley, B., Cotte, M., Church, J., Mehta, A. (2013b). The photodegradation of cadmium yellow paints in Henri Matisse's Le Bonheur de vivre (1905–1906). *Applied Physics A*, 111(1), 59–68.

Mazel, V., Richardin, P., Debois, D., Touboul, D., Cotte, M., Brunelle, A., Walter, P., Laprévote, O. (2007). Identification of ritual blood in African artifacts using TOF-SIMS and synchrotron radiation microspectroscopies. *Analytical Chemistry*, 79, 9253–9260.

Meirer, F., Cabana, J., Liu, Y., Mehta, A., Andrews, J.C., Pianetta, P. (2011). Three-dimensional imaging of chemical phase transformations at the nanoscale with full-field transmission X-ray microscopy. *Journal of Synchrotron Radiation*, 18(5), 773–781.

Meirer, F., Liu, Y., Pouyet, E., Fayard, B., Cotte, M., Sanchez, C., Andrews, J.C., Mehta, A., Sciau, P. (2013). Full-field XANES analysis of Roman ceramics to estimate firing conditions – A novel probe to study hierarchical heterogeneous materials. *Journal of Analytical Atomic Spectrometry*, 28(12), 1870–1883.

Michelin, A., Drouet, E., Foy, E., Dynes, J., Neff, D., Dillmann, P. (2013). Investigation at the nanometre scale on the corrosion mechanisms of archaeological ferrous artefacts by STXM. *Journal of Analytical Atomic Spectrometry*, 28(1), 59–66.

Miliani, C., Monico, L., Melo, M.J., Fantacci, S., Angelin, E.M., Romani, A., Janssens, K. (2018). Photochemistry of artists' dyes and pigments: Towards better understanding and prevention of colour change in works of art. *Angewandte Chemie International Edition*, 57(25), 7324–7334.

Monico, L., Janssens, K.H., Miliani, C., Brunetti, B.G., Vagnini, M., Vanmeert, F., Falkenberg, G., Abakumov, A.M., Lu, Y., Tian, H. et al. (2013a). The degradation process of lead chromate in paintings by Vincent van Gogh studied by means of spectromicroscopic methods. Part III: Synthesis, characterization and detection of different crystal forms of the chrome yellow pigment. *Analytical Chemistry*, 85(2), 851–859.

Monico, L., Janssens, K., Miliani, C., Van der Snickt, G., Brunetti, B.G., Cestelli Guidi, M., Radepont, M., Cotte, M. (2013b). The degradation process of lead chromate in paintings by Vincent van Gogh studied by means of spectromicroscopic methods. Part IV: Artificial ageing of model samples of co-precipitates of lead chromate and lead sulfate. *Analytical Chemistry*, 85(2), 860–867.

Monico, L., Janssens, K., Alfeld, M., Cotte, M., Vanmeert, F., Ryan, C.G., Falkenberg, G., Howard, D.L., Brunetti, B.G., Miliani, C. (2015a). Full spectral XANES imaging using the Maia detector array as a new tool for the study of the alteration process of chrome yellow pigments in paintings by Vincent van Gogh. *Journal of Analytical Atomic Spectrometry*, 30(3), 613–626.

Monico, L., Janssens, K., Cotte, M., Romani, A., Sorace, L., Grazia, C., Brunetti, B.G., Miliani, C. (2015b). Synchrotron-based X-ray spectromicroscopy and electron paramagnetic resonance spectroscopy to investigate the redox properties of lead chromate pigments under the effect of visible light. *Journal of Analytical Atomic Spectrometry*, 30(7), 1500–1510.

Monico, L., Janssens, K., Hendriks, E., Vanmeert, F., Van der Snickt, G., Cotte, M., Falkenberg, G., Brunetti, B.G., Miliani, C. (2015c). Evidence for degradation of the chrome yellows in Van Gogh's sunflowers: A study using noninvasive in situ methods and synchrotron-radiation-based X-ray techniques. *Angewandte Chemie*, 127(47), 14129–14133.

Monico, L., Janssens, K., Cotte, M., Sorace, L., Vanmeert, F., Brunetti, B.G., Miliani, C. (2016). Chromium speciation methods and infrared spectroscopy for studying the chemical reactivity of lead chromate-based pigments in oil medium. *Microchemical Journal*, 124, 272–282.

Monico, L., Chieli, A., De Meyer, S., Cotte, M., de Nolf, W., Falkenberg, G., Janssens, K., Romani, A., Miliani, C. (2018). Role of the relative humidity and the Cd/Zn stoichiometry in the photooxidation process of cadmium yellows ($CdS/Cd_{1-x}Zn_xS$) in oil paintings. *Chemistry – A European Journal*, 24(45), 11584–11593.

Monico, L., Sorace, L., Cotte, M., de Nolf, W., Janssens, K., Romani, A., Miliani, C. (2019). Disclosing the binding medium effects and the pigment solubility in the (photo) reduction process of chrome yellows ($PbCrO_4/PbCr_{1-x}S_xO_4$). *ACS Omega*, 4(4), 6607–6619.

Monico, L., Cartechini, L., Rosi, F., Chieli, A., Grazia, C., De Meyer, S., Nuyts, G., Vanmeert, F., Janssens, K., Cotte, M. (2020a). Probing the chemistry of CdS paints in The Scream by in situ noninvasive spectroscopies and synchrotron radiation X-ray techniques. *Science Advances*, 6(20), eaay3514.

Monico, L., Cartechini, L., Rosi, F., De Nolf, W., Cotte, M., Vivani, R., Maurich, C., Miliani, C. (2020b). Synchrotron radiation Ca K-edge 2D-XANES spectroscopy for studying the stratigraphic distribution of calcium-based consolidants applied in limestones. *Scientific Reports*, 10(1), 1–14.

Monico, L., Cotte, M., Vanmeert, F., Amidani, L., Janssens, K., Nuyts, G., Garrevoet, J., Falkenberg, G., Glatzel, P., Romani, A. et al. (2020c). Damages induced by synchrotron radiation-based X-ray microanalysis in chrome yellow paints and related Cr-compounds: Assessment, quantification, and mitigation strategies. *Analytical Chemistry*, 92(20), 14164–14173.

Pattammattel, A., Tappero, R., Ge, M., Chu, Y., Huang, X., Gao, Y., Yan, H. (2020). High-sensitivity nanoscale chemical imaging with hard x-ray nano-XANES. *Science Advances*, 6(37), eabb3615.

Plahter, U. and Topalova-Casadiego, B. (2011). The scream by Edvard Munch: Painting techniques and colouring materials, in studying old master paintings technology and practice. In *The National Gallery Technical Bulletin 30th Anniversary Conference Postprints*, Spring, M. (ed.). Archetype, London.

Pouyet, E., Cotte, M., Fayard, B., Salomé, M., Meirer, F., Mehta, A., Uffelman, E., Hull, A., Vanmeert, F., Kieffer, J. (2015a). 2D X-ray and FTIR micro-analysis of the degradation of cadmium yellow pigment in paintings of Henri Matisse. *Applied Physics A*, 121(3), 967–980.

Pouyet, E., Fayard, B., Salomé, M., Taniguchi, Y., Sette, F., Cotte, M. (2015b). Thin-sections of painting fragments: Opportunities for combined synchrotron-based micro-spectroscopic techniques. *Heritage Science*, 3(1), 3.

Radepont, M., de Nolf, W., Janssens, K., Van der Snickt, G., Coquinot, Y., Klaassen, L., Cotte, M. (2011). The use of microscopic X-ray diffraction for the study of HgS and its degradation products corderoite (α-$Hg_3S_2Cl_2$), kenhsuite (γ-$Hg_3S_2Cl_2$) and calomel (Hg_2Cl_2) in historical paintings. *Journal of Analytical Atomic Spectrometry*, 26(5), 959–968.

Rapantzikos, K. and Balas, C. (2005). Hyperspectral imaging: Potential in non-destructive analysis of palimpsests. In *IEEE International Conference on Image Processing 2005*, Genoa, II–618.

Ryan, C.G., Siddons, D.P., Kirkham, R., Dunn, P.A., Kuczewski, A., Moorhead, G., Geronimo, G.D., Paterson, D.J., de Jonge, M.D., Hough, R.M. et al. (2010). The new Maia detector system: Methods for high definition trace element imaging of natural material. *AIP Conference Proceedings*, 1221(1), 9–17.

Ryan, C., Kirkham, R., De Jonge, M., Siddons, D.P., van der Ent, A., Pages, A., Boesenberg, U., Kuczewski, A., Dunn, P., Jensen, M. (2018). The Maia detector and event mode. *Synchrotron Radiation News*, 31(6), 21–27.

Sakurai, K., Iida, A., Takahashi, M., Gohshi, Y. (1988). Chemical state mapping by X-ray fluorescence using absorption edge shifts. *Japanese Journal of Applied Physics*, 27(9A), L1768.

Salomé, M., Cotte, M., Baker, R., Barrett, R., Benseny-Cases, N., Berruyer, G., Bugnazet, D., Castillo-Michel, H., Cornu, C., Fayard, B. et al. (2013). The ID21 scanning X-ray microscope at ESRF. *Journal of Physics: Conference Series*, 425(19), 192001.

Samain, L., Silversmit, G., Sanyova, J., Vekemans, B., Salomon, H., Gilbert, B., Grandjean, F., Long, G.J., Hermann, R.P., Vincze, L. et al. (2011). Fading of modern Prussian blue pigments in linseed oil medium. *Journal of Analytical Atomic Spectrometry*, 26(5), 930–941.

Sciau, P. and Goudeau, P. (2015). Ceramics in art and archaeology: A review of the materials science aspects. *The European Physical Journal B*, 88(5), 1–11.

Sciau, P. and Wang, T. (2019). Full-field transmission X-ray microspectroscopy (FF-XANES) applied to cultural heritage materials: The case of ancient ceramics. In *Synchrotron Radiation: Useful and Interesting Applications*, Joseph, D. (ed.). IntechOpen, London.

Sciau, P., Leon, Y., Goudeau, P., Fakra, S.C., Webb, S., Mehta, A. (2011). Reverse engineering the ancient ceramic technology based on X-ray fluorescence spectromicroscopy. *Journal of Analytical Atomic Spectrometry*, 26(5), 969–976.

Sole, V.A., Papillon, E., Cotte, M., Walter, P., Susini, J. (2007). A multiplatform code for the analysis of energy-dispersive X-ray fluorescence spectra. *Spectrochimica Acta – Part B Atomic Spectroscopy*, 62(1), 63–68.

Tajima, S., Asahi, R., Isheim, D., Seidman, D., Itoh, T., Hasegawa, M., Ohishi, K. (2014). Atom-probe tomographic study of interfaces of Cu2ZnSnS4 photovoltaic cells. *Applied Physics Letters*, 105(9), 093901.

Van der Snickt, G., Dik, J., Cotte, M., Janssens, K., Jaroszewicz, J., De Nolf, W., Groenewegen, J., Van der Loeff, L. (2009). Characterization of a degraded cadmium yellow (CdS) pigment in an oil painting by means of synchrotron radiation based X-ray techniques. *Analytical Chemistry*, 81(7), 2600–2610.

Van der Snickt, G., Janssens, K., Dik, J., De Nolf, W., Vanmeert, F., Jaroszewicz, J., Cotte, M., Falkenberg, G., Van der Loeff, L. (2012). Combined use of synchrotron radiation based micro-X-ray fluorescence, micro-X-ray diffraction, micro-X-ray absorption near-edge, and micro-Fourier transform infrared spectroscopies for revealing an alternative degradation pathway of the pigment cadmium yellow in a painting by Van Gogh. *Analytical Chemistry*, 84(23), 10221–10228.

Vanmeert, F., Hendriks, E., Van der Snickt, G., Monico, L., Dik, J., Janssens, K. (2018). Chemical mapping by macroscopic X-ray powder diffraction (MA-XRPD) of Van Gogh's sunflowers: Identification of areas with higher degradation risk. *Angewandte Chemie International Edition*, 57(25), 7418–7422.

Wang, T., Zhu, T.Q., Feng, Z.Y., Fayard, B., Pouyet, E., Cotte, M., De Nolf, W., Salome, M., Sciau, P. (2016). Synchrotron radiation-based multi-analytical approach for studying underglaze color: The microstructure of Chinese Qinghua blue decors (Ming dynasty). *Analytica Chimica Acta*, 928, 20–31.

Wilke, M., Hahn, O., Woodland, A.B., Rickers, K. (2009). The oxidation state of iron determined by Fe K-edge XANES-application to iron gall ink in historical manuscripts. *Journal of Analytical Atomic Spectrometry*, 24(10), 1364–1372.

4

Structural Cartography and Tomography by Diffraction/ Diffusion: A Local Selective Analysis of Cultural Heritage Materials

Jean-Louis HODEAU[1], Michèle ALVAREZ-MURGA[1], Michel ANNE[1], Pierre-Olivier AUTRAN[2], Nils BLANC[1], Pierre BLEUET[3], Nathalie BOUDET[1], Pierre BORDET[1], Sophie CERSOY[4], Catherine DEJOIE[2], Eric DOORYHÉE[1], Florian KERGOULAY[1], Olivier LEYNAUD[1], Pauline MARTINETTO[1], Alain PRAT[1] and Philippe WALTER[5]

[1] *Institut Néel CNRS/UGA, Grenoble, France*

[2] *European Synchrotron Radiation Facility, Grenoble, France*

[3] *CEA-LETI, Grenoble, France*

[4] *Centre de Recherche sur la Conservation, Paris, France*

[5] *Laboratoire d'archéologie moléculaire structurale, Sorbonne University, Paris, France*

Cultural heritage materials are often extremely heterogeneous: they consist of a complex assembly of different phases (organic, mineral, etc.) in very varied crystalline states (amorphous, powders, polycrystals, etc.). Knowledge of these phases and their spatial distribution is important to understand and compare cultural artifacts. Here, we illustrate with some examples, methods of 2D and 3D mappings that use chemical and structural information to image them, in order to easily extract this information. The "diffraction-tomography" method is based on the computer processing of a large number of measurements obtained by illuminating a sample with a thin

Synchrotron Radiation, Cultural Heritage, Biomineralization, coordinated by Catherine DEJOIE, Pauline MARTINETTO and Nobumichi TAMURA.

X-ray beam from multiple angles. These structural mappings/tomographies use X-rays for single- or multi-method measurements (diffraction/scattering/fluorescence/absorption), the variations of which make it possible to construct structural and/or chemical images of the samples/artifacts studied. Through recombination, which uses signal differences, this tomography improves the sensitivity of the measurements and makes it possible to extract very weak signals coming from minor phases. This selectivity is essential for heritage materials because very often their specificities/originalities are linked to preparations involving slight variations in its components (pigments, microstructures, amorphous phases, binders, etc.). The non-destructive characteristics of diffraction mapping/tomography experiments are important for material science and especially for cultural heritage samples, which must be studied non-destructively. In this chapter, we illustrate this with some case studies of heritage artifacts, both ancient and recent, published in the literature. The chemical and mineral maps obtained describe the characteristics of these artifacts to differentiate the different structural phases, as well as the granularities and textures, information that can be linked to the artist's know-how.

4.1. Introduction

The study of cultural heritage materials is an interesting challenge for crystallography and materials science. Indeed, their specificities, which are both precious and unique, mean that their study requires the use of analytical methods that are, if possible, non-invasive (i.e. without sample removal), or at least non-destructive (i.e. the samples taken must be preserved). Furthermore, heritage materials are archetypes of heterogeneous materials, where often only a few minor components are essential to express their nature and characteristics. For example, studies of polychrome artifacts not only show that they are composed of varied mixtures of different pigments but also show that they have various chemical compositions, grain sizes, overlapping of crystallized or non-crystallized layers, varnishes, as well as phases modified by time.

Chemical information and information on short-range structural environments can be obtained by various fluorescence or absorption spectroscopies, which are non-invasive methods with very high chemical sensitivity (Solé et al. 2007; Cotte et al. 2010; Alfeld and Janssens 2015). An example of this is the study of the degradation process of lead chromate, giving rise to a darkening of the original yellow paint in "Sunflowers" by Vincent van Gogh. This study was carried out using fluorescence (XRF) and absorption (XANES) synchrotron radiation spectroscopies and by some complementary methods, both on artificially aged model samples and on original samples of layers of this painting (Monico et al. 2011a, 2011b). Such chemical information can tell us about the origins, preparations and how pigments were used, i.e. the artist's working methods. Another example of this type of study is chemical mapping, using synchrotron radiation X-ray fluorescence to visualize covered paintings. It revealed, for instance, the head of a woman in the painting "Patch of Grass" by Vincent van Gogh (Dik et al. 2008).

The complexity of these mixtures often makes their characterization very difficult. This is why a combination of spectroscopic analysis methods with structural analysis methods, which provide information on the phases present and their microstructure, is necessary. This structural information and the compositions of polycrystalline materials are obtained by X-ray diffraction, discovered in 1912 by M. von Laue and by W.H. & W.L. Bragg, who were the first to observe: (i) that the interaction of matter with this X-ray light was linked to the repetition of the atomic order existing in crystals; and (ii) that X-rays reveal the atomic arrangement of materials. This developed into the field of crystallography, with further discoveries revealing compositions and atomic arrangements not only in crystals but also in powders, and even in poorly ordered or non-crystallized materials (Hodeau and Guinebretière 2007). Recent developments have also allowed crystallography to open up to modern materials such as thin films, nanomaterials and those containing mixtures of phases and/or amorphous contributions and/or defects, an important characteristic of heritage materials.

Now, the contribution of powerful sources of synchrotron radiation means that diffraction measurements can be collected with significantly higher intensity, allowing for the use of effective filtering methods to remove experiment noise and to suppress noise from the environment of the object being studied (Hodeau et al. 1998; Dejoie et al. 2018a). This possibility of performing "clean" measurements is extremely important for studies of heritage artifacts or samples, which almost all consist of powders containing very low proportions of essential compounds such as crystallized and/or amorphous and/or degraded pigments (Dooryhee et al. 2004; Welcomme et al. 2007; Reguer et al. 2009; Van der Snickt et al. 2009; Monico et al. 2011b; Dejoie et al. 2014, 2018b; Cersoy et al. 2016; Cotte et al. 2017).

To extract unmixed and accurate information on the 3D contents of samples or artifacts, many measurements of diffracted X-rays must be taken, with the object studied placed in varying orientations with respect to the incident X-ray beam. With current developments, (i) intense light sources such as synchrotron radiation and micro-beams, as well as (ii) analysis tools and (iii) calculation methods, such selective analyses of complex materials have become possible at micrometric scales. In fact, the contribution of these powerful sources means that measurements can be collected quickly with high intensity to obtain multiple specific pieces of information about the object studied. In parallel with these advances, the development of 2D multi-detectors (which in the 1990s were only used by bio-crystallographers) means that diffraction measurements have become even faster. The contribution of new 2D or 1D "pixel detectors" can be added to this, which have no intrinsic noise, allowing a significant advance in measurement sensitivity. Naturally, the large amount of data collected requires more resources and computing speed, which are now available in all major research centers and laboratories. These three major advances allow researchers to develop new methods for selecting important information for the studies of heritage materials. These experiments very often use different modes of interaction (diffraction, diffusion, fluorescence, absorption) where the experiments are associated with, more commonly referred to as "mapping" and/or "tomography".

4.2. 2D mapping using diffraction

4.2.1. *First imaging experiments*

Mapping studies correspond to series of measurements taken during translation (x,y) of the object to be studied, which directly give 2D images. These measurements use X-ray fluorescence as an analysis method to measure the spatial distribution of chemical elements in a sample. They can also use X-ray diffraction, which makes it possible to determine the spatial distribution of the crystallographic structures of the powders and microcrystals in the artifacts. The first diffraction maps, carried out in situ on thin metal layers and environment materials, made it possible to discover and identify the structures/microstructures and the proportions of the phases present in the object studied (Tamura et al. 2002a). Combining absorption and diffraction methods enables the analysis of complex materials. However, it is only in the last 10–20 years that diffraction has been used as a tool for quantitative mapping and imaging. This is due to the difficulty in quantitatively interpreting diffraction patterns, as well as due to the fact that X-ray diffraction studies are often limited by: (i) too large source sizes; (ii) a beam penetration depth linked to the X-ray wavelength; and (iii) an angular alignment often difficult to achieve on a "real" object. Advances in X-ray source dimensions, nature, intensity, optics, detection and analysis software have made such analyses now feasible for multiple applications. For example, in the field of biology, Al-Jawad et al. (2007) studied the change in texture and structure of hydroxyapatite in a cross-section of tooth enamel. Variations in the intensities of the diffraction ring (002) due to preferential orientation were analyzed and texture mapping and the distribution of the lattice parameters were obtained through refinement using the Rietveld method (Rietveld 1969). Likewise, in the field of geosciences, Manceau et al. (2002) used fluorescence, diffraction and X-ray absorption mapping to study nickel sequestration in heterogeneous ferromanganese nodules in the soil and to show that nickel substitutes to manganese in a lithiophorite phase. In the field of cultural heritage materials, Dooryhee et al. (2005) studied a Roman fresco painting and demonstrated the selectivity and potential of 2D diffraction mapping. We will present this last case study in more detail, carried out directly on a fragment of wall painting from a

Roman villa from the end of the 2nd century CE, from excavations carried out by I. Bardiès in Metz, France.

4.2.2. *2D mapping of a "raw" heritage artifact*

This 2D structural mapping experiment, which analyzes the positions, intensities and widths of diffraction lines, was one of the first quantitative studies of a "raw" artifact. The experiment was carried out at the European Synchrotron Radiation Facility (ESRF) on the ID11 beamline using a large fluorescent screen coupled with a 2D CCD camera (Figure 4.1(a)). Since the sample of fresco painting was thick, the X-ray beam was not able to pass through it; therefore, this study was carried out in reflection mode with the object slightly inclined. This inclination was optimized to balance (i) the penetration depth of the beam and (ii) the dimension of its impact on the surface of the analyzed area, thus determining the size of each measured voxel (x,y). As a result, for each measurement zone (x,y) on the fresco painting, the images collected contain a set of diffraction arcs, some broader and/or punctuated (Figure 4.1(b)). The fine arcs are due to powders; the punctuated arcs are due to small crystals (notably calcite crystals, as well as quartz contained in the support, which often only generates one to three diffraction peaks). The arcs are coaxial and their 2θ angles correspond to the Bragg angles of the structural diffraction lines of the various crystalline materials comprising the fresco (support and/or pigments). After correcting any response/flatness errors in the detection system (Hammersley et al. 1996) (Figure 4.1(c)), these 2θ angles were used to determine (albeit with difficulty) the different crystalline phases present. Indeed, as the fresco supports are not strictly planar and as the surface studied has a slight inclination relative to the incident beam, depending on the zone (x,y) analyzed, there is a significant point-to-point variation in the distance between the analysis zone and the detector. This results in a variation and error in the "calculated" Bragg angles, based on the size of the arcs/lines. Consequently, the agreement of these 2θ angles with the inter-reticular distances *d(hkl)* does not match well with those of the known phases for substrates and/or pigments… However, like any research, a problem raised often contains the clue to the solution (… or a new discovery/application, which was the case for anomalous scattering, initially seen as anomalous scattering). The most intense diffraction peaks seemed to come from quartz crystals, in the same region of the support, but their *d(hkl)* did not correspond to the inter-reticular distances of the different diffraction lines of quartz! This disagreement guided the authors to find, for each voxel (x,y), the exact distance between the analysis zone and detector. Using this new exact distance for all arcs in this same area, the values of angle 2θ (and *d(hkl)*) could be calibrated and, incidentally, the roughness of the sample was determined (Figure 4.1(d)). The data generated a calibrated 2D diffraction image for each voxel (x,y) of the object. These 2D images were then transformed into 1D diffractograms, giving the intensity of the diffraction as a function of the Bragg angle 2θ or as a function of the inter-reticular distances *d(hkl)* corresponding to the lines *(hkl)* of the different phases existing in the studied area/voxel (x,y) or for a scan line (Figure 4.1(e)). These lines with their $2\theta/d(hkl)$ angles and their corresponding intensities provide detail on the contributions of each phase present. This identification can be done for each voxel (x,y), either by the individual study of all the individual lines (ROI) or by global studies using the sum of the different phases/structures and the Rietveld method (Rietveld 1969; Roisnel and Rodriguez-Carvajal 2000) (Figure 4.1(f)). The relative intensities of all the lines provide information on the proportions of the different structures, with some variations due to texture. The shape of the lines informs us about the size of the grains, their strains and also imperfections.

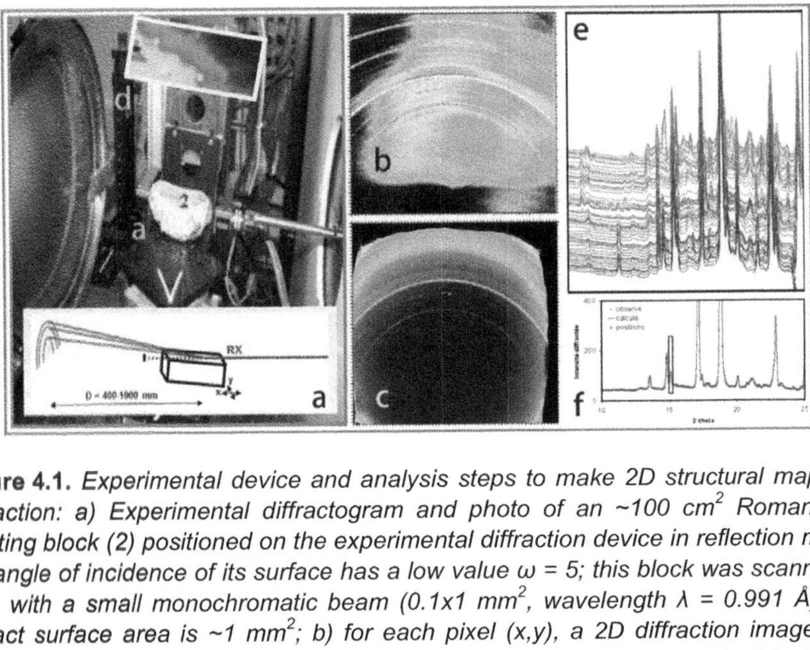

Figure 4.1. *Experimental device and analysis steps to make 2D structural maps by diffraction: a) Experimental diffractogram and photo of an ~100 cm^2 Roman wall painting block (2) positioned on the experimental diffraction device in reflection mode: the angle of incidence of its surface has a low value ω = 5; this block was scanned in (x,y) with a small monochromatic beam (0.1x1 mm^2, wavelength λ = 0.991 Å), the impact surface area is ~1 mm^2; b) for each pixel (x,y), a 2D diffraction image was generated; the width and intensity distribution of the diffraction arcs relate to the microstructure and texture of the structural minerals; c) using data from relevant references including the contribution of quartz present in the object, the intensity and position of all the diffraction arcs were corrected; d) the artifact surface was determined using the correction to the quartz contribution; e) an azimuthal integration of the diffraction arc intensity makes it possible to construct the diffractograms as a function of the angle 2Θ for each pixel (x,y), for a scanning line and to compare them with calculated diffraction patterns. This determines the respective mass fractions of the mineral phases present; f) example of a diffractogram calculated using the Rietveld method for a pixel (x,y) (Dooryhee et al. 2005). For a color version of this figure, see www.iste.co.uk/dejoie/synchrotron.zip*

Using the proportions of the different structural phases found in each voxel (x,y), it is possible to represent their quantity on diffraction maps of the area studied, for each phase, and to superimpose them onto the photograph of the Gallo-Roman wall painting fragment (Figure 4.2). A large amount of calcite ($CaCO_3$) was detected: calcite is an essential element of plaster, which also contains grains of quartz (SiO_2). By focusing on the lines of the calcite diffraction patterns, we observe two subsets of peaks that correspond to two different distribution maps. Map 4.2b shows a uniform (but very punctuated) presence of calcite throughout the area studied: it corresponds to the substrate and the plaster (with large crystals not all in a diffracting position). However, by making 2D reconstructions from very intense $CaCO_3$ diffraction lines alone, with strong preferential orientation, we obtain a different distribution resembling the contours of the nose and the left cheek (Figure 4.2(c)): this corresponds to a thin layer of calcite used as a white pigment. The intensities of the diffraction lines in this layer vary, indicating the presence of texture. The large width of these diffraction peaks suggests that this calcite was finely ground before its application. The diffraction maps also show the existence of two iron-containing phases: hematite Fe_2O_3 and goethite FeO(OH). Hematite is present on the surface (Figure 4.2(d)) and gives the artifact its red-ocher color and dominant patterns, while the shading of the hair and darker areas of the lip were from goethite (Figure 4.2(e)). This artifact also contains cuprorivaite ($CaCu(Si_4O_{10})$), particularly in the hair and eyelids (Figure 4.2(f)). This pigment is a major constituent of Egyptian blue, frequently used in Antiquity (Le Fur 2002). On the bridge of the nose and the corner of the lips, to the right of the face, the diffraction patterns show lines corresponding to lead carbonate and lead magnesium ($PbMg(CO_3)_2$) (Figure 4.2(g)). There are also a few scattered areas/voxels (x,y) that contain lead oxide diffraction lines. Such an observation suggests the initial presence of a lead pigment such as massicot yellow, which would have reacted with calcium carbonate containing Mg (magnesium is a common impurity in lime) to produce lead magnesium carbonate. This last observation shows that the structural selectivity of diffraction maps can also suggest transformations of materials/pigment, which are important to obtain in-depth knowledge about the artifacts. In Figure 4.2(i), all the identification maps have been superimposed to form an overall diffraction map for each phase. Finally, with these pigment location and quantification results, each elementary pixel (x,y) is assigned a color respecting the mass fraction and the hue of each pigment phase (Figure 4.2(j)).

With fluorescence being excellent for understanding chemical compositions, an elementary chemical analysis using proton-induced X-ray emission (PIXE) was also carried out on this wall painting fragment on the AGLAE accelerator at the Louvre

(Calligaro et al. 1998). X-ray fluorescence radiation emitted from the upper 10–20 μm of the artifact's surface revealed its chemical composition, and PIXE maps clearly showed the presence of calcium (Ca), iron (Fe), copper (Cu) and lead (Pb). These maps showed the abundance of Fe over large areas of the face but could not distinguish hematite Fe_2O_3 from goethite FeO(OH) because, unlike diffraction, fluorescence cannot differentiate between minerals with common chemical elements. Fluorescence also shows the presence of copper, where diffraction maps found cuprorivaite, and also the presence of lead, where lead carbonates and oxides were observed by diffraction. The superposition of these different elementary PIXE maps is represented in Figure 4.2(h) using the same representation mode.

This case study shows that the chemical and structural maps of the measured mineral phases are strongly correlated and make it possible to accurately describe the main characteristics of the painting. In order to perfectly correlate these chemical elementary maps and the phase structure maps for the same area/voxel (x,y) of the artifact, the fluorescence and diffraction signals can be recorded simultaneously with detections in parallel (this was not the case in this analysis, but this will be the case in most of the examples of 2D mapping and 3D tomography in the remainder of this chapter).

This first example of 2D mapping also demonstrates that it is possible to locate pigments mixed intimately with the substrate. Indeed, fresco-type art is applied directly to the wall, which is covered with a mortar (lime and sand) mixed with a fine marble powder. The pigments are applied to the wall when it is still wet and it is during drying that the carbonation of the slaked lime occurs, forming a layer of calcite crystals around the pigments: the colors are therefore intimately mixed with the plaster (Vitruvius *De Architectura*, around 15 BCE; Pliny the Elder *Naturalis Historia*, circa 77 CE; Duran et al. 2010a).

Studying only the texture of the artifact layers, it is possible to determine whether a component was used as the substrate or as pigments. We also see from this example of 2D mapping that X-ray diffraction can also be a non-invasive tool for mapping the distribution of phases and microstructures on irregular surfaces. In this example from 2005, the spatial resolution was 1 mm^2, but with new X-ray sources and experimental beamlines, the resolution can adapt to the scale and heterogeneity of the material, ranging from mm to a few 0.1 μm.

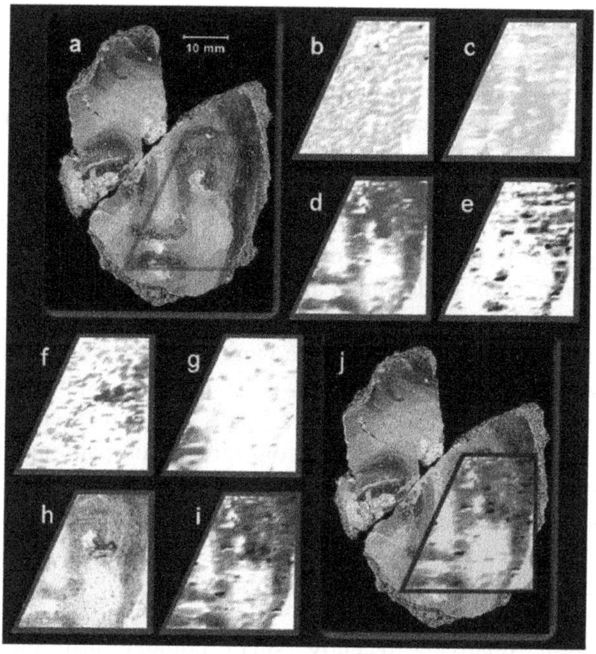

Figure 4.2. *2D structural maps by diffraction of a fresco-type Roman wall painting: a) analyzed fragment of the Roman wall painting (Metz museum, France). The brown frame indicates the analyzed area. The maps of the different phases observed by diffraction (with a scan of 60×60 voxels) are as follows: b) underlying preparation layer of calcite → Ca; c) surface layer of fine powdered calcite pigment → Ca; d) red-ochre pigment of hematite → Fe; e) darker red goethite pigment → Fe; f) Egyptian blue pigment (cuprorivaite) → Cu; g) transformed lead carbonate → Pb; h) superposition of PIXE fluorescence analysis maps revealing the elements iron (Fe), lead (Pb) and copper (Cu) with a similar color code; i) superposition of all diffraction maps of the different phases with the same color code as for the individual maps (b–g); j) superposition of all diffraction maps of the different phases with a "suggestive" color code emphasizing that the image interpreted by diffraction reveals the main characteristics of a face (Dooryhee et al. 2005). For a color version of this figure, see www.iste.co.uk/dejoie/synchrotron.zip*

The wealth of information obtained from these mappings leads to their application on heritage materials of various shapes and natures. Additionally, conditioned by the characteristics of the artifacts and to respect their preservation, the configurations of the experiments may vary. Below, we will present some of the different analysis configurations and illustrate their complementarity.

4.2.3. *Complementarity of 2D mappings performed on samples and/or directly on artworks: the example of Lead White*

In order to best preserve cultural heritage artifacts, 2D mapping experiments were initially conducted with microscopic measurements from small samples using powerful sources such as synchrotron radiation (Sciau et al. 2006; Welcomme et al. 2006, 2007). Specifically, they were performed on paint cross-sections to describe the pigment degradation pathways (Cotte et al. 2008; Van der Snickt et al. 2009, 2012; Monico et al. 2011b; Radepont et al. 2011; Bergmann et al. 2012; Gonzalez et al. 2016, 2017). However, it is relatively difficult to take micro-samples from paintings to analyze, so in theory, valuable works of art are better studied using in situ non-invasive sampling techniques. Nevertheless, such experiments prove to be very difficult but are possible on some artifacts using high-energy X-rays. For example, such analyses have been carried out directly on paintings to produce 2D maps, some revealing hidden paintings (De Nolf et al. 2011). However, for safety and preservation reasons, such direct measurements made with a synchrotron beam are very rare; in fact, these artworks are almost never moved to a laboratory equipped with a source of synchrotron radiation.

For this reason, relatively recently, small portable instruments have been developed that can be easily taken to museums like the Louvre or the Rijksmuseum: they allow further analysis directly on the artifacts. Many of these sources and portable devices have been designed to take direct X-ray absorption and diffraction measurements of heritage artifacts (Chiari et al. 2008; Gianoncelli et al. 2008; Sarrazin et al. 2008; Duran et al. 2009, 2010b; De Viguerie et al. 2009; Eveno et al. 2011; Van der Snickt et al. 2011; Walter et al. 2011; Janssens et al. 2013; Hirayama et al. 2018; Vanmeert et al. 2018a; De Meyer et al. 2019; Poline et al. 2023). For example, diffraction can easily identify lead compounds, which generate a variety of colored pigments, or the pigment *lapis lazuli*, which is difficult to identify using fluorescence alone. With such portable devices, around a hundred works of art have been studied in situ in museums and on monuments. Such devices were notably used to analyze paintings by Rembrandt, Leonardo da Vinci, Mantegna, van Gogh, Vermeer, etc. (Eveno et al. 2010; Nakai and Abe 2012; Hirayama et al. 2018; Vanmeert et al. 2018a, 2018b; De Meyer et al. 2019). In fact, such measurements have mainly been taken of flat artifacts such as paintings, as well as of three-dimensional artifacts such as ceramics, vases, bronzes, altarpieces or statues. For example, Martinetto et al. (2019, 2021) analyzed brocades with measurements taken directly on statues from the Duchy of Savoy from the late Middle Ages (Figure 4.3). For this, they developed the MobiDiff portable device as part of the CDP "PatrimAlp" program (PatrimAlp 2018; Poline et al. 2023).

Figure 4.3. *Example of statues originating from the Duchy of Savoy, dating back to the late Middle Ages, with applied brocades. Information was obtained through analyzing the painting directly and also from samples taken of the artifact: a) the "Vierge de Pitié avec saint Jean et sainte Marie-Madeleine" from the Château de Montrottier (Haute Savoie) was studied by taking samples from areas containing brocades and also direct analysis of the painting; b) a photograph of such a direct analysis system with a portable diffraction (Cu-Ka source) and fluorescence device (source Pd-Kα)(Martinetto et al. 2021; Poline et al. 2023). For a color version of this figure, see www.iste.co.uk/dejoie/synchrotron.zip*

It should also be noted that one of these portable devices, CheMin, was one of the 10 Curiosity robot instruments, which was sent into space to take the first diffraction measurements on Mars! (Bish et al. 2014). The various works and references cited in this paragraph highlight the complementarity of these macroscopic measurements (of relatively moderate precision), made with such a portable instrument directly on the works, with the microscopic measurements (of high precision) made with synchrotron radiation of samples collected from the same areas of interest. A more detailed illustration of the contribution of these two analysis configurations is the study of "lead white"-type pigments, presented in the following paragraph.

Although analyses of Roman paintings or cosmetics, in particular from Pompeii and Gaul, have shown the importance of using materials commonly found in nature such as gypsum or calcite (see the previous example of Roman mapping, Dooryhee et al. (2005)), "lead white"-type pigments have been used and synthesized since Antiquity for cosmetic and artistic purposes. Ancient texts describe different recipes and preparation processes as well as various production sites; for example, Pliny the Elder specified the geographical origins and explained that the best quality white Psimithium or ceruse (also known as white lead) came from the island of Rhodes (Pliny the Elder, 77 CE). This type of pigment has also been one of the white pigments widely used in oil paints. Also, because of its significant use, it has been the subject of numerous studies. Welcomme et al. (2006) notably showed that, during the Hellenistic period, the white pigments used were mainly lead white, which contained a mixture of cerussite and hydrocerusite phases, whose synthesis initially involved altering metallic lead with vinegar vapors (Martinetto et al. 2002; Walter 2015). Such studies of lead whites are therefore important for defining the various lead pigments, which could be indicative of a particular craftmanship (washing, decanting, grinding, etc.), a country or period (Stols-Witlox et al. 2012).

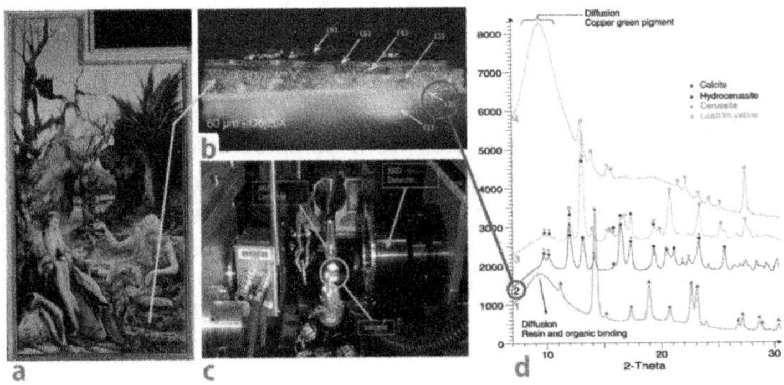

Figure 4.4. *XRD and XRF analyses of the ID22 beamline of the ESRF, from microsamples taken from the panel painted by Matthias Grunewald "Visite de Saint Antoine à Saint Paul l'Hermite" from the Isenheim Altarpiece; a) sampling area of "Visite de Saint Antoine à Saint Paul l'Hermite"; b) optical microscope observation of the cross-section of this MGN7 micro-sample, ×20 magnification; the different layers are indicated by the arrows (1), (2), (3), (4), (5), (6), which go from the preparation undercoat to the varnish. Layer (1): white preparation undercoat; layer (2): primer coat; layers (3), (4), (5): colored layers; layer (6): varnish. c) experimental device from the ID22 line used to simultaneously carry out micro-XRD and micro-XRF measurements on this sample inserted in a resin block; the diffraction analysis was carried out in transmission mode with a monochromatic micro-beam (2.9x1.5 μm^2 having an energy of 18 keV); d) μXRD diffractograms corresponding to the four layers (1), (2), (3) and (4): layer (1): calcium carbonate preparation sub-layer; layer (2): primer layer rich in lead carbonates (hydrocerussite and cerussite); layers (3) and (4): copper pigments (green) mixed with lead tin yellow and white of the two lead carbonates (Welcomme et al. 2007). For a color version of this figure, see www.iste.co.uk/dejoie/synchrotron.zip*

A detailed characterization of these lead white pigments requires identification of variations in their mineral phase composition and their granularity. Early studies were carried out on micro-samples or micro-cross-sections of heritage artifacts. For example, in 2006 and 2007, Welcomme et al. studied the white pigments in paint samples taken from two pieces painted by Matthias Grunewald, an early Renaissance artist and contemporary of Albrecht Dürer. The paintings analyzed were the "Isenheim Altarpiece", painted between 1512 and 1516, kept at the Unterlinden Museum in Colmar, and the "Crucifixion", painted around 1500–1508, kept at the Kunstmuseum in Basel. Very small samples (100 µm × 400 µm × 400 µm) were analyzed, embedded in a support resin and containing successive layers of paints (Figure 4.4). Quantitative X-ray diffraction of these micro-samples made it possible to identify the phases and to establish the different composition ratios between the two phases present: hydrocerusite $2PbCO_3$-$Pb(OH)_2$ and cerussite $PbCO_3$ (Welcomme et al. 2007). The series of diffraction analyses identified and located these different mineral/chemical phases in the sample stratigraphy and quantitatively analyzed these lead whites in each layer. The variation in the hydrocerusite/cerussite ratio of the area sampled on the artifact indicated the know-how of a particular period and school.

Many other studies have observed such variability in the ratio of these two lead carbonates (cerussite and hydrocerussite) in lead white pigments from other artworks (Walter 2015; Gonzalez et al. 2017; De Meyer et al. 2019). In fact, this lead white, which was the most widely used artificial pigment from Antiquity until the end of the 19th century, often has a composition of its two lead carbonates that varies from one painting to another, and also within different areas of the same artwork, as is the case, for example, in the painting the "Virgin and Child with Saint Anne" by Leonardo da Vinci. These variations can be linked to the supply of pigments as well as to the artist's choice because the appearance and covering power of hydrocerussite and cerussite differ (Walter 2015). Other studies carried out using synchrotron diffraction on micro-samples of around 30 paintings by major artists over a period ranging from the 14th to the 19th century, confirm such variations in the cerussite/hydrocerusite ratios and seem to indicate a significant increase in the size of crystallites over these five centuries (Gonzalez et al. 2017). Depending on the proportions of hydrocerusite and cerussite, lead white can be classified into different subtypes, each with different optical properties. Recently, after an analysis of Vermeer's "Girl with a Pearl Earring", De Meyer et al. (2019) were able to differentiate four subtypes. Their study also illustrated the complementarity of 2D mappings, made from microscopic measurements of samples and macroscopic ones done directly on the painting: these two types of measurements were very useful to these authors in that they supported this differentiation of the lead whites. This latest study also illustrates the specificities of two possible collection geometries on paintings.

Indeed, for "paintings", the X-ray beam passes through the artwork without significantly absorbing its diffracted signal, allowing the realization of mappings that can be measured by experiments with different diffraction geometries: transmission and reflection. These two analysis geometries are defined by the position of the artwork relative to the directions of the incident X-ray beam and the detector. Direct 2D maps measured in reflection mode must be created with a low angle of incidence of the X-ray beam to collect the diffracted beams. This angle must not be too small to balance the X-ray beam penetration depth and the dimension of each voxel analyzed. These constraints impose a relatively low beam angle of incidence which, generally, will not be able to completely reach deep layers of the artifact. Indeed, during its long journey in the surface layers, the X-ray beam absorbs depending on the chemical elements composing the pigments (high for lead or mercury, low for calcium, sulfur, oxygen or carbon). 2D maps measured in reflection mode therefore reinforce surface information, which can mask data from deeper layers. On the contrary, direct 2D maps measured in transmission mode can only be generated if the incident and diffracted X-ray beams can easily penetrate the artifact; these measurements therefore extract information about all layers of the painting, which can in turn mask the surface pictorial information. These two diffraction geometries in reflection mode and in transmission mode are therefore complementary.

This complementarity was illustrated in the cross-experiments carried out by De Meyer et al. (2019) on Vermeer's "Girl with a Pearl Earring" (Mauritshuis Museum, La Haye, The Netherlands) in particular using a portable device for easy scanning (x, y) of X-ray diffraction measurements (project "Girl in the Spotlight" De Meyer et al. 2019; Vandivere et al. 2020). Two large 2D diffraction maps were thus collected on this 17th century painting. To analyze the lead whites in this painting, these authors used both sampling methods with diffraction, reflection and transmission analyses to specify the variations in the hydrocerusite/cerussite ratios according to their position in the stratigraphy. These macroscopic diffraction analyses were also combined with some microscopic analyses carried out with the PETRA III synchrotron source on associated samples to confirm and increase the precision of the compositions in the different layers containing lead (De Meyer et al. 2019). This is how these authors were able to identify four different lead white subtypes used for the face of the "Girl with a Pearl Earring". This result indicates that Vermeer selected various lead whites to create opaque and/or transparent layers of paint. The background sublayer (i) contains a white lead pigment rich in hydrocerusite, embedded in a chalk matrix. The light/shiny parts of the face contain two types of lead white pigment containing a hydrocerusite/cerussite mixture, whose proportions differ depending on the depth: one rich in hydrocerusite, associated with vermilion in the upper layer (ii) and the another even richer in hydrocerusite and poor in cerussite in the lower layer (iii). The shadow areas (iv) contain a thin layer of more transparent pigment rich in cerussite (De Meyer et al. 2019).

The differences in the way the various lead whites are visually perceived (light/bright/opaque/dark) are linked both to the hydrocerusite/cerussite proportions and to the large/small size of their particles, which influence the opacity/transparency of the pigment. Obtaining these different proportions is achieved during the preparation of the pigments. Studies have shown that the chemical behavior of Pb^{2+} ions in aqueous solutions in the presence of calcite varies depending on the concentrations and pH, and that it influences the precipitation of carbonates in the form of cerussite or hydrocerusite (Shahwan et al. 2005). The crystal growth characteristics are also different for these two phases and give either nano-crystalline precipitates or polyhedral crystals (Sánchez-Navas et al. 2013). Hydrocerusite crystallizes in the form of very fine platelets, while cerussite forms small, elongated, ovoid-shaped crystals (Martinetto et al. 2002). It is these variations in the shapes of crystals depending on the chemistry of their preparation that cause variations in the visual perceptions of lead whites. In fact, the particle size can be chosen by washing/decanting the pigments or by washing/grinding the pigments in vinegar or vinegar vapors, which promotes the recrystallization of small cerussite particles (Stols-Witlox et al. 2012; Walter 2015; Gonzalez et al. 2017; De Meyer et al. 2019).

4.3. 3D mapping by tomography

4.3.1. *Method history*

The discovery of tomography earned Cormack and Hounsfield the Nobel Prize in 1979. G. Hounsfield had the idea of an experiment to determine the content of an object from X-ray images measured at different angles and built a device capable of taking such measurements to reconstruct an image of the interior of an object and visualize it layer by layer (Hounsfield 1973). Independently, A. M. Cormack established the mathematical theory describing this method. These studies require the collection of a series of images taken as the studied object undergoes translation y and rotation ω (or when the object is a body, rotation of the entire source/detector assembly). This reconstruction of the object – its 3D volume and its 2D slices – is achieved mathematically from this series of projections, which is why tomography is often called computed tomography (CT) (Kak and Slaney 1988, 2001; Maire and Withers 2014). The interest of this imaging technique in the medical field led to the development of tomography, often called CT scan or scanner.

These tomography reconstructions can use various physical interactions; however, the most common tomography is absorption tomography, long used in medical imaging with X-rays, as well as for the study of materials such as alloys. Like most studies using X-rays, tomography has benefited from advances in X-ray sources, X-ray optics, detectors and computing methods. For such analyses, the samples can be used in their natural form/state, without the need to polish or cut them into thin slices. In addition, they can also be analyzed in a non-destructive way, which is very important for Heritage Materials.

4.3.2. *3D tomography with different modes of physical interactions*

With absorption tomography, it is possible to visualize density variations in various materials such as trabecular bones (Salome et al. 1999), objects with voids/cracks or with small differences in contrast between the materials that compose it (Kak and Slaney 1988, 2001; Wu et al. 2017). Another method widely used for the study of materials is X-ray fluorescence, which gives the 3D spatial distribution of chemical elements in a sample. The high chemical sensitivity of fluorescence means that it can extract trace elements from an object (Brunetti and Golosio 2001; Schroer et al. 2001; Golosio et al. 2003; Kim et al. 2006; La Riviere and Vargas 2006). These experiments can be performed with high sensitivity, down to less than 0.01% by volume. We can also utilize the information provided by X-ray absorption near edge structure (XANES) at energies close to X-ray absorption thresholds, which give insights into the chemical state of the elements (Nelson et al. 2011; Matsui et al. 2017; Sanchez et al. 2017). A Compton-type interaction can also be used to measure density, particularly to image low-Z materials (Brunetti et al. 2004; Golosio et al. 2004). By using the coherence of synchrotron sources or XFEL (X-ray free-electron laser), we can extract and use the very small phase variations expressed in the real part of the refractive index, making it possible to amplify and visualize very small contrast differences in the object studied (Snigirev et al. 1995; Cloetens et al. 1997, 1999). Using this interaction offers new imaging possibilities: phase contrast radiography, phase contrast micro-tomography and holo-tomography, methods now widely used for non-destructive studies of paleontological specimens (Tafforeau et al. 2006). More recently, quantitative nano-imaging of heterogeneous materials is possible by combining tomography with 3D X-ray ptychography. This imaging, which makes it possible to simultaneously map the complex refractive index of the object in relation to the coherent X-ray source, has a very high resolving power and sensitivity and can also be used at high energy (Da Silva and Menzel 2015; Da Silva et al. 2019). It is also possible to use the Small-Angle X-ray Scattering (SAXS) signal from the object to extract its characteristic dimensions and certain short-range orders (Schroer et al. 2006; Kulmann et al. 2008). These SAXS tomography analyses make it possible, for example, to analyze brain tumor tissues (Jensen et al. 2011) or trabecular bones of the vertebrae (Liebi et al. 2015, 2018).

However, to extract precise structural information about the different phases constituting an object, one of the richest methods is wide angle diffraction. Indeed, only diffraction tomography makes it possible to finely extract the 3D spatial distribution of the different structures of the phases that make up an object or a sample thereof.

4.3.3. *Different types of 3D diffraction tomography*

As we will see in the examples below, cultural heritage materials are in almost all cases very heterogeneous materials containing both polycrystals, powders and/or amorphous materials, in varied quantities, compositions and structures. If we want to visualize and quantify the distribution of each phase, we must do so using structural images of diffraction-tomography. However, visualizing diffraction information in 2D/3D images using tomography requires mitigating the sensitivity of the diffraction line intensities to the object orientation. Its sensitivity is even more pronounced when the object is composed of crystals, but fortunately it is low if the object is composed of powders or amorphous materials. Ultimately, a variety of diffraction/diffusion methods, using synchrotron X-ray radiation in most experiments have been developed and allow us to construct structural maps for each type of material.

In the case of well-crystallized materials, where the grains only diffract if they have a suitable orientation, we can overcome this by using the Laue method, which exploits the wide-angle diffraction of a polychromatic beam and allows us to "see" the contribution of crystallized grains over a wide range of orientations. Such images can be obtained on polymer phantom samples, crystallized thin films, cements or alloys (Hall et al. 1998; Larson et al. 2002; Tamura et al. 2002b; Budai et al. 2003; Jupe et al. 2004; Beale et al. 2007; Lazzari et al. 2009, 2012; Cernik et al. 2011; Jacques et al. 2011; Hofmann et al. 2012). To facilitate these tomographic energy-dispersive diffraction imaging (TEDDI) studies, a strategy and analysis process was developed by Lazzari et al. (2009). They tested its potential on a phantom object and compared the relative strengths and limitations of this method with the classic angular dispersion method (which is discussed later in this review). The reconstruction accuracy of the TEDDI method is very good, even if its sensitivity is lower than that obtained by monochromatic mode diffraction analyses and despite an absorption correction being necessary for the diffraction of low-energy X-rays (Lazzari et al. 2009, 2012).

In the case of very well-crystallized materials, tomography can also use another imaging approach: topography, a method using diffraction to find defects in "perfect" crystals. Topography can be combined with tomography to produce a topo-tomography, known as diffraction contrast tomography (DCT) (Ludwig et al. 2008).

This provides 3D images of the sizes and orientations of the crystals in an object; it can visualize in 3D the orientations and microstructures of the grains making up an alloy as well as ceramic compounds containing crystals with limited mutual disorientations (Johnson et al. 2008; King et al. 2008, 2010; Ludwig et al. 2008).

Still in the case of well-crystallized materials, a completely different method has also been developed. It uses the direction of the paths of the diffracted beams of each grain to find their position and thus reconstruct a 3D image. This analysis is done using a 2D detector and arrays of conical slits corresponding to the 2Θ diffraction angles of well-known structures – this method cannot therefore be applied to materials with unknown structures (Nielsen et al. 2000; Poulsen et al. 2004).

When the object of study contains nano-crystallized or polycrystallized materials such as powders, it is easier to differentiate the different structures because the orientations of the grains composing it are often distributed homogeneously, which reduces the reconstruction errors. In this case, the diffraction experiments of a monochromatic beam generate almost continuous diffraction rings, whose relevant parameters are the different diameters of these rings – corresponding to the Bragg diffraction angles 2Θ signatures of each object phase – and make it possible to easily quantify the heterogeneities of this object or artifact. In this situation, common in heritage artifacts, the diffraction intensities are less sensitive to the orientation of the object and the experiments are easier to carry out: the structural information only varies with the 2Θ angle and can be collected on a 1D detector (and also 2D, by converting the collected intensities into a single 1D information). The principle of this method, called diffraction/scattering computed tomography (DSCT), involves acquiring diffraction patterns in transmission mode with a 2D detector while translating (perpendicular to the beam axis) and rotating the object illuminated by a small X-ray beam (Alvarez-Murga et al. 2012b). In artifacts containing polycrystallized, nano-crystallized or amorphous powders, the DSCT method can extract all the information and variations in the diffracted lines and 2Θ angles, important characteristics related to the structures of the materials comprising this object.

4.4. Diffraction/scattering computed tomography: DSCT

4.4.1. *Initial DSCT experiments*

Given the strong potential of imaging in medicine, the first selective structural DSCT images were obtained on tissues and model biological objects (Harding et al. 1987). One of the first DSCT experiments was carried out on an educational object, a rib of beef!

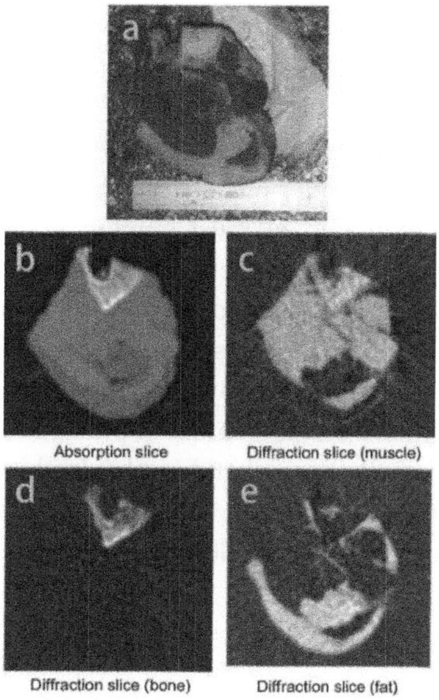

Figure 4.5. *One of the first observations of diffraction/scattering computed tomography (DSCT): a) object analyzed: a prime rib! b) 2D section extracted by the absorption computed tomography method (μCT); (c–e) 2D sections extracted by the DSCT method using the angular range signal 2θ where c) the intensities scattered by the muscle are significant; d) the intensities scattered by the bone are significant; e) the intensities scattered by the fat are significant (Kleuker et al. 1998). For a color version of this figure, see www.iste.co.uk/dejoie/synchrotron.zip*

As shown in Figure 4.5, this DSCT experiment made it possible to separate meat from bones, thanks to the different structural characteristics of their constituents (Kleuker et al. 1998). Meat, muscles and fat have different scattering/diffraction patterns, close to those of poorly crystallized or amorphous phases, while bones have patterns closer to those of polycrystalline phases with a certain microstructure.

These structural images are produced from a selection of a plane or a section of a 3D map of an object and require the collection of thousands (or tens of thousands) of diffraction patterns. These are collected by scans along the sample y (and/or z) axis and with the object or artifact rotated along ω (it is not always easy to carry out this experiment on heritage objects!). Then, it is through tomographic reconstruction that we can visualize the image of the desired plane or cross-section. This reconstruction is based on the variations (in the case of a heterogeneous material) in the diffracted/scattered signal of the volume irradiated by the beam for a sample orientation (y_i,ω_i) relative to another position (y_j,ω_j). Such a reconstruction of DSCT images makes it possible to restore the distribution either of the scattered/diffracted intensity, or of a desired structural parameter (Bragg angle, texture, phase, proportion of phases, particle size, etc.) for each point/*voxel* (x,y,z) on the sample. This absorption tomographic reconstruction principle can only be used for diffraction/scattering on an amorphous material or one containing powders after appropriate azimuthal integration. Indeed, each illuminated volume/voxel must contain a sufficiently high number of randomly oriented crystallites, because otherwise the signal can vary depending on the crystal orientations of its large, crystallized grains. For example, for a point/volume with a voxel size of 1 μm, the sample must be composed of grains of a few tens of nanometers. To better average the effects of randomly oriented crystallites, as well as to reduce the acquisition time, we can also use a pink X-ray beam (i.e. not perfectly monochromatic but with a small energy bandwidth). In this case, the Debye–Scherrer rings become thicker, which unfortunately reduces the possibility of detecting minor phases. This pink beam analysis is excellent for samples/artifacts with only a few phases with distinct diffractograms.

4.4.2. *Methodology of DSCT analyses and reconstructions*

DSCT experiments begin with (i) ***a series of data acquisitions*** on a sample at different locations along a line and at multiple projection angles ω over a range of angles (0, 180°) (Figure 4.6(a),(b)). When the sample contains large grains, the 2D diffraction images also contain very intense single-crystal diffraction spots which make analysis difficult (this is the case, for example, in fresco-type paints which contain large grains of quartz and calcite). An easy way to solve this is to apply a saturation threshold to the 2D diffraction images before integrating them; this removes most of the signals from these large grains, but it also removes information about the material (see the two roles of calcite in the study of a Roman fresco painting). Another way involves dividing the contents of the 2D image into two: a powder-type pattern and a crystal-type pattern (Voltolini 2010). Next, (ii) ***azimuth integration*** of the set of Debye–Scherrer diffusion rings makes it possible to convert this data into one-dimensional information (Hammersley et al. 1996; De Nolf and Janssens 2010) (Figure 4.6(c)). The next step is (iii) ***the construction of sinograms*** $S(y,\omega)$, which corresponds to calculating and representing the integrated intensity of the total diffractogram as a function of y and ω (Figure 4.6(d)) and the integrated intensities of the structurally significant data of this diffractogram; in most cases, this corresponds to the diffracted intensities of a particular phase such as a significant Bragg angle, or to a selective structural parameter for each phase (these can be extracted with Rietveld or PDF-type refinements from these diffractograms). We can create as many selective sinograms as phases (Figure 4.6(e)). The selection of selective contributions in these diffractograms is facilitated by various software such as *PyMca* (Solé et al. 2007), *ImageJ* (Magelhaes et al. 2004), *XRDUA* (De Nolf and Janssens 2010) or *EasyDD* (Sochi 2013). It should be noted that the precision of the sinograms and of the reconstruction is linked to the number of steps in y and ω: in fact, for each diffraction measurement point (y_i, ω_j), the analyzed volume is elongated resulting in low resolution along the beam path; the resolution is improved by combining it with diffraction data from the other inclined volumes ω_j, ω_k. We can then obtain by (iv) ***direct analysis***, a more visual representation of the data using a mathematical process based on Fourier transforms and linear integrations (Kak and Slaney 1988), which allows us to reconstruct the selective images (x,y) of the phases contained in the sample (Figure 4.6(f)). These image reconstructions can be done in particular with software packages *XRDUA* (De Nolf and Janssens 2010), *DGT3D* (Bleuet 2008) or *EasyDD* (Sochi 2013).

4.4.3. *Selectivity of DSCT analyses*

Figure 4.6(a)–(f) which illustrates the process of this DSCT analysis shows images of reconstructions obtained on a sample of fullerene (C60) transformed into crystallized or amorphous diamond under high pressure at room temperature (Figure 4.6(a)) (Regueiro et al. 1992). This example allows us to highlight the selectivity of the DSCT method because this sample theoretically consists of only carbon with phases having similar compositions and densities. As with all tomography studies, the scattering/diffraction data obtained are combined to form ***sinograms*** to selectively image the spatial distribution of each phase (crystallized or amorphous) within the sample. For this sample, the absorption and fluorescence tomography images are similar, which is expected, but surprising, but the DSCT images show multiple contrast variations where five phases can be identified! (Figure 4.6(f)). Inside the glass capillary, there are cubic diamond microcrystals embedded in an sp^3 amorphous carbon matrix. In addition, crystallized ferrite impurities are observed on the surface of the sample (αFe) as well as an isolated grain of the impurity calcite ($CaCO_3$). These two additional and unexpected identifications highlight the sensitivity of this method. The ferrite is likely contamination from the razor blade used to extract the sample, while the calcite is probably a dust particle in contact with the pressure cell. The results of this DSCT analysis also highlight the sensitivity of diffraction/scattering tomography to separate crystalline phases from amorphous phases within a sample.

In this study (Bleuet et al. 2008), then in the complete 3D reconstruction of the same carbon sample (Alvarez-Murga et al. 2012a), a (v) ***reverse analysis*** approach was taken, which better illustrates and quantifies the structural selectivity of this method, by making it possible to recover all the diffraction/scattering diffractograms containing selective local structural information (Figure 4.7). The principle of this ***reverse analysis*** is to extract all the scattering/diffraction patterns as a function of 2Θ (and significant atomic interplanar distances $d_{(hkl)}$) of a selected area corresponding to one or a set of voxels (x,y) of the reconstructed image. This extraction is carried out from images reconstructed by ***direct analysis***, without a priori knowledge of the structures of the phases themselves. It allows us to obtain a direct and easy visualization of the phases present, select them and extract the most significant diffractograms of the minor phases. The size of the analyzed area can be adapted to extract local information, from a single voxel (~µm^3) to the total volume of the sample (~mm^3) (Alvarez-Murga et al. 2011, 2012b).

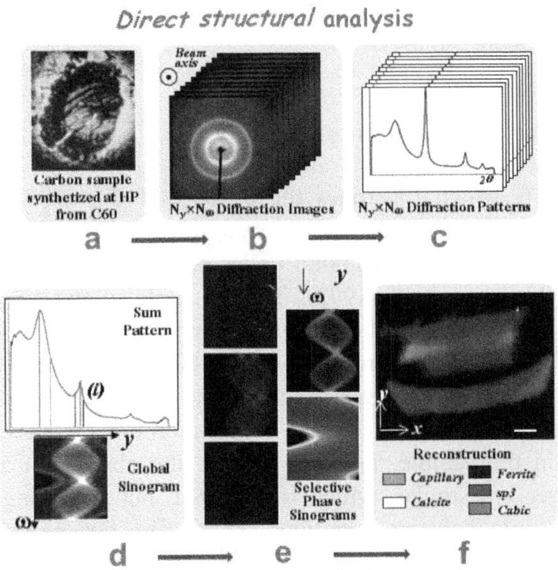

Figure 4.6. *The DSCT method: reconstruction by direct analysis. For a color version of this figure, seewww.iste.co.uk/dejoie/synchrotron.zip*

NOTE ON FIGURE 4.6.– a) Photo of a C60 sample transformed under HP, a section of which has been reconstructed; b, c) carrying out the reconstruction of this transverse section: a 2D diffusion/diffraction image is collected for each sample position (y,ω), then they are converted into 1D(y,ω) diffractograms varying in intensity according to the angle 2Θ; d) we can reconstruct, from this set of data (y,ω,Θ), a 1D "sum" diffractogram as a function of Θ and a "global" 2D sinogram; this "global" sinogram $S_g(y,\omega)$ is constructed by integrating for each sample position (y,ω) the intensity of the "sum" diffractogram over all the diffraction angles 2Θ, this total scattering intensity is plotted as a function of (y,ω); e) selective sinograms $S_i(y,\omega)$ can be extracted using the intensities of the Bragg angle ranges alone (colored 2Θ angles), corresponding to a given scattering contribution or to a diffraction peak; this allows us to extract a relevant sinogram $S_i(y,\omega)$ (colored) of the corresponding individual phase; f) from all of these selective sinograms, a mathematical process reconstructs, by direct analysis, the selective (x,y) images (colored here with the same colors) of the different phases contained in the sample (Bleuet et al. 2008).

As the contribution of the surrounding areas is removed in these extracted diffractograms, their selectivity is greatly improved and this makes it possible to detect and locate ultra-diluted phases in the object studied. Figure 4.7(b)–(f), showing results of this *reverse analysis*, gives the scattering/diffraction diffractograms corresponding to the five phases highlighted in Figure 4.6. For example, Figure 4.7(d) corresponds to one of the minor phases with a volume/total volume ratio less than 10^{-3}: its quality allows its crystallographic analysis, corresponding to the calcite grain visualized by the DSCT analysis on only 4 voxels of the 8,000 voxels measured (Figure 4.6(f)).

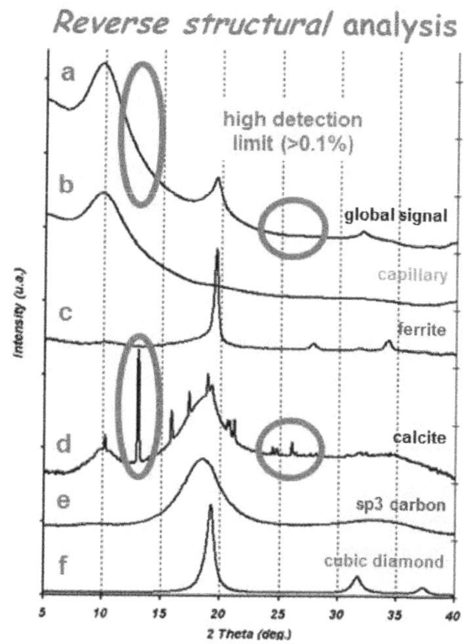

Figure 4.7. *The DSCT method: reconstruction by reverse analysis. For a color version of this figure, see www.iste.co.uk/dejoie/synchrotron.zip*

NOTE ON FIGURE 4.7.– a) reconstruction by reverse analysis giving the global diffractogram; b–f) reconstructions by reverse analysis giving the selective diffractograms extracted from the zones/voxels where each phase is in the majority: b) capillary; c) ferrite impurity; d) calcite ultra-minor impurity present in only 4 of the 8,000 voxels measured; e) amorphous phase sp^3; and f) cubic crystallized diamond; the selectivity of this extraction makes it easy to determine the lattice parameters and the structure of the impurity $CaCO_3$ (R-3c, a=0.4991 nm, c=1.7062 nm) (Alvarez-Murga et al. 2012b).

It should be noted that the detection sensitivity of DSCT is at least an order of magnitude higher than that of conventional diffraction: it allows non-destructive structural analyses with a sensitivity better than 0.1% by mass. This advantage of the DSCT method relies on the redundancy and variations in the data by using all the diffraction/scattering images where the beam intercepts the minor phase for all the different orientations of the sample: there are snapshots where the minor phase signal is less diluted by the matrix information and where the contribution of this matrix varies for each snapshot, while the contribution of the minor phase is almost constant.

4.4.4. *Selectivity of "multimodal" DSCT analyses*

Following the pioneering work of Harding et al. (1987), Kleuker et al. (1998) and Stock et al. (2008) on model biological objects, the selectivity of the DSCT method has been confirmed for other complex materials such as pigments (Bleuet et al. 2008). Using the DSCT method, Bleuet et al. (2008) obtained the structure and local composition of a heritage test sample: a polycrystalline powder containing quartz-chalcedony and various iron pigments, all inside a glass capillary. For this study, the measurements were collected with an 18 keV synchrotron X-ray beam on the former ID22 beamline (now ID16) of the ESRF (Grenoble, France). This DSCT experiment was carried out in parallel with fluorescence and absorption measurements to complement the structural results with chemical and density information, thus leading to a ***multimodal*** analysis (Figure 4.8(a)). The setup used allowed these simultaneous measurements while moving the sample (along y) and rotating it (about ω). As the focused X-ray beam has a size close to 2.3×1.6 μm^2, the chosen reconstruction step of the 2D section corresponds to a voxel (x,y) of 2.3×2.3 μm^2 and the layer analyzed has a thickness of 1.6 μm. With this setup, ***multimodal*** tomographic analysis was carried out on a powder containing, in a 300 μm capillary, a mixture of chalcedony and iron pigments, particularly hematite (αFe_2O_3). Chalcedony is composed of quartz crystallites, arranged as microfibers, with a diameter of less than 100 nm. 2D structural and chemical images of this cross-section were measured and reconstructed (Figure 4.8b). Although chalcedony is by far the dominant phase, we observe at least three types of iron grains: hematite and siderite, and a mixture of hematite and phyllosilicate phases including greenalite. The ***reverse analysis*** of the data, we can extract a selective diffraction pattern for each grain (for extremely tiny volumes of powder, as small as the size of the voxel (x,y)) (Figure 4.8(c)), which enables recognition of possible unknown phases and subsequent simulation of the data or refinement of their structure (Figure 4.8(d)).

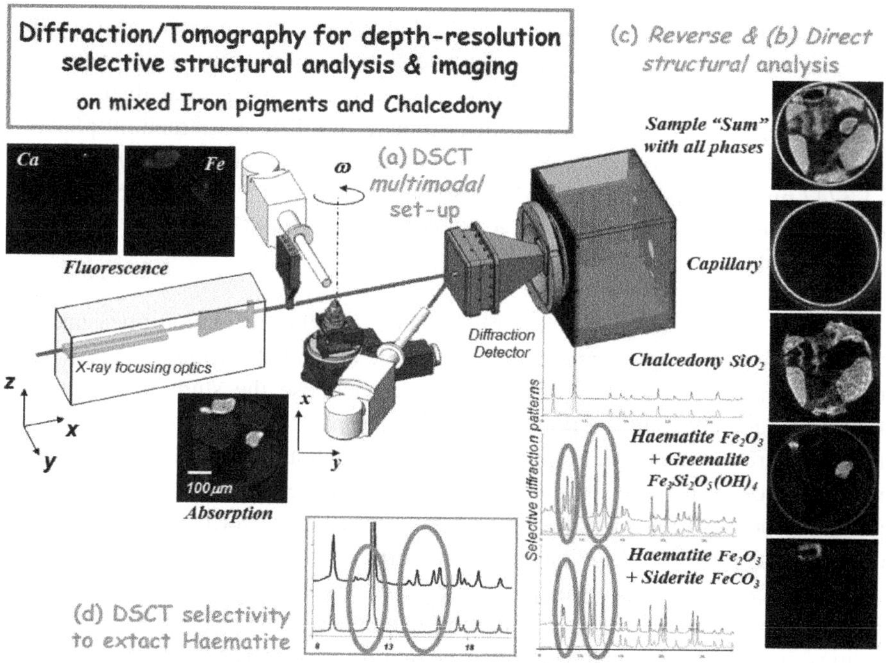

Figure 4.8. *Multimodal DSCT diffraction–tomography experimental device: a) an X-ray beam is focused on the sample positioned on a stage allowing it to be translated (y) and rotated (ω). The fluorescence signal (chemical analysis), transmitted intensity (absorption) and diffracted radiation (phase analysis) are measured simultaneously; b) images of absorption, fluorescence (left) and "sum" diffraction (top right) and several selective images (right) of the same sample, containing chalcedony powder + various iron pigments, are obtained by selecting one or more of its components; c) experimental (black) and calculated (red) diffracted intensities extracted by an "reverse analysis" reveal the phases present: chalcedony and pigments: hematite + greenalite, and hematite + siderite (the most intense lines of which are surrounded by blue ellipses); d) enlargement of the "sum" diffractogram (in black) for angles 2Θ containing the most intense diffraction lines of hematite and simulation (in red) of the chalcedony diffractogram; its comparison (blue ellipses) with the selective diffractograms of iron pigments underlines the sensitivity of the method (according to Bleuet et al. 2008). For a color version of this figure, see www.iste.co.uk/dejoie/synchrotron.zip*

This combination of three analysis methods (scattering, fluorescence, absorption), during non-invasive tomography studies at high spatial resolution, enables reconstructions of cross-sections to visualize the known phases as well as the unidentified phases in the nanomaterials and polycrystalline materials, while associating density cross-sections and/or chemical species. For example, in Figure 4.8, we see that absorption, fluorescence and DSCT are combined to discriminate the mixture of various iron pigments and chalcedony. Absorption images show a contrast between areas with different attenuation coefficients, mainly given by Si (Z=14) and Fe (Z=26). Fluorescence is sensitive to the general distribution of elemental Fe, while DSCT can also discriminate the different structural phases of iron pigments with a common Fe atom: hematite (Fe_2O_3, R-3c, a =0.50355 nm, c =1.3747 nm), siderite ($FeCO_3$, R-3c, a =0.46916 nm, c =1.5379 nm) or monoclinic greenalite ($Fe_3Si_2O_5(OH)_4$, C 2/m, a =0.5390 nm, b =0.9336 nm, c =1.4166 nm, β=90.01°) (Bleuet et al. 2008).

These 3D reconstructions require at least 10,000 diffraction measurements (y,z,ω). These images are very sensitive to sample/artifact absorption as a function of the rotation angle ω, so it is also necessary to choose the study areas carefully because the geometry and maximum sample size must be considered, particularly when there are pigments containing heavy atoms such as lead or mercury. It is also necessary to carefully evaluate the size of the voxel in relation to the total volume to be studied, because the measurement times can become enormous. We can select from different types of experiments: those carried out with a mobile device (with long measurement times and macroscopic voxel sizes (De Meyer et al. 2019)) or those done on large synchrotron radiation centers (with very high X-ray intensity densities, particularly on future 4th-generation centers (ESRF-EBS 2020) and therefore with much shorter measurement times and the possibility of microscopic or even nanoscopic voxel sizes).

4.4.5. *The wide range of DSCT applications*

These "multimodal" DSCT analyses are very useful when studying heritage materials as well as in fields such as materials science, chemistry, geology, environmental sciences or medicine. They open new avenues for non-destructive study in various areas of application: soils, sediments, trace elements and pollutants, metallurgy and industry, where understanding the properties of materials at the microstructural level is paramount (Beale et al. 2014; Maire and Withers 2014). Diffraction tomography has proven particularly useful for studying dilute distributions of minor phases such as those encountered in catalysts (Jacques et al. 2011; O Brien et al. 2012), automotive paint pigments (De Nolf and Janssens 2010), carbon-based materials (Alvarez-Murga et al. 2011, 2012a, 2012b; Autran et al.

2021), as well as complex materials such as cements (Artioli et al. 2010; Valentini et al. 2011; Voltolini et al. 2013; Artioli et al. 2014), materials that protect the limestone of monuments (Vanmeert et al. 2013), filament composite materials (Stock and Almer 2012), speciation of uranium and nuclear fuel (Palancher et al. 2011; Bonnin et al. 2014), as well as to visualize the heterogeneities of radionuclides in mining site soil in Central Asia (Lind et al. 2013). DSCT applications also concern bio-mineralized tissues (Leemreize et al. 2014a; Xiao and Stock 2014), like shell fibers (Leemreize et al. 2013, 2014b), hydroxyapatite bone phantoms (Frolich et al. 2016), transformations/wear of zirconia-based dental prostheses (Mochales et al. 2011) or more directly the enamel, dentine and texture of the teeth (Egan et al. 2013b; Forien et al. 2015).

This wide range of DSCT applications is due to the selectivity of its method as well as to the resolution of the reconstructed images. This resolution is linked to the size of the X-ray beam on the object studied: in the 2000s, the first experiments were carried out with medium resolutions of several μm to now with resolutions up to 10 nm with 4th-generation synchrotron sources (ESRF-EBS 2020). However, very high-resolution analyses require many measurement points and therefore significant calculation resources and a high X-ray intensity. The power of current or future sources, detectors and computing resources means that such analyses became very fast, so kinetic experiments could be carried out by DSCT (Vamvakeros et al. 2016). It should be noted that the DSCT method can also be used with a neutron source, with the advantage of low absorption for large objects, but with a low resolution (a few mm) due to the low neutron flux and difficulty in producing submillimeter beams. It should also be noted that the DSCT method can also be used with a laboratory X-ray source (Cersoy et al. 2015) or with portable macroscopic sources (De Meyer et al. 2019), again with less good spatial resolution.

4.4.6. *Analysis tools for DSCT imaging*

These numerous applications have been facilitated by the continued developments in software, exploiting the diverse needs of structural tomography and make these analyses much more user-friendly for non-specialist users. They rely on the wide variety of powder diffraction software, developed by the crystallography community that allows us to obtain an initial data analysis in a relatively user-friendly manner. However, most of these tools require information on the phases that the object studied could contain, and this knowledge prerequisite is not always possible in the case of heritage materials. Also, for tomography analyses, researchers in this field first used methods based on the selective and partial exploitation of diffraction/scattering information (distribution, positions and shapes of the lines, etc.) coming from the various polycrystals and phases in their sample (Bleuet 2008;

Lazzari et al 2009; Sochi 2013; De Nolf 2014; Frolich and Birkedal 2015). Among these tools, note the development by De Nolf et al. (2014) of the software *XRDUA*, which transforms 2D powder diffraction data (known or unknown) by various reconstruction algorithms and makes it possible to perform various data reductions required to convert the series of powder diffraction patterns into tomography data/images. Also, the development of analysis tools that allow the use of weak signals includes scattering from disordered phases or amorphous phases (Jacques et al. 2013; Birkbak et al. 2017).

Researchers have combined several complementary analysis tools with the DSCT method: (i) pair distribution function (PDF) analyses of amorphous phases (Toby and Egami 1992; Proffen and Billinge 1999; Jacques et al. 2013; Bordet 2018); (ii) analyses of the total profile of the diffractograms with the Rietveld method (Rietveld 1969; Roisnel and Rodriguez-Carvajal 2000; Palancher et al. 2011; Frølich and Birkedal 2015; Frølich et al. 2016); (iii) analyses of particular structural parameters of diffraction lines (profiles, positions, etc.) (Alvarez-Murga et al. 2012a; Vamvakeros et al. 2018); (iv) quantitative characterization of microstructures (Korsunsky et al. 2011); (v) use the entire diffractogram (lines and diffuse background) to simultaneously determine the shapes of polycrystals, nano-crystals and amorphous phases (Birkbak et al. 2015, 2017): structural tomography is sensitive to local size variations in the diffracting/scattering, the texture or the defective parts of the sample, and allows simultaneous 3D mapping of these various properties and/or defects. The diversity and complexity of "real" materials, such as heritage materials, mean that they contain mixtures of polycrystalline powders, amorphous phases, as well as crystals, all interacting differently with the X-ray beam. As relatively large crystals do not have a continuous response as the sample is rotated, software had to be developed to overcome this (i) either to use this specific characteristic (Sorensen et al. 2012; Voltolini et al. 2013); (ii) or to filter them out (Alvarez-Murga et al. 2012b; Voltolini et al. 2013; Vamvakeros et al. 2015). With the help of all these data processing tools, diffraction tomography analyses have become as important as absorption and X-ray fluorescence imaging in several fields, including biomaterials, environmental sciences, paleontology and cultural heritage.

4.4.7. DSCT analyses dedicated to heritage materials

DSCT analyses are effective for the study of heritage materials containing multiple phases and even minor phases such as pigments, or poorly crystallized phases such as binders. Very often, the analyses carried out on these materials are "multimodal" – combined with absorption and fluorescence measurements – because these two measurements are easy to integrate on a DSCT experimental setup. It is extremely important to carry out these measurements in a non-destructive way on micro-samples of an artifact or in a non-invasive way directly on small artifacts. This makes it possible to extract, using structural and chemical maps, the stratigraphy of the paint layers to locate, identify and characterize the successions of fine layers of crystallized, amorphous and/or degraded pigments (Walter 2013; Gonzalez et al. 2019). An example of these studies is the discoloration of a lead-based red pigment, used since antiquity and still on many paintings; this discoloration was better understood thanks to the identification of the various lead phases on a microscopic sample of a van Gogh painting (Vanmeert et al. 2015). Likewise, a study on the carbon blacks of Pompeii, using DSCT experiments and PDF analyses, made it possible to identify minor phases and to isolate the amorphous diffractograms of the carbon blacks in order to compare them with references of known carbon blacks (Cersoy et al. 2016).

Another example of a DSCT study on carbon blacks is that carried out on writings on pieces of papyrus from Egypt by Champollion (Champollion Museum, Vif, France). These experiments were carried out on the ID22 powder diffraction beamline as well as on the CRG-D2AM beamline at the ESRF. As we can see in the following images, extracted from the ID22 data, they extract the signal from the black inks from all of the mixed signals corresponding to the various compounds on the papyrus support (Figure 4.9). Metallic elements have also been identified there, but they are not correlated with the ink: pollution or unintentional contamination:

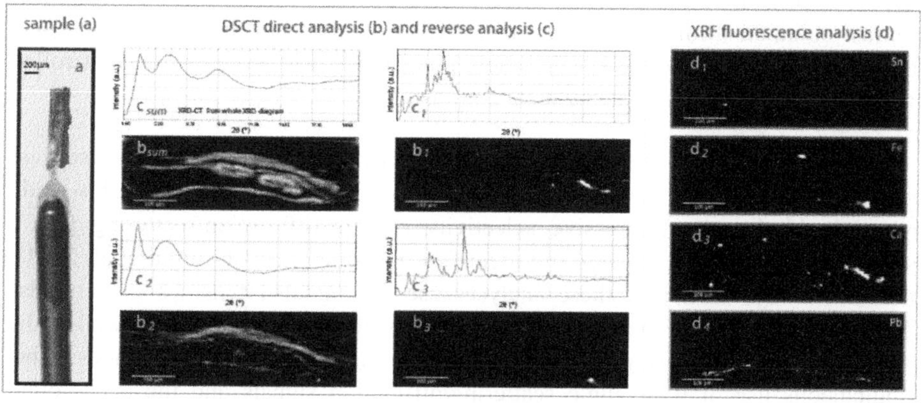

Figure 4.9. *Example of structural and chemical mapping of a sample of papyrus from ancient Egypt. For a color version of this figure, see www.iste.co.uk/dejoie/synchrotron.zip*

NOTE ON FIGURE 4.9.– a) Photo of a piece of papyrus analyzed from the collection of the Champollion Museum in Vif, France – age, 3,500 years old; dimensions of the fragment, ~1.2 x 0.4 x 0.15 mm^3; experiment carried out on the ID11 line of the ESRF at an energy of 65 keV with a 2 µm beam and a 2 µm gap (90 x 260 steps and 154 projections); b_{sum}) DSCT map → "global" diffractogram → papyrus + ink; b_1) DSCT map → selective diffractogram with a line at 2.2° → calcite impurity ?; b_2) DSCT map → diffractogram with line (002) of carbon at 1.9° → papyrus + ink; b_3) DSCT map → selective diffractogram a line at 7.0° → iron oxide impurity?; c_{sum}) reverse DSCT reconstruction → "global" diffractogram with all the lines → papyrus + ink + impurities; c_1) reverse DSCT reconstruction → selective diffractogram with the piece containing Ca → calcite impurity ?; c_2) reverse DSCT reconstruction → diffractogram with top layer → papyrus + ink → almost exclusively carbon; c_3) reverse DSCT reconstruction → selective diffractogram with the piece containing Fe → iron oxide impurity?; d_1) fluorescence map → Sn → impurity ?; d_2) fluorescence map → Fe → iron oxide impurity; d_3) fluorescence map → Ca → calcite impurity; d_4) fluorescence map → Pb → the bottom layer contains some lead (ink?); in this example, we can notice the complementarity of the fluorescence/diffraction analyses and the differentiation of the upper and lower layers of the papyrus (Autran et al. 2021).

The specific signal of the ink extracted by DSCT turns out to be very close to that of carbon black: we therefore have all the elements of a carbon ink, in agreement with what is described in the bibliography on Ancient Egypt, but which had never been directly identified on an artifact (Autran et al. 2021). The structural contrast obtained from this diffraction/scattering signal from the carbon black, i.e. from the black pigment itself, offers new possibilities for "reading" ancient rolled and/or fragile papyri, without chemical contrasts.

All the components of these black inks on papyrus and of the papyrus itself are poorly crystallized and composed mainly of carbon-based materials, so this study indeed highlights the impressive selectivity of this diffraction/scattering DSCT imaging method, which isolates and identifies the components (Autran et al. 2021). In such heritage materials, these analyses use the signals/intensities of particular structural parameters of the diffraction peaks (positions, profiles, etc.), as well as the entire diffraction/scattering patterns by combining Rietveld/PDF methods. This simultaneous use of wide-angle scattering/diffraction signals and the background signal from the refinements makes it possible to map both the crystallized phases (pigments) and the amorphous phases (binders). However, the two global Rietveld/PDF methods can present difficulties when there are many phases present, or when there are reconstruction artifacts due to the presence of large, crystallized grains or absorbing heavy metals, etc., plus the sheer number of diffractograms to process, which can exceed several tens of thousands. Thus, Bordet et al. (2021) proposed another approach, consisting of processing DSCT data using a "multivariate" statistical analysis, which makes it possible to extract the principal components of the whole data as well as the evolution of their proportions in function of the voxel (x, y) measured. For this, they used the multivariate curve resolution–alternative least squares method (MCR-ALS) (Jaumot et al. 2015a). This method has been previously used for mass spectrometry imaging data analyses (Jaumot et al. 2015b), X-ray absorption analyses as a function of time (Cassinelli et al. 2014), absorption and X-ray fluorescence analyses of complex diluted samples (Egan et al. 2013a) and for fluorescence analyses of the inks of Marie-Antoinette's letters to extract falsified passages (Pottier et al. 2017).

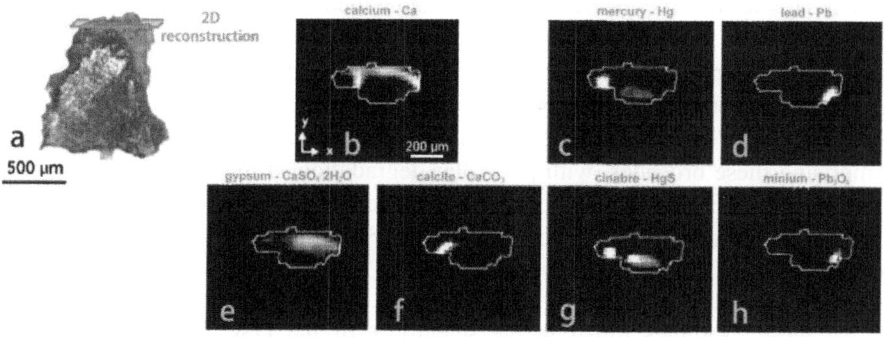

Figure 4.10. *Example showing the mapping of an applied brocade sample taken from a medieval sculpture: a) Photo of a sample taken from the statue of St John from the medieval statuary group from the Château de Montrottier – Haute Savoie, France (Figure 3); the maps presented here correspond to the 2D section located at the top of the sample, the measurements were collected on the CRG-D2AM line of the ESRF with an energy of 20 keV and a beam of 30 x40µm^2: b) fluorescence map → Ca → calcium; c) fluorescence map → Hg → mercury; d) fluorescence map → Pb → lead; e) DSCT diffraction map → gypsum → $CaSO_4$, $2H_2O$; f) DSCT diffraction map → calcite → $CaCO_3$; g) DSCT diffraction map → cinabre → HgS; h) DSCT diffraction map → minium → Pb_3O_4; We can also note the complementarity of fluorescence/diffraction analyses: for example, Ca → gypsum and/or calcite (according to Bordet et al. 2021)*

The studies by Bordet et al. (2021) show that "multivariate" DSCT analyses also give very good results with heritage materials. They applied it to the multiple diffraction measurements necessary to reconstruct their structural maps of applied brocades samples from baroque statues originating from the Duchy of Savoy (Bordet et al. 2021). The decorations, mainly done to statues around the late Middle Ages, aimed to imitate the appearance of the rich textiles/silks embroidered with gold or silver, worn by the highest society classes. They are very heterogeneous and made up of a complex superposition of layers of pigments, binders and metal foils (Lelong et al. 2015a, 2015b). On several applied brocades of wooden statuary groups undergoing restoration (Figure 4.3), samples with volume less than 1 mm^3 were taken and then studied by multimodal DSCT with the ESRF CRG-D2AM beamline. Several virtual sections (Figure 4.10) were collected on these samples, each section requiring the measurement of 10,000 diffraction/fluorescence snapshots (Bordet et al. 2021). Since the phases of these samples were unknown, they determined the proportions of each phase, then, using these initial values made tomographic reconstructions to refine the concentration of each voxel (xy). With this multivariate statistical analysis, they quantified the phases corresponding to the main components in their samples and they were able to determine their concentrations. This method is much faster and more robust than a sequential analysis by refinement of diffraction profiles, so its application to heritage artifacts is very promising. For example, this study made it possible to identify the stratigraphy of the brocades and it verified the association of "binder – support – pigments – metal foil – gilding" materials comprising these brocades, with particular degradation of the tin foil. These results provide information on Savoyard workshops and clarify degradation mechanisms to improve the effectiveness of restorations (Kergourlay et al. 2019; Bordet et al. 2021).

All these examples illustrate the selectivity of the DSCT method, which, by using an intense source of X-ray radiation, obtains accurate, high-resolution 3D structural and chemical maps. However, moving an artifact from a museum to a large laboratory such as synchrotron light centers is generally difficult, which is why precise mapping in DSCT mode is often carried out on micro-samples (Vanmeert et al. 2013; Bordet et al. 2021).

As illustrated in the previous examples, the limited access to heritage artifacts can be overcome by taking samples directly from the artifact for synchrotron studies and associating them with measurements made with portable devices which, in turn, can be installed where the artifact is located. These instruments, like those presented in section 4.1.3 for 2D maps, are being adapted to 3D mapping. Naturally, the power of their sources is quite low, and even if their DSCT maps are carried out at low resolution, they complement the precise measurements made by synchrotron radiation on samples. These two types of DSCT experimental measurements, macroscopic and microscopic, were combined to study the easel painting "*Christ with Singing and Music-Making Angels*" from the end of the 15th century by Hans Memling, revealing a palette of around 10 pigments, water-based gilding, a combination of chalk and gypsum, etc. and characterized the degradation products (Van der Snickt et al. 2011). Two other examples of such a combination of macroscopic and microscopic DSCT analyses are the study of the formation of parasitic phases during the degradation of the yellow orpiment pigment on a 17th century still life painting by Martinus Nellius (Simoen et al. 2019), and the study of brocade stratigraphy of baroque statues from Savoy (Bordet et al. 2021; Martinetto et al. 2021). Fluorescence and diffraction analyses were also used in conjunction in these examples because chemical analysis/imaging could not sufficiently interpret the chemical transformations linked to deterioration.

4.5. Conclusion

The association of structural images/maps with visual images of artifacts allows us to correlate the sensations of our vision more easily with the heterogeneous materials that compose them. Structural mapping/tomography studies have shown their potential to discriminate all phases (natural chemical, polycrystalline, amorphous) with high sensitivities (a contrast of 0.1% is possible with the support material of the heritage object). This great selectivity of structural mapping/tomography is based on the redundancy and variations in the thousands of data points collected where the beam intercepts a minor phase at different “artifact–beam” orientations. 2D mapping analyses can be non-invasive and performed directly on important artifacts, and/or non-destructive on a sample of the artifact, an essential requirement for works of art. From virtual sections reconstructed by DSCT tomography, it is possible to produce 3D maps to identify the different layers of a decoration or painting, either directly on small artifacts or paintings, or on millimeter or micrometric samples of such objects. Chemical mapping/tomography studies are complementary to these structural maps, which are also very sensitive and can easily be carried out in parallel on the same observation areas. Such “multimodal” 2D mapping or 3D DSCT measurements differentiate and determine the location of pigments/binders/supports according to the artifact stratigraphy, including those with chemical elements in common, by playing on certain characteristics (structure, chemical composition, particle size, constraints, textures, etc.). The complexity and compositional variations of the pigments of artifacts may be linked to the different sources of raw materials, but many of these studies also show structural/chemical/morphological variations which are a signature of the preparation, history of the material and its aging. As we saw in the example of studies of lead whites, we see that an artifact is associated with a particular skillset which, itself, is linked to particular uses and preparations of materials. This structural and chemical information observed on artifacts allows us to clarify the “paint palette” concept corresponding to an artist at a given period/workshop (Walter 2015; Walter and de Viguerie 2018): ultimately, they reveal the undertakings, attempts, intentions and methods of the artist.

4.6. Acknowledgments

This work benefited from French governmental support managed by the ANR under the program *Investissements d'Avenir*: ANR-15-IDEX-02 (CDP Patrimalp, IDEX UGA).

4.7. References

Alfeld, M. and Janssens, K. (2015). Strategies for processing mega-pixel X-ray fluorescence hyperspectral data: A case study on a version of Caravaggio's painting "Supper at Emmaus". *Journal of Analytical Atomic Spectrometry*, 30(3), 777–789.

Al-Jawad, M., Steuwer, A., Kilcoyne, S.H., Shore, R.C., Cywinski, R., Wood, D.J. (2007). 2D mapping of texture and lattice parameters of dental enamel. *Biomaterials*, 28(18), 2908–2914.

Alvarez-Murga, M., Bleuet, P., Marques, L., Lepoittevin, C., Boudet, N., Garbarino, G., Mezouar, M., Hodeau, J.L. (2011). Microstructural mapping of C60 phase transformation into disordered graphite at high pressure, using X-ray diffraction micro-tomography. *Journal of Applied Crystallography*, 44(1), 163–171.

Alvarez-Murga, M., Bleuet, P., Garbarino, G., Salamat, A., Mezouar, M., Hodeau, J.L. (2012a). "Compressed graphite" formed during C60 to diamond transformation as revealed by scattering computed tomography. *Physical Review Letters*, 109(2), 025502.

Alvarez-Murga, M., Bleuet, P., Hodeau, J.L. (2012b). Diffraction/scattering computed tomography for three-dimensional characterization of multi-phase crystalline and amorphous materials. *Journal of Applied Crystallography*, 45(6), 1109–1124.

Artioli, G., Cerulli, T., Cruciani, G., Dalconi, M.C., Ferrari, G., Parisatto, M., Rack, A., Tucoulou, R. (2010). X-ray diffraction micro-tomography (XRDCT), a novel tool for non-invasive mapping of phase development in cement materials. *Analytical and Bioanalyical Chem*istry, 397(6), 2131–2136.

Artioli, G., Valentini, L., Dalconi, M.C., Parisatto, M., Voltolini, M., Russo, V., Ferrari, G. (2014). Imaging of nano-seeded nucleation in cement pastes by X-ray diffraction tomography. *International Journal of Materials Research*, 105(7), 628–631.

Autran, P.O., Dejoie, P., Bordet, P., Hodeau, J.L., Dugand, C., Gervason, M., Anne, M., Martinetto, P. (2021). Revealing the nature of black pigments used on ancient Egyptian papyri from Champollion collection. *ACS Analytical Chemistry*, 93(2), 1135–1142.

Beale, A.M., Jacques, S.D.M., Bergwerff, J.A., Barnes, P., Weckhuysen, B.M. (2007). Tomographic energy dispersive diffraction imaging as a tool to profile in three dimensions the distribution and composition of metal oxide species in catalyst bodies. *Angewandte Chemie International Edition*, 46(46), 8832–8835.

Beale, A.M., Jacques, S.D.M., Gibson, E.K., Di Michiel, M. (2014). Progress towards five dimensional diffraction imaging of functional materials under process conditions. *Coordination Chemistry Reviews*, 277, 208–223.

Bergmann, U., Manning, P.L., Wogelius, R.A. (2012). Chemical mapping of paleontological and archeological artifacts with synchrotron X-rays. *Annual Review of Analytical Chemistry*, 5, 361–389.

Birkbak, M.E., Leemreize, H., Frolich, S., Stock, S.R., Birkedal, H. (2015). Diffraction computed tomography reveals the inner structure of complex biominerals. *Nanoscale*, 7(44), 18402–18410.

Birkbak, M.E., Nielsen, I.G., Frolich, S., Stock, S.R., Kenesei, P., Almer, J.D., Birkedal, H. (2017). Concurrent determination of nanocrystal shape and amorphous phases in complex materials by diffraction scattering computed tomography. *Journal of Applied Crystallography,* 50(1), 192–197.

Bish, D., Blake, D., Vaniman, D., Sarrazin, P., Bristow, T., Achilles, C., Dera, P., Chipera, S., Crisp, J., Downs, R.T. et al. (2014). The first X-ray diffraction measurements on Mars. *IUCrJ*, 1(6), 514–522.

Bleuet, P. (2008). Dgt3D Software. European Synchrotron Radiation Facility, Grenoble.

Bleuet, P., Welcomme, E., Dooryhee, E., Susini, J., Hodeau, J.L., Walter, P. (2008). Probing the structure of heterogeneous diluted materials by diffraction tomography. *Nature Materials*, 7(6), 468–472.

Bonnin, A., Wright, J.P., Tucoulou, R., Palancher, H. (2014). Impurity precipitation in atomized particles evidenced by nano X-ray diffraction computed tomography. *Applied Physics Letters,* 105(8), 084103.

Bordet, P. (2018). Application of the pair distribution function analysis for the study of cultural heritage materials. *Comptes Rendus de Physique,* 19(7), 561–574.

Bordet, P., Kergourlay, F., Pinto, A., Blanc, N., Martinetto, P. (2021). Applying multivariate analysis to X-ray diffraction computed tomography: The study of medieval applied brocades. *Journal of Analytical Atomic Spectrometry*, 36, 1724–1734.

Brunetti, A. and Golosio, B. (2001). Software for X-ray fluorescence and scattering tomographic reconstruction. *Computer Physics Communications*, 141(3), 412–425.

Brunetti, A., Cesareo, R., Rao, V.D., Golosio, B. (2004). Geometrical factor influence on Compton profile measurement for biological samples. *Nuclear Instruments and Methods A*, 526(3), 584–592.

Budai, J.D., Yang, W., Tamura, N., Chung, J-S., Tischler, J.Z., Larson, B.C., Ice, G.E., Park C., Norton, D.P. (2003). X-ray micro-diffraction study of growth modes and crystallographic tilts in oxide films on metal substrates. *Nature Materials*, 2(7), 487–492.

Calligaro, T., Dran, J.C., Hamon, H., Moignard, B., Salomon, J. (1998). An external milli-beam for archaeometric applications on the AGLAE IBA facility of the Louvre museum. *Nuclear Instruments and Methods in Physics Research Section B: Beam Interactions with Materials and Atoms*, 136, 339–343.

Cassinelli, W.H., Martins, L., Passos, A.R., Pulcinelli, S.H., Santilli, C.V., Rochet, A., Briois, V. (2014). Multivariate curve resolution analysis applied to time-resolved synchrotron X-ray Absorption Spectroscopy monitoring of the activation of copper alumina catalyst. *Catalysis Today*, 229, 114–122.

Cernik, R.J., Hansson, C.T., Martin, C.M., Preuss, M., Attalah, M., Korsunsky, A.M., Belnoue, J.P., Jun, T.S., Barnes, P., Jacques, S. (2011). A synchrotron tomographic energy-dispersive diffraction imaging study of the aerospace alloy Ti 6246. *Journal of Applied Crystallography*, 44(1), 150–157.

Cersoy, S., Leynaud, O., Álvarez-Murga, M., Martinetto, P., Bordet, P., Boudet, B., Chalmin, E., Castets, G., Hodeau. J.L. (2015). Laboratory implementation of X-ray diffraction/scattering computed tomography. *Journal of Applied Crystallography*, 48(1), 159–165.

Cersoy, S., Martinetto, P., Bordet, P., Hodeau, J.L., Van Elslande, E., Walter, P. (2016). Identifying and quantifying amorphous and crystalline content in complex powdered samples: Application to archaeological carbon blacks. *Journal of Applied Crystallography*, 49(2), 585–593.

Chiari, G. (2008). Saving art *in situ*: A conservation scientist explains how borrowing gadgets from Mars rovers helps preserve culture on Earth. *Nature*, 453(7192), 159.

Cloetens, P., Pateyron-Salomé, M., Buffière, J., Peix, G., Baruchel, J., Peyrin, F., Schlenker, M. (1997). Observation of microstructure and damage in materials by phase sensitive radiography and tomography. *Journal of Applied Physics*, 81(9), 5878–5886.

Cloetens, P., Ludwig, W., Baruchel, J., Van Dyck, D., Vand Landuyt, J., Guigay, J.P., Schlenker, M. (1999). Holotomography: Quantitative phase tomography with micrometer resolution using hard synchrotron radiation X rays. *Applied Physics Letters*, 75(19), 2912–2914.

Cormack, A.M. (1924–1998). Cormack Allan Macleod (1924–1998). Encyclopædia Universalis [Online]. Available at: https://www.universalis.fr/encyclopedie/allan-macleod-cormack/.

Cotte, M., Susini, J., Sole, A., Taniguchi, Y., Chillida, J., Checroun, E., Walter, P. (2008). Applications of synchrotron-based micro-imaging techniques to the chemical analysis of ancient paintings. *Journal of Analytical Atomic Spectrometry (JAAS)*, 23(6) 820–828.

Cotte, M., Susini, J., Dik, J., Janssens, K. (2010). Synchrotron-based X-ray absorption spectroscopy for art conservation: Looking back and looking forward. *Account of Chemical Research*, 43(6), 705–714.

Cotte, M., Pouyet, M., Salomé, M., Rivard, C., De Nolf, W., Castillo-Michel, H., Fabris, T., Monico, L., Janssens, K., Wang, T. et al. (2017). The ID21 X-ray and infrared microscopy beamline at the ESRF status and recent applications to artistic materials. *Journal of Analytical Atomic Spectrometry (JAAS)*, 32(3), 477–493.

Da Silva, J.C. and Menzel, A. (2015). Elementary signals in ptychography. *Optics Express*, 23(26), 33812–33821.

Da Silva J.C., Guilloud C., Hignette O., Jarnias C., Ponchut C., Ruat M., Labiche J.C., Pacureanu A., Yang Y., Salomé M. et al. (2019). Overcoming the challenges of high-energy X-ray ptychography. *Journal of Synchrotron Radiation*, 26(5), 1751–1762.

De Meyer, S., Vanmeert, F., Vertongen, R., Van Loon, A., Gonzalez, V., Delaney, J., Dooley, K., Dik J., Van der Snickt, G., Vandivere, A. et al. (2019). Macroscopic X-ray powder diffraction imaging reveals Vermeer's discriminating use of lead white pigments in *Girl with a Pearl Earring*. *Science Advances – Chemical Physics*, 5(8), 2216–2229.

De Nolf, W. and Janssens, K. (2010). Micro X-ray diffraction and fluorescence tomography for the study of multilayered automotive paints. *Surface and Interface Analysis*, 42(5), 411–418.

De Nolf, W., Dik, J., Van der Snickt, G., Wallert, A., Janssens, K. (2011). High energy X-ray powder diffraction for the imaging of (hidden) paintings. *Journal of Analytical Atomic Spectrometry (JAAS)*, 26(5), 910–916.

De Nolf, W., Vanmeert, F., Janssens, K. (2014). XRDUA: Crystalline phase distribution maps by two-dimensional scanning and tomographic (micro) X-ray powder diffraction. *Journal of Applied Crystallography*, 47(3), 1107–1117.

De Viguerie, L., Duran, A., Bouquillon, A., Sole, V.A., Castaing, J., Walter, P. (2009). Quantitative X-ray fluorescence analysis of an Egyptian faience pendant and comparison with PIXE. *Analytical and Bioanalytical Chemistry*, 395, 2219–2225.

Dejoie, C., Martinetto, P., Tamura, N., Kunz, M., Porcher, F., Bordet, P., Brown, R., Dooryhee, E., Anne, M., McCusker, L. (2014). Crystal structure of an indigo@silicalite hybrid related to the ancient maya blue pigment. *Journal of Physical Chemistry C*, 118(48), 28032–8042.

Dejoie, C., Coduri, M., Petitdemange, S., Giacobbe, C., Covacci, E., Grimaldi, O., Autran, P.O., Mogodi, M.W., Jung, D.S., Fitch, A.N. (2018a). Combining a nine-crystal multi-analyser stage with a two-dimensional detector for high-resolution powder X-ray diffraction. *Journal of Applied Crystallography*, 51(6), 1721–1733.

Dejoie, C., Autran, P.O., Bordet, P., Fitch, A.N., Martinetto, P., Sciau, P., Tamura, N., Wright, J. (2018b). X-ray diffraction and heterogeneous materials: An adaptive crystallography approach. *Comptes Rendus Physique*, 19(7), 553–560.

Dik, J., Janssens, K., Van Der Snickt, G., Van der Loeff, L., Rickers, K., Cotte, M. (2008). Visualization of a lost painting by Vincent van Gogh using synchrotron radiation based X-ray fluorescence elemental mapping. *Analytical Chemistry*, 80(16), 6436–6442.

Dooryhee, E., Martinetto, P., Walter, P., Anne, M. (2004). Synchrotron X-ray analyses in art & archaeology. *Radiation Physics & Chemistry*, 71(3–4), 863–868.

Dooryhee, E., Anne, M., Bardies, I., Hodeau, J.L., Martinetto, P., Rondot, S., Salomon, J., Vaughan, G.B.M., Walter, P. (2005). Non-destructive synchrotron X-ray diffraction mapping of a Roman painting. *Applied Physics A-Materials Science & Processing*, 81(4), 663–667.

Duran, A., Perez-Rodriguez, J.L., Espejo, T., Franquelo, M.L., Castaing, J., Walter, P. (2009). Characterization of illuminated manuscripts by laboratory-made portable XRD and micro-XRD systems. *Analytical and Bioanalytical Chemistry*, 395(7), 1997–2004.

Duran, A., Jiménez de Haro, M.C., Perez-Rodriguez, J.L., Franquelo, M.L., Herrera, L.K., Justo, A. (2010a). Determination of pigments and binders in Pompeian wall paintings using synchrotron radiation – high resolution X-ray powder diffraction and conventional spectroscopy – chromatography. *Archaeometry,* 52(2), 286–307.

Duran, A., Castaing, J., Walter, P. (2010b). X-ray diffraction studies of Pompeian wall paintings using synchrotron radiation and dedicated laboratory-made systems. *Applied Physics A: Materials Science and Processing*, 99, 333–340.

Egan, C.K., Jacques, S.D.M., Cernik, R.J. (2013a). Multivariate analysis of hyperspectral hard X-ray images. *X-Ray Spectrometry*, 42(3), 151–157.

Egan, C.K., Jacques, S.D.M., Di Michiel, M., Cai, B., Zandbergen, M.W., Lee, P.D., Beale, A.M., Cernik, R.J. (2013b). Non-invasive imaging of the crystalline structure within a human tooth. *Acta Biomaterialia,* 9(9), 8337–8345.

ESRF-EBS (2020). [Online]. Available at: http://www.esrf.eu/home/news/general/content-news/general/opening-of-esrf-ebs-a-new-generation-of-synchrotron.html.

Eveno, M., Duran, A., Castaing, J. (2010). A portable X-ray diffraction apparatus for *in situ* analyses of masters' paintings. *Applied Physics A, Materials Science & Processing*, 100, 577–584.

Eveno, M., Moignard, B., Castaing, J. (2011). Portable apparatus for in situ X-ray diffraction and fluorescence analyses of artworks. *Microscopy and Microanalysis* 17, 667–673.

Forien, J.B., Fleck, C., Cloetens, P., Duda, G., Fratzl, P., Zolotoyabko, E., Zaslansky, P. (2015). Compressive residual strains in mineral nanoparticles as a possible origin of enhanced crack resistance in human tooth dentin. *Nano Letters*, 15(6), 3729–3734.

Frolich, S. and Birkedal, H. (2015). *MultiRef*: Software platform for Rietveld refinement of multiple powder diffractograms from *in situ*, scanning or diffraction tomography experiments. *Journal of Applied Crystallography*, 48(6), 2019–2025.

Frolich, S., Leemreize, H., Jakus, A., Xiao, X., Shah, R., Birkedal, H., Almer, J.D., Stock, S.R. (2016). Diffraction tomography and Rietveld refinement of a hydroxy-apatite bone phantom. *Journal of Applied Crystallography*, 49(1), 103–109.

Gianoncelli, A., Castaing, J., Ortega, L., Dooryhee, E., Salomon, J., Walter, P., Hodeau, J.L., Bordet, P. (2008). A portable instrument for in situ determination of the chemical and phase compositions of cultural heritage objects. *X-Ray Spectrometry*, 37(4), 418–423.

Golosio, B., Simionovici, A., Somogyi, A., Chukalina, M., Brunetti, A. (2003). Internal elemental microanalysis combining X-ray fluorescence, Compton and transmission tomography. *Journal of Applied Physics*, 94(1), 145–156.

Golosio, B., Brunetti, A., Cesareo, R. (2004). Algorithmic techniques for quantitative Compton tomography. *Nuclear Instruments & Methods B*, 213, 108–111.

Gonzalez, V., Calligaro, T., Wallez, G., Eveno, M., Toussaint, K., Menu, M. (2016). Composition and microstructure of the lead white pigment in masters paintings using HR Synchrotron XRD. *Microchemical Journal*, 125, 43–49.

Gonzalez, V., Wallez, G., Calligaro, T., Cotte, M., De Nolf, W., Eveno, M., Ravaud, E., Menu, M. (2017). Synchrotron-based high angle resolution and high lateral resolution X-ray diffraction: Revealing lead white pigment qualities in old masters paintings. *Analytical Chemistry*, 89(24), 13203–13211.

Gonzalez, V., Cotte, M., Vanmeert, F., De Nolf, W., Janssens, K. (2019). X-ray diffraction mapping for cultural heritage science: A review of experimental configurations and applications. *Chemistry European Journal*, 25, 1–19 .

Hall, C., Barnes, P., Cockcroft, J.K., Colston, S.L., Häusermann. D., Jacques. S.D.M., Jupe, A.C., Kunz, M. (1998). Synchrotron radiation energy-dispersive diffraction tomography. *Nuclear Instruments and Methods in Physics Research, Section B*, 140(1–2), 253–257.

Hammersley, A.P., Svensson, S.O., Hanfland, M., Fitch, A.N., Hausermann, D. (1996). Two-dimensional detector software: From real detector to idealized image or two-theta scan. *High Pressure Research*, 14(4–6), 235–248.

Harding, G., Kosanetsky, J., Neitzel, U. (1987). X-ray-diffraction-computed-tomography. *Medical Physics*, 14(4), 515–525.

Hirayama, A., Abe, Y., van Loon, A., De Keyser, N., Noble, P., Vanmeert, F., Janssens, K., Tantrakarn, K., Taniguchi, K., Nakai, I. (2018). Development of a new portable X-ray powder diffractometer and its demonstration to on-site analysis of two selected old master paintings from the Rijksmuseum. *Microchemical Journal*, 138, 266–272.

Hodeau, J.L. and Guinebretière, R. (2007). Crystallography: Past and present. *Applied Physics A - Materials Science & Processing*, 89(4), 813–823.

Hodeau, J.L., Bordet, P., Anne, M., Prat, A., Fitch, A.N., Dooryhee, E., Vaughan, G., Freund, A. (1998) Nine crystal multi-analyzer stage for high resolution powder diffraction between 6 and 40keV. *SPIE Proceedings*, 3448, 353–361.

Hofmann, F., Abbey, B., Connor, L., Baimpas, N., Song, X., Keegan, S., Korsunsky, A.M. (2012). Imaging of grain-level orientation and strain in thicker metallic poly-crystals by high energy transmission micro-beam Laue (HETL) diffraction techniques. *International Journal of Materials Research*, 103(2), 192–199.

Hounsfield, G.N. (1973). Computerized transverse axial scanning (tomography): Part I. Description of system. *British Journal of Radiology*, 46(552), 1016–1022.

Jacques, S.D.M., Di Michiel, M., Beale, A.M., Sochi, T., O'Brien, M.G., Espinosa-Alonso, L., Weckhuysen, B.M., Barnes, P. (2011). Dynamic X-ray diffraction computed tomography reveals real-time insight into catalyst active phase evolution. *Angewandte Chemie International Edition*, 50(43), 10148–10152.

Jacques, S.D.M., Di Michiel, M., Kimber, S.A.J., Yang, X., Cernik, R.J., Beale, A.M., Billinge, S.J.L. (2013). Pair distribution function computed tomography. *Nature Communications*, 4(43), 2536.

Janssens, K., Alfeld, M., Van der Snickt, G., De Nolf, W., Vanmeert, F., Radepont, M., Monico, L., Dik, J., Cotte, M., Falkenberg, G. et al. (2013). The use of synchrotron radiation for the characterization of artists' pigments and paintings. *Annual Review of Analytical Chemistry*, 6, 399–425.

Jaumot, J., De Juan, A., Tauler, R. (2015a). MCR-ALS GUI 2.0: New features and application. *Chemometrics and Intelligent Laboratory Systems*, 140, 1–12.

Jaumot, J., De Juan, A., Tauler, R. (2015b). Potential use of multivariate curve resolution for the analysis of mass spectrometry images. *Analyst*, 140(3), 837–846.

Jensen, T.H., Bech, M., Bunk, O., Thomsen, M., Menzel, A., Bouchet, A., Le Duc, G., Feidenhans'l, R., Pfeiffer, F. (2011). Brain tumor imaging using small-angle X-ray scattering tomography. *Physics in Medecine & Biology*, 56(6), 1717–1726.

Johnson, G., King, A., Honnicke, M.G., Marrow, J., Ludwig, W. (2008). X-ray diffraction contrast tomography: A novel technique for three-dimensional grain mapping of poly-crystals. II. The combined case. *Journal of Applied Crystallography*, 41(2), 310–318.

Jupe, A.C., Stock, S.R., Lee, P.L., Naik, N.N., Kurtis, K.E., Wilkinson, A.P. (2004). Phase composition depth profiles using spatially resolved energy dispersive X-ray diffraction. *Journal of Applied Crystallography*, 37, 967–976.

Kak, A.C. and Slaney, M. (1988). *Principles of Computerized Tomographic Imaging*. IEEE Press, New York.

Kak, A.C. and Slaney, M. (2001). *Principles of Computerized Tomographic Imaging*. SIAM, Philadelphia.

Kergourlay, F., Martinetto, P., Blanc, N., Boudet, N., Arnaud, S., Hodeau, J.L., Bordet, P., Lelong, F., Champdavoine, S., Fabre, F. (2019). *Analyse multivariée pour l'etude de polychromies médiévales par XRF/XRD-CT*. Colloque RX & Matière, Nancy.

Kim, S.A., Punshon, T., Lanzirotti, A., Li, L., Alonso, J.M., Ecker, J.R., Kaplan, J., Guerinot, M.L. (2006). Localization of iron in Arabidopsis seed requires the vacuolar membrane transporter VIT1. *Science*, 314(5803), 1295–1298.

King, A., Johnson, G., Engelberg, D., Ludwig, W., Marrow, J. (2008). Observations of intergranular stress corrosion cracking in a grain-mapped poly-crystal. *Science*, 321(5887), 382–385.

King, A., Herbig, M., Ludwig, W., Reischig, P., Lauridsen, E.M., Marrow, T., Buffière, J. (2010). Non-destructive analysis of micro texture and grain boundary character from X-ray diffraction contrast tomography. *Nuclear Instruments and Methods in Physics Research B*, 268(3–4), 291–296.

Kleuker, U., Suortti, P., Weyrich, W., Spanne, P. (1998). Feasibility study of X-ray diffraction computed tomography for medical imaging. *Physics in Medicine and Biology*, 43(10), 2911–2923.

Korsunsky, A.M., Baimpas, N., Song, X., Belnoue, J., Hoffmann, F., Abbey, B., Xie, M., Andrieux, J., Buslaps, T., Neo, T.K. (2011). Strain tomography of polycrystalline zirconia dental prostheses by synchrotron X-ray diffraction. *Acta Materialia*, 59(6), 2501–2513.

Kuhlmann, M., Feldkamp, J.M., Roth, S.V., Schroer, C.G. (2008). Development of XAS micro-tomography and related methods. *IEEE Nuclear Science Symposium Conference*, 8, 570–573.

La Riviere, P.J. and Vargas, P.A. (2006). Monotonic penalized-likelihood image reconstruction for X-ray fluorescence computed tomography. *IEEE Transactions on Medical Imaging*, 25(9), 1117–1129.

Larson, B.C., Yang, W., Ice, G.E., Budai, J.D., Tischler, J.Z. (2002). Three-dimensional X-ray structural microscopy with sub-micrometer resolution. *Nature*, 415(6874), 887–890.

Lazzari, O. Jacques, S., Sochi, T., Barnes, P. (2009). Reconstructive color X-ray diffraction imaging: A novel TEDDI imaging method. *Analyst*, 134(9), 1802–1807.

Lazzari, O., Egan, C.K., Jacques, S.D.M., Sochi, T., Di Michiel, M., Cernik, R.J., Barnes, P. (2012). A new approach to synchrotron energy-dispersive X-ray diffraction computed tomography. *Journal of Synchrotron Radiation*, 19(4), 471–477.

Le Fur, D. (2002). Les pigments dans la peinture égyptienne. In *Pigments et colorants de l'Antiquité et du Moyen Âge*. CNRS Editions, Paris, 181–188.

Leemreize, H., Almer, J.D., Stock, S.R., Birkedal, H. (2013). Three-dimensional distribution of polymorphs and magnesium in a calcified underwater attachment system by diffraction tomography. *Journal of the Royal Society Interface*, 10(86), 20130319.

Leemreize, H., Birkbak, M., Frolich, S., Kenesei, P., Almer, J.D., Stock, S.R., Birkedal, H. (2014a). Diffraction computed tomography reveals the inner structure of complex biominerals. *Proceedings of SPIE*, 9212, 92120C.

Leemreize, H., Eltzholtz, J.R., Birkedal, H. (2014b). Lattice macro and microstrain fluctuations in the calcified byssus of Anomia simplex. *European Journal of Mineralogy*, 26(4), 517–522.

Lelong, F., Champdavoine, S., Pouyet, E., Guiblain, T. (2015a). Caractérisation de décors dits "brocarts appliqués" dans la polychromie des sculptures, datées des années 1480–1530, rattachées à l'ancien duché de Savoie. *La rubrique des patrimoines de Savoie*, 35, 24–25.

Lelong, F., Champdavoine, S., Lefèvre, M., Peurichard, S., Pouyet, E., Terpent C. (2015b). Identification de décors en brocarts appliqués dans la polychromie de sculptures rattachées à l'ancien duché de Savoie. *La rubrique des patrimoines de Savoie*, 36, 8–9.

Liebi, M., Georgiadis, M., Menzel, A., Schneider, P., Kohlbrecher, J., Bunk, O., Guizar-Sicairos, M. (2015). Nanostructure surveys of macroscopic specimens by small-angle scattering tensor tomography. *Nature*, 527(7578), 349–352.

Liebi, M., Georgiadis, M., Kohlbrecher, J., Holler, M., Raabe, J., Usov, I., Menzel, A., Schneider, P., Bunk, O., Guizar-Sicairos, M. (2018). Small-angle X-ray scattering tensor tomography: Model of the three-dimensional reciprocal-space map, reconstruction algorithm and angular sampling requirements. *Acta Crystallographica Section A: Foundations and Advances*, 74(1), 12–24.

Lind, O.C., De Nolf, W., Janssens, K., Salbu, B. (2013). Micro-analytical characterization of radioactive heterogeneities in samples from Central Asian TENORM sites. *Journal of Environmental Radioactivity*, 123, 256–261.

Ludwig, W., Schmidt, S., Lauridsen, E.M., Poulsen, H.F. (2008). X-ray diffraction contrast tomography: A novel technique for three-dimensional grain mapping of poly-crystals: I. Direct beam case. *Journal of Applied Crystallography*, 41(2), 302–309.

Magelhaes, P.J., Ram S.J., Abramoff, M.D. (2004). Image processing with ImageJ. *Biophotonics International*, 11(7), 36–42.

Maire, E. and Withers, P.J. (2014). Quantitative X-ray tomography. *International Material Review*, 59(1), 1–43.

Manceau, A., Tamura, N., Marcus, M.A., MacDowell, A.A., Celestre, R.S., Sublett, R.E., Sposito, G., Padmore, H.A. (2002). Deciphering Ni sequestration in soil ferromanganese nodules by combining X-ray fluorescence, absorption, and diffraction at micrometer scales of resolution. *American Mineralogist*, 87(10), 1494–1499.

Martinetto, P., Anne, M., Dooryhee, E., Walter, P., Tsoucaris, G. (2002). Synthetic hydrocerussite by powder X-ray diffraction. *Acta Crystallographica C*, 58(6), i82–i84.

Martinetto, P., Blanc, N., Boudet, N., Arnaud, S., Hodeau, J.L., Bordet, P., Lelong, F., Champdavoine, S., Fabre, F. (2019). Analyse multivariée pour l'etude de polychromies médiévales par XRF/XRD-CT. *Colloque RX & Matière*, Nancy.

Martinetto, P., Blanc, N., Bordet, P., Champdavoine, S., Fabre, F., Guiblain, T., Hodeau, J.L., Lelong, F., Leynaud, O., Prat, A. et al. (2021). Non-invasive X-ray investigations of medieval sculptures: New insights on "applied tin-relief brocade" technique. *Journal of Cultural Heritage*, 47(2), 89–99.

Matsui, H., Ishiguro, N., Uruga, T., Sekizawa, O., Higashi, K., Maejima, N., Tada, M. (2017). Operando 3D visualization of migration and degradation of a platinum cathode catalyst in a polymer electrolyte fuel cell. *Angewandte Chemie*, 56(32), 9371–9375.

Mendoza Cuevas, A., Bernardini, A., Gianoncelli, A., Tuniza, C. (2015). Energy dispersive X-ray diffraction and fluorescence portable system for cultural heritage applications. *X-Ray Spectrometry*, 44, 105–115.

Mochales, C., Maerten, A., Rack, A., Cloetens, P., Mueller, W.D., Zaslansky, P., Fleck, C. (2011). Monoclinic phase transformations of zirconia-based dental prostheses, induced by clinically practiced surface manipulations. *Acta Biomaterialia*, 7(7), 2994–3002.

Monico, L., Van der Snickt, G., Janssens, K., De Nolf, W., Miliani, C., Verbeeck, J., Tian, H., Tan, H., Dik, J., Radepont, M. et al. (2011a). Degradation process of lead chromate in paintings by Vincent van Gogh studied by means of synchrotron X-ray spectromicroscopy and related methods. 1. Artificially aged model samples. *Analytical Chemistry*, 83(4), 1214–1223.

Monico, L., Van der Snickt, G., Janssens, K., De Nolf, W., Miliani, C., Dik, J., Radepont, M., Hendriks, E., Geldof, M., Cotte, M. (2011b). Degradation process of lead chromate in paintings by Vincent van Gogh studied by means of synchrotron X-ray spectromicroscopy and related methods. 2. Original paint layer samples. *Analytical Chemistry*, 83(4), 1224–1231.

Musée Champollion (2020). Musée situé à Vif dans la demeure familiale des Champollion présentant une valorisation culturelle et scientifique de ses collections [Online]. Available at: https://www.culture.gouv.fr/Regions/Drac-Auvergne-Rhone-Alpes/Focus/Le-musee-Champollion-a-Vif-devient-musee-de-France.

Nakai, I. and Abe, Y. (2012). Portable X-ray powder diffractometer for the analysis of art and archaeological materials. *Applied Physics A, Materials Science & Processing*, 106(2), 279–293.

Nelson, G.J., Harris, W.M., Izzo, J.R., Grew, K.N., Chiu, W.K.S., Chu, Y.S., Yi, J., Andrews, J.C., Liu, Y., Pianetta, P. (2011). Three-dimensional mapping of nickel oxidation states using full field X-ray absorption near edge structure nano-tomography. *Applied Physics Letters*, 98(17), 173109-3.

Nielsen, S.F., Wolf, A., Poulsen, H.F., Ohler, M., Lienert, U., Owen, R.A. (2000). A conical slit for three-dimensional XRD mapping. *Journal of Synchrotron Radiation*, 7, 103–109.

O'Brien, M.G., Jacques, S.D.M., Di Michiel, M., Barnes, P., Weckhuysen, B.M., Beale, A.M. (2012). Active phase evolution in single Ni/Al2O3 methanation catalyst bodies studied in real time using combined mu-XRDCT and mu-absorption-CT. *Chemical Science*, 3(2), 509–523.

Palancher, H., Tucoulou, R., Bleuet, P., Bonnin, A., Welcomme, E., Cloetens, P. (2011). Hard X-ray diffraction scanning tomography with sub-micrometre spatial resolution: Application to an annealed gamma-U0.85Mo0.15 particle. *Journal of Applied Crystallography,* 44(5), 63–70.

PatrimAlp (2018). Sciences du patrimoine. Regards croisés sur l'œuvre. Université Grenoble Alpes [Online]. Available at: https://patrimalp.univ-grenoble-alpes.fr/.

Pliny the Elder (1991). *Natural History: A Selection*. Translated by J. Healy. Penguin Classics, London.

Poline, V., Bordet, P., Leynaud, O., Prat, A., Bruyere, R., Blanc, N., Lelong, F., Martinetto, P. (2023). A mobile instrument for joint X-ray fluorescence and diffraction measurements on complex-shape Cultural Heritage objects. *The European Physical Journal Plus*, 138, 239.

Pottier, F., Aristide-Hastir, I., Lheureux R., Michelin, A., Andraud, C., Genachte-Le Bail, A., Mirabaud, S., Lemaigre-Gaffier, P. and Rabin I. (2017). REX II. Recherches sur l'extraction et l'exploitation des tracés sous-jacents dans les manuscrits anciens : application aux lettres de Marie-Antoinette II. *Fondation des Sciences du Patrimoine* [Online]. Available at: https://www.sciences-patrimoine.org/projet/rex-2-marie-antoinette/.

Poulsen, H.F., Jensen, D.J., Vaughan, G.B.M. (2004). Three-dimensional X-ray diffraction microscopy using high-energy X-rays. *MRS Bulletin*, 29(3), 166–169.

Proffen, T. and Billinge, S.J.L. (1999). PDFFIT, a program for full profile structural refinement of the atomic pair distribution function. *Journal of Applied Crystallography*, 32(3), 572–575.

Radepont, M., De Nolf, W., Janssens, K., Van der Snickt, G., Coquinot, Y., Klaassen, L., Cotte, M. (2011). The use of microscopic X-ray diffraction for the study of HgS and its degradation products corderoite (α-$Hg_3S_2Cl_2$), kenhsuite (γ-$Hg_3S_2Cl_2$) and calomel (Hg_2Cl_2) in historical paintings. *Journal of Analytical Atomic Spectrometry (JAAS)*, 26(5), 959–968.

Regueiro, M.N., Monceau, P., Hodeau, J.L. (1992). Crushing C60 to diamond at room-temperature. *Nature*, 355(6357), 237–239.

Reguer, S., Mirambet, F., Dooryhee, E., Hodeau, J.L., Dillmann, P., Lagarde, P. (2009). Structural evidence for the desalination of akaganeite in the preservation of iron archaeological objects, using synchrotron X-ray powder diffraction and absorption spectroscopy. *Corrosion Science,* 51(12), 2795–2802.

Rietveld H.M. (1969). A profile refinement method for nuclear and magnetic structures. *Journal of Applied Crystallography*, 2(2), 65–71.

Roisnel, T. and Rodriguez-Carvajal, J. (2000). WinPLOTR: A Windows tool for powder diffraction pattern analysis. *Materials Science Forum*, 378(1), 118–123.

Salome, M., Peyrin, F., Cloetens, P., Odet, C., Laval-Jeantet, A.M., Baruchel, J., Spanne, P. (1999). A synchrotron radiation micro-tomography system for the analysis of trabecular bone sample. *Medical Physics*, 26(10), 2194–2204.

Sanchez, D.F., Simionovici, A.S., Lemelle, L., Cuartero, V., Mathon, O., Pascarelli, S., Bonnin, A., Shapiro, R., Konhauser, K., Grolimund, D. et al. (2017). 2D/3D microanalysis by energy dispersive X-ray absorption spectroscopy tomography. *Scientific Reports*, 7, 16453.

Sánchez-Navas, A., López-Cruz, O., Velilla, N., Vidal, I. (2013). Crystal growth of lead carbonates: Influence of the medium and relationship between structure and habit. *Journal of Crystal Growth*, 376, 1–10.

Sarrazin, P., Chiari, G., Gailhanou, M. (2008). A portable non-invasive XRD/XRF instrument for the study of Art objects. *Advances in X-ray Analysis*, 52, 175–186.

Schroer, C.G. (2001). Reconstructing X-ray fluorescence micro-tomograms. *Applied Physics Letters*, 79(12), 1912–1914.

Schroer, C.G., Kuhlmann, M., Roth, S.V., Gehrke, R., Stribeck, N., Almendarez-Camarillo, A., Lengeler, B. (2006). Mapping the local nanostructure inside a specimen by tomographic small-angle X-ray scattering. *Applied Physics Letters,* 88(16), 164102.

Sciau, P., Goudeau, P., Tamura, N., Dooryhee, E. (2006). Micro scanning X-ray diffraction study of Gallo-Roman Terra Sigillate ceramics. *Applied Physics A - Materials Science & Processing*, 83(2), 219–224.

Shahwan, T., Zunbul, B., Akar, D. (2005). Study of the scavenging behavior and structural changes accompanying the interaction of aqueous Pb2+ and Sr2+ ions with calcite. *Geochemical Journal*, 39(4), 317–326.

Simoen, J., De Meyer, S., Vanmeert, F., de Keyser, N., Avranovich, E., Van der Snickt, G., Van Loon, A., Keune, K., Janssens, K. (2019). Combined micro- and macro-scale X-ray powder diffraction mapping of degraded Orpiment paint in a 17th century still life painting by Martinus Nellius. *Heritage Science*, 7(83), 1–12.

Snigirev, A., Snigireva, I., Kohn, V., Kuznetsov, S., Schelokov, I. (1995). On the possibilities of X-ray phase contrast micro-imaging by coherent high-energy synchrotron radiation. *Review of Scientific Instruments*, 66(12), 5486–5492.

Sochi, T. (2013). EasyDD software: New program with new approach for spectral data analysis. *Measurement,* 46(9), 3502–3507.

Solé, V.A., Papillon, E., Cotte, M., Walter, P., Susini, J. (2007). A multiplatform code for the analysis of energy-dispersive X-ray fluorescence spectra. *Spectrochimica Acta Part B*: *Atomic Spectroscopy*, 62, 63–68.

Sorensen, H.O., Schmidt, S., Wright, J.P., Vaughan, G.B.M., Techert, S., Garman, E.F., Oddershede, J., Davaasambuu, J., Paithankar, K.S., Gundlach, C. et al. (2012). Multigrain crystallography. *Zeitschrift für Kristallographie – Crystalline Materials*, 227(1), 63–78.

Stock, S.R. and Almer, J.D. (2012). Diffraction microcomputed tomography of an Al-matrix SiC-monofilament composite. *Journal of Applied Crystallography,* 45(6), 1077–1083.

Stock, S.R., De Carlo, F., Almer, J.D. (2008). High energy X-ray scattering tomography applied to bone. *Journal of Structural Biology*, 161(2), 144–150.

Stols-Witlox, M.J.N., Megens, L., Carlyle, L. (2012). "To prepare white excellent...": Reconstructions investigating the influence of washing, grinding and decanting of stack-process lead white on pigment composition and particle size. In *The Artist's Process: Technology and Interpretation*, Eyb-Green, S., Townsend, J.H., Clarke, M., Nadolny, J., Kroustallis, S. (eds). Archetype Publications Ltd, London, 112–129.

Tafforeau, P., Boistel, R., Boller, E., Bravin, A., Brunet, M., Chaimanee, Y., Cloetens, P., Feist, M., Hoszowska, J., Jaeger, J.J. et al. (2006). Applications of X-ray synchrotron micro-tomography for non-destructive 3D studies of paleontological specimens. *Applied Physics A*, 83, 195–202.

Tamura, N., Spolenak, R., Valek, B.C., Manceau, A., Meier Chang, N., Celestre, R.S., MacDowell. A.A., Padmore, H.A., Patel, J.R. (2002a). Submicron X-ray diffraction and its applications to problems in materials and environmental science. *Review of Scientific Instruments*, 73(3), 1369–1372.

Tamura, N., MacDowell, A.A., Celestre, R.S., Padmore, H.A., Valek, B., Bravman, J.C., Spolenak, R., Brown, W.L., Marieb, T., Fujimoto, H. et al. (2002b). High spatial resolution grain orientation and strain mapping in thin films using polychromatic submicron X-ray diffraction. *Applied Physics Letters*, 80(20), 3724–3726.

Toby, B.H. and Egami, T. (1992). Accuracy of pair distribution function analysis applied to crystalline and non-crystalline materials. *Acta Crystallographica - Section A,* 48(3), 336–346.

Valentini, L., Dalconi, M.C., Parisatto, M., Cruciani, G., Artioli, G. (2011). Towards three-dimensional quantitative reconstruction of cement microstructure by X-ray diffraction micro-tomography. *Journal of Applied Crystallography*, 44(2), 272–280.

Vamvakeros, A., Jacques, S.D.M., Di Michiel, M., Middelkoop, V., Egan, C.K., Cernik, R.J., Beale, A.M. (2015). Removing multiple outliers and single-crystal artefacts from X-ray diffraction computed tomography data. *Journal of Applied Crystallography*, 48(6), 1943–1955.

Vamvakeros, A., Jacques, S.D.M., Di Michiel, M., Senecal, P., Middelkoop, V., Cernik, R., Beale, A.M. (2016). Interlaced X-ray diffraction computed tomography. *Journal of Applied Crystallography*, 49(2), 485–496.

Vamvakeros, A., Jacques, S.D.M., Di Michiel, M., Matras, D., Middelkoop, V., Ismagilov, I.Z., Matus, E.V., Kuznetsov, V.V., Drnec, J., Senecal, P. et al. (2018). 5D operando tomographic diffraction imaging of a catalyst bed. *Nature Communications*, 9, 4751.

Van der Snickt, G., Dik, J., Cotte, M., Janssens, K., Jaroszewicz, J., De Nolf, W., Groenewegen, J., Van der Loeff, L. (2009). Characterization of a degraded cadmium yellow (CdS) pigment in an oil painting by means of synchrotron radiation based X-ray techniques. *Analytical Chemistry*, 81(7), 2600–2610.

Van der Snickt, G., Miliani, C., Janssens, K., Brunetti, B.G., Romani, A., Rosi, F., Walter, P., Castaing, J., De Nolf, W., Klaassen, L. et al. (2011). Material analyses of 'Christ with singing and music-making Angels', a late 15th-C panel painting attributed to Hans Memling and assistants: Part I. non-invasive *in situ* investigations. *Journal of Analytical Atomic Spectrometry (JAAS)*, 26(11), 2216–2229.

Van der Snickt, G., Janssens, K., Dik, J., De Nolf, W., Vanmeert, F., Jaroszewicz, J., Cotte, M., Falkenberg, G., Van der Loeff, L. (2012). Characterization of a degraded cadmium yellow (CdS) pigment in an oil painting by means of synchrotron radiation based X-ray techniques. *Analytical Chemistry*, 84(23), 10221–10228.

Vandivere, A., Wadum, J., Leonhardt, E. (2020). The girl in the spotlight: Vermeer at work, his materials and techniques in "Girl with a Pearl Earring". *Heritage Science*, 8, 20.

Vanmeert, F., Mudronja, D., Fazinic, S., Janssens, K., Tibljas, D. (2013). Semi-quantitative analysis of the formation of a calcium oxalate protective layer for monumental limestone using combined micro-XRF and micro-XRPD. *X-Ray Spectrometry*, 42(4), 256–261.

Vanmeert, F., Van der Snickt, G., Janssens, K. (2015). Plumbonacrite identified by X-ray powder diffraction tomography as a missing link during degradation of red lead in a Van Gogh painting. *Angewandte Chemie – International Edition*, 54(12), 3607–3610.

Vanmeert, F., De Nolf, W., De Meyer, S., Dik, J., Janssens, K. (2018a). Macroscopic X-ray powder diffraction scanning, a new method for highly selective chemical imaging of works of art: Instrument optimization. *Analytical Chemistry*, 90(11), 6436–6444.

Vanmeert, F., Hendriks, E., Van der Snickt, G., Monico, L., Dik, J., Janssens, K. (2018b). Chemical mapping by macroscopic X-ray powder diffraction of Van Gogh's *Sunflowers*: Identification of areas with higher degradation risk. *Angewandte Chemie – International Edition*, 57(25), 7418–7422.

Vitruvius (2009). *De architectura VII.* Translated by R. Schofield. Penguin Classics, London.

Voltolini, M. (2010). XRD data cleaner/splitter plugin for Image J [Online]. Available at: http://marcov.x10.mx/index_files/DCtomo_splitter/.

Voltolini, M., Dalconi, M.C., Artioli, G., Parisatto, M., Valentini, L., Russo, V., Bonnin, A., Tucoulou, R. (2013). Understanding cement hydration at the microscale: New opportunities from 'pencil-beam' synchrotron X-ray diffraction tomography. *Journal of Applied Crystallography*, 46(1), 142–152.

Walter, P. (2013). Chemical analysis and painted colors: The mystery of Leonardo's Sfumato. *European Review*, 21(2), 175–189.

Walter, P. (2015). *Sur la palette de l'artiste. La physicochimie dans la création artistique*. Collège de France, Paris.

Walter, P. and de Viguerie, L. (2018). Materials science challenges in paintings. *Nature Materials*, 17(2), 106–109.

Walter, P., de Viguerie, L., Castaing, J. (2011). Appareils portables pour l'analyse des œuvres d'art aux rayons X. *Images de la physique*, 79–85.

Welcomme, E., Walter, P., Van Eslande, E., Tsoucaris, G. (2006). Investigation of white pigments used as make-up during the Greco-Roman period. *Applied Physics A – Materials Science & Processing*, 83(4), 551–556.

Welcomme, E., Walter, P., Bleuet, P., Hodeau, J.L., Dooryhee, E., Martinetto, P., Menu, M. (2007). Classification of lead white pigments using synchrotron radiation micro X-ray diffraction. *Applied Physics A-Materials Science & Processing*, 89(4), 825–832.

Wu, S.C., Xiao, T.Q. Withers, P.J. (2017). The imaging of failure in structural materials by synchrotron radiation X-ray micro-tomography. *Engineering Fracture Mechanics*, 182, 127–156

Xiao, X. and Stock, S.R. (2014). X-ray computed tomography. *Biomineralization Sourcebook: Characterization of Biominerals and Biomimetic Materials*, DiMasi, E. and Gower, L.B. (eds). CRC Press, Boca Raton, 233–247.

5

Contribution of Synchrotron Radiation to the Study of Glazed Ancient Ceramics

Philippe SCIAU and Chantal BROUCA-CABARRECQ
CEMES, Toulouse, France

5.1. Introduction

Ceramic objects have fairly short life spans. Once broken or chipped, these objects are typically abandoned, and unlike metals or glass, are difficult to recycle. Since they degrade little in soil in most cases, they are often discovered in abundance during archaeological excavations[1]. In addition, the shapes of ceramic objects and their decorations are excellent cultural markers, which follow societal developments and serve as precise and reliable chronological benchmarks for archaeologists. Due to this key role played by ceramic objects, their study is a scientific discipline in its own right within archeology: ceramology. The importance of ceramics in the study of ancient civilizations naturally led scientists from so-called "hard" sciences to take an interest and, since the birth of archaeometry[2] in the 1960s, they have been the focus of numerous works, whether to date them (thermoluminescence, etc.) or to study their life cycle (raw materials, production

1 The first manmade terracotta ceramics date back to the Paleolithic period, but it was only really since the Neolithic period that their use developed.
2 Archeometry: scientific discipline using methods borrowed from physics, geology, chemistry or biochemistry to study archaeological sites or objects.

Synchrotron Radiation, Cultural Heritage, Biomineralization,
coordinated by Catherine DEJOIE, Pauline MARTINETTO and Nobumichi TAMURA.

processes, use, state of alteration, restoration and conservation) (Tite 2008; Maritan 2019).

The first real physicochemical studies of ancient ceramics, however, predate the creation of archaeometry as a scientific discipline. At the end of the 19th century, petrographic analyses of thin sections of clay were carried out as part of provenance studies, i.e. studies whose objective was to identify where they were made (Bamps 1883). Gravimetric methods were also among the first to be used (Jenkins 1908). Chemical analyses of heavy metals were carried out a few decades later to understand more about their provenance (Oakley 1933). It was also at this time that the first attempts to analyze mineral composition by X-ray diffraction were made, but it was not until the end of the 1950s that relevant information was obtained. After this, many techniques (Raman spectroscopy, electron microscopy), among the most advanced at the time, were used to analyze ancient clay and coatings[3], and then the decorative layers of ancient ceramics. Among these advanced techniques, analysis methods based on synchrotron radiation, particularly with the arrival of X-ray beams with sizes around a micron (or micro-beams), have proven to be particularly well suited to the study of the chemistry and microstructure of these complex materials, as we will show in this chapter.

5.2. The problem

Ceramic objects are usually made up of a body, a coating (slip, enamel, glaze, etc.) and decorations, which can be embossing, part of the coating or painted under or on the latter (Figure 5.1). Several different coatings can be applied to the same object, as in the example where a white slip has been applied between the glaze and the body of the vase. Most natural clays generally contain enough iron to color them, and this process saves white clay resources, particularly for the production of common ceramics. Honey lead glaze (75% by mass of lead oxide PbO) of the example given in Figure 5.1 reacted with the components of the slip during firing to form the crystals observed at the interface on the scanning electron microscopy (SEM) image (Figure 5.1(c)). Given the size of these crystals, spatial resolution of around a micrometer is necessary for a detailed study of their structure and chemical composition.

The thicknesses of the coatings are generally between a few tens and a few hundred micrometers, and in many cases, as we will see a little later, the depth or distance from the surface is a parameter to consider in structural studies.

3 Generic term designating the surface layer(s) partially or totally covering the body of the ceramic.

The chemical composition and distribution of the crystals they contain vary significantly with depth, so it is often essential to consider this parameter and to carry out stratigraphic studies on cross-sections.

Furthermore, due to the dimensions involved and the intrinsic heterogeneity of these layers, it is necessary to carry out stratigraphic studies both over sufficiently large areas and with a spatial resolution, at least a micrometer or sub-micrometer in depth.

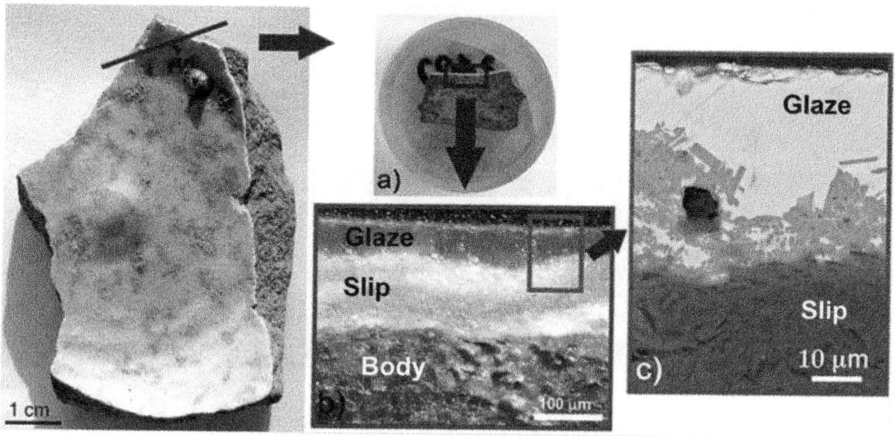

Figure 5.1. *18th century bowl base (Cox, Haute-Garonne, France) with an indication of the fragment taken for the stratigraphic analysis. a) Sample cross-section; b) optical image of the different layers (honey glaze, white slip and body); and c) detail observed by electron microscopy of the glaze/slip interface (backscattered electrons). For a color version of this figure, see www.iste.co.uk/dejoie/synchrotron.zip*

Obtaining elemental composition maps meeting these criteria can be easily achieved using standard laboratory equipment such as a scanning electron microscope equipped with an X-ray fluorescence detector (SEM-EDS) (Sciau and Godet 2021). Figure 5.2 shows an example of this type of analysis. The spatial resolution of the EDS maps, around a micrometer, clearly highlights the difference in composition and texture between the slip and the clay. The distributions of quartz crystals (bright grains on the silicon map) and iron oxide crystals (shiny grains in the iron map) are also observed. Raman spectrometry makes it possible to verify that grains rich in silicon and those rich in iron correspond respectively to quartz and hematite crystals.

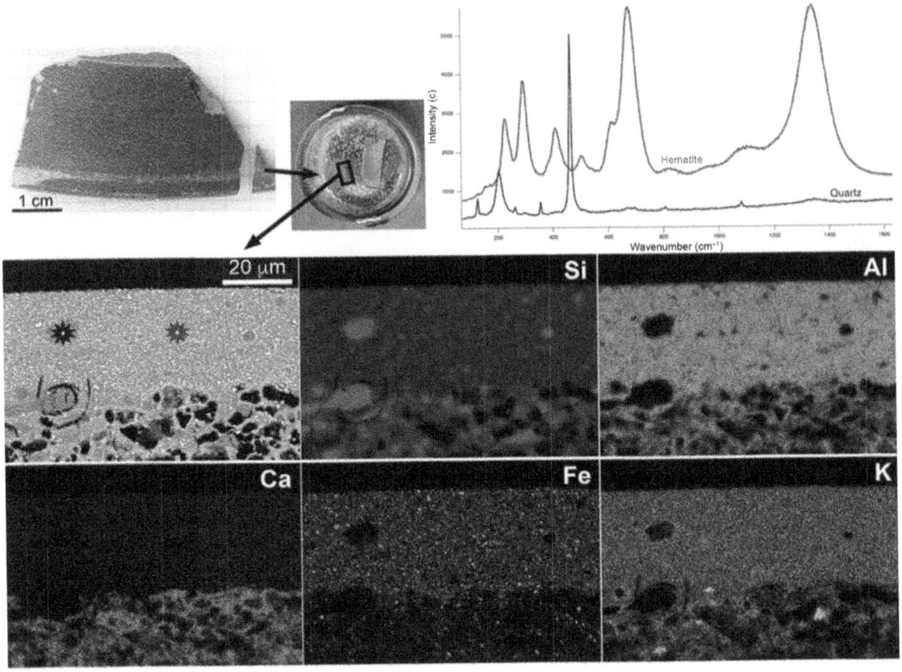

Figure 5.2. *Example of an analysis of a South Gaul sigillata glaze from the 1st century CE by SEM-EDS and Raman spectroscopy. The difference in composition between the paste and the slip is clearly demonstrated by the maps of the distribution of chemical elements (Ca, Fe and K), while Raman spectroscopy makes it possible to specify the nature of certain crystals; large quartz crystal (blue dot) and hematite crystals (red dot). For a color version of this figure, see www.iste.co.uk/dejoie/synchrotron.zip*

Analyzing the mineral or crystal composition at the required scale is, however, a little trickier. Laboratory diffractometers do not yet have the required spatial resolution and other techniques such as Raman spectroscopy, which allows measurements at the micron scale, are too selective to provide an exhaustive description of the mineral composition. As we will see a little later, the crystalline part of the *sigillata* slips (ceramics characteristic of the Roman period) is mainly made up of quartz, hematite and corundum. Quartz and hematite are easily identifiable by Raman spectroscopy (Figure 5.2). On the other hand, it is almost impossible to obtain a spectrum of corundum from a glaze, despite being primarily in the crystalline phase (Leon et al. 2010a). The same is true for many other phases such as spinel ($MgAl_2O_4$) or hercynite ($FeAl_2O_4$) present respectively in the slips of Italian *sigillata* and the black varnishes of Attic vases from the Greek period.

X-ray diffraction is still one of the most efficient techniques for analyzing mineral compositions and studying the crystal structures of the different phases. However, as its spatial resolution is directly linked to the size of the incident beam and more precisely to that of its point of impact on the sample, this requires the use of micrometric (or even sub-micrometric) beams to obtain the required resolution, currently only accessible with synchrotron sources. In contrast to elemental analyses based on X-ray absorption, whose physical phenomenon operates at the atomic scale, even encompassing its close neighbors for XANES, and thus independent, except in intensity, of the size of the incident beam here, the diffraction pattern is heavily dependent on the ratio between the size of the beam and that of the crystallites. Crystallites with a size similar to the beam will behave like single crystals and their study will require the use of single-crystal diffraction techniques. On the other hand, crystallites with sizes much smaller than the beam will require the use of powder diffraction techniques. In the following examples, we will see how to use and combine these different techniques to the most comprehensive description possible of the mineral composition, while maintaining good spatial resolution. Synchrotron beamlines used for diffraction are also, as a general rule, equipped with X-ray fluorescence detectors, thus allowing the collection of both chemical and structural data. In certain cases, however, it is more interesting to use spectroscopy beamlines and to use several beamlines for a complete analysis. This is particularly the case for studies of valence states and atomic environments for which much better results are obtained by working on dedicated beamlines.

5.3. Spectroscopic techniques

As mentioned previously, although elemental analyses at the required scale can easily be carried out in the laboratory (SEM-EDS), synchrotron studies are essential for speciation and selective phase studies. The significant flux of photons and the ability of modifying their energy also allow for very short acquisition times and the detection of elements in low concentration.

5.3.1. *Microbeam absorption spectroscopy*

In this setup, the beam, usually monochromatic, is focused on the sample and the spatial resolution is directly linked to the size of the impact point. As in SEM-EDS, the analysis can be point-based or cover a larger area. In the first case, the sample is fixed and the volume analyzed corresponds to the interaction zone between the photons of the incident beam and the atoms in the sample (interaction volume). Area analysis is achieved by moving the sample perpendicular to the beam to scan the area of interest. Unlike analyses using SEM-EDS, the area is scanned by moving the sample and not the beam (fixed source). Different detectors placed before or after the sample make it possible to collect emitted X-ray fluorescence or transmitted X-rays (for transparent samples).

The first example presented concerns the study of pre-*sigillata* glazes (Sciau et al. 2011). This type of ceramics was produced in Gaul around the middle of the 1st century BCE before the production of real *sigillata*, which only started at the beginning of the 1st century CE. The shapes of these productions are strongly inspired by Italian *sigillata* and characterized by an unglazed red or black glaze for early productions, which, in some cases, can be vitrified. Unlike real *sigillata,* which were fired in kilns that isolated the pieces from flames and combustion gases, the kilns for pre-*sigillata* wares had direct flames. This type of kiln allows us, by adjusting the draft and smoking, to vary the cooking atmosphere (reducing/oxidizing), both during the heating and cooling. These variations in atmosphere make it possible, by using iron-rich clays, to obtain different colored glazes (red or black) and body (reddish or gray).

The ceramic fragments in Figure 5.3 come from a plate with reddish clay and a partially vitrified black glaze. The analyses were carried out at the Stanford Synchrotron Radiation Lightsource (SSRL, Stanford, California, USA) using a spectroscopy and imaging beamline with a beam size of 2.5x2 μm^2 on the sample. The sample, a cross-section similar to those in Figures 5.1 and 5.2, was oriented so that the smallest dimension (2 μm) coincided with the depth. Under these conditions, the resolution of the maps is sufficient to highlight the differences in composition between the glaze and the body, which can be seen in Figure 5.3. We find characteristics similar to those of true *sigillata*, a glaze richer in iron and potassium in a clay rich in calcium. This type of resolution is, however, not sufficient to study the interfaces (paste/coating) or the distributions of crystals (quartz, iron oxides, etc.), particularly in the coatings. On the other hand, the possibility of varying the energy of the incident photons without modifying the point of impact on the sample makes it possible to punctually analyze the fine structures of the absorption edges of selected elements (XANES, X-ray Absorption Near Edge Structure) and thus obtain both structural (coordination, metal-ligand distances) and electronic (oxidation state, spin state) information.

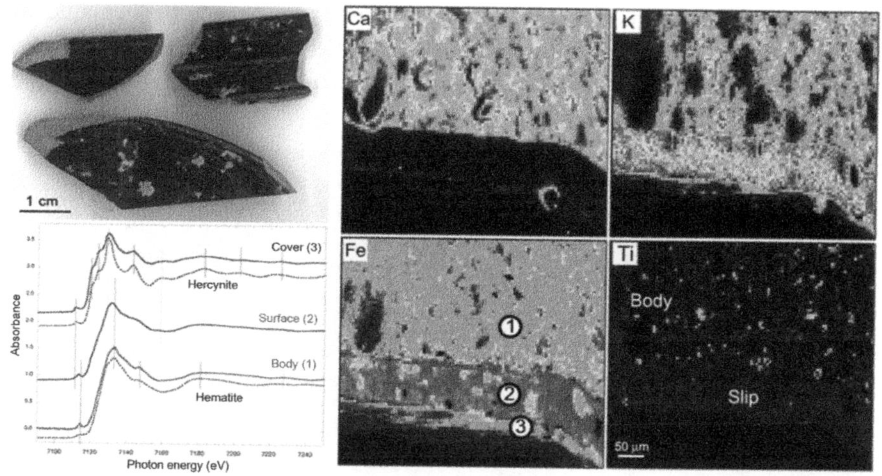

Figure 5.3. *Fragment of South Gaul pre-sigillata (1st century BCE) with black coating. Chemical maps of Ca, K, Fe and Ti and XANES spectra at the Fe threshold of the three points identified on the iron map. Study carried out at the Stanford Synchrotron Radiation Lightsource (SSRL USA) on line 2.3. For a color version of this figure, see www.iste.co.uk/dejoie/synchrotron.zip*

Given the role of iron in this type of ceramic, the XANES spectra were recorded at the K edge (7,112 eV) of this element (Figure 5.3), the $L_{2,3}$ edges being at too low an energy (700–730 eV) for the capabilities of the beamline used. The spectrum of point (1) on the clay is similar to that of iron (III) oxide with the hematite structure (Fe_2O_3), while that of point (2) presents the characteristics of hercynite, a mixed iron (II) and aluminum oxide ($FeAl_2O_4$). The energy positions of the pre-peak and edge of the surface spectrum (3) are intermediate to those of the glaze (2) and the clay (1). These correspond to a mixture of Fe(II)/Fe(III). The acquisition of XANES spectra takes some time and if only the degree of oxidation is targeted, it may be interesting to directly map this data by simply recording the iron fluorescence map at four energies around its absorption edge. Two energies for which the fluorescence is not (7,100 eV) or slightly dependent (7,138 eV) on the degree of oxidation to consider variations in iron concentration and to normalize the signal, and two intermediate energies (7,122 eV and 7,134 eV) selected to have the highest possible positive and negative contrast between the response of Fe(II) and Fe(III). After normalizing and correcting (dead times, drift) the maps, the four fluorescence values of each pixel were adjusted, taking into account the background noise and the hematite and hercynite spectra as references for Fe(III) and Fe(II) respectively. The result obtained is presented in Figure 5.4 in the form of a two-color map.

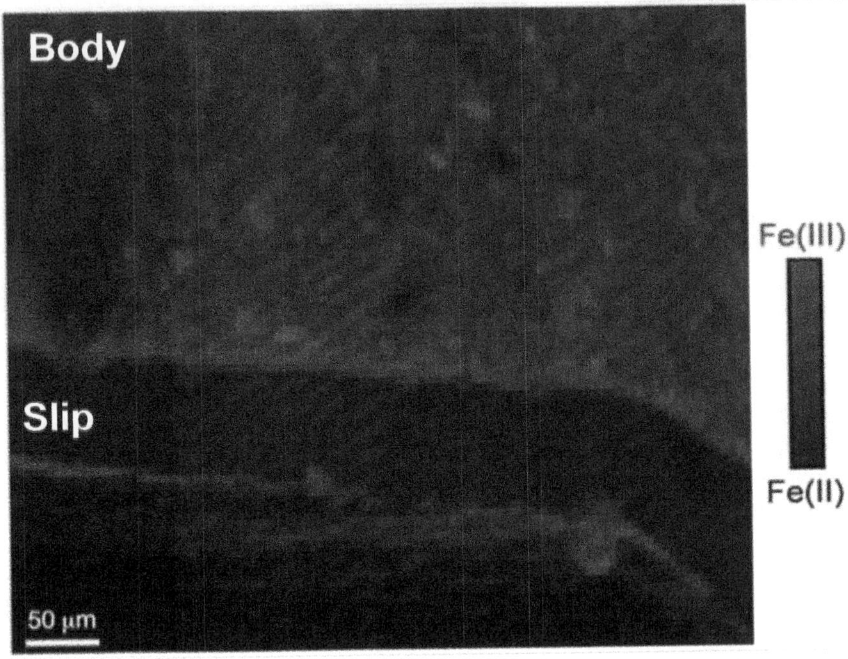

Figure 5.4. *Two-color mapping of the Fe(II)/Fe(III) distribution of the pre-sigillata in Figure 5.3. For a color version of this figure, see www.iste.co.uk/dejoie/synchrotron.zip*

The difference in Fe(II)/Fe(III) distribution between the slip and the body confirms the results of the point XANES analyses. The slip almost exclusively contains Fe(II), with the exception of the surface, which is slightly oxidized. In the body, the iron is mainly in the form of Fe(III) with some areas a little richer in Fe(II). This distribution is characteristic of a ceramic fired in a direct flame kiln with a rise in temperature under a reducing atmosphere, during the vitrification of the coating, and cooling in air. The body, unvitrified and porous, was re-oxidized during cooling, resulting in the transformation of a large part of the Fe(II) into Fe(III).

Re-oxidation also affected the first few micrometers of the slip surface (or the superficial layers of the slip).

Studies of the Fe(II)/Fe(III) distribution have been carried out on other types of pre-*sigillata*. As expected, the red slips contain mainly Fe(III) and the black ones contain Fe(II), but significant differences are observed within each group (Figure 5.5). Thus, the pre-*sigillata* in Figure 5.5(d) is quite different from that in Figures 5.4 and 5.5(c). The slip shows no, or very little, surface re-oxidation and a large part of the iron in the body is still in the form of Fe(II). The atmosphere around this ceramic, during the cooling phase, was much less oxidizing than that of the two other pre-*sigillata* with a black slip (Figures 5.4 and 5.5(c)). The fumes were still present around the vessel at the start of cooling and it was only later, at a relatively low temperature, that the air around the vase began to be rich enough in oxygen to modify the iron valence. This type of firing can be obtained by increasing the production of smoke and closing the kiln at the end of the heating phase to trap the smoke (damper, chimney, vents, etc.) to maintain a reducing atmosphere around the vessels for as long as possible. For the other two pre-*sigillata*, the change in atmosphere took place much more quickly and the re-oxidation began as soon as the temperature began to drop, at a temperature high enough to allow oxidation of the iron oxides in contact with the atmosphere (coating surface and body) and not protected by the vitreous matrix (coating).

The ceramic fragment in Figure 5.5(b) has a red coating on one side and a black coating on the other. The two coatings have similar chemical compositions, however, with a very slightly higher level of calcium and iron for the red slip (1.5% versus 1% for CaO and 12.9% versus 9.8% for Fe_2O_3) compensated by a slightly lower silicon content (41.4% compared to 46.7% for SiO_2). Given the heterogeneity of these coatings, these slight variations must be interpreted with caution; it is difficult to conclude, as proposed for Attic vases, that the apparent excess of calcium in the red coating is due to its greater porosity (Figure 5.6).

The difference in porosity between the two layers is indeed responsible for their different behavior during cooling. The re-oxidation may have occurred more deeply in the most porous layer. However, this re-oxidation is only partial; certain areas of the slip still predominantly contain Fe(II). The other layer, although mainly composed of Fe(II), shows stronger surface oxidation than the other black coatings, with even some areas showing deep oxidation.

Figure 5.5. *Maps of the Fe(II)/Fe(III) distribution for different types of pre-sigillata. For a color version of this figure, see www.iste.co.uk/dejoie/synchrotron.zip*

The valence of iron has also been studied in *sigillata* with a vitrified red slip (Sciau et al. 2011). These ceramics were fired in kilns designed to protect the vessels from flames and combustion gases in order to maintain an oxidizing atmosphere around them during the heating phase and thus achieve red and shiny coatings, i.e. well vitrified. The majority of *sigillata* kilns unearthed during excavations are in such a poor state of conservation that it is difficult to determine the effectiveness of the equipment in channeling the flames and insulating the vessels. The study of Fe(II) residues, particularly at the heart of the coating, is one way of addressing this question by directly examining the materials produced. During vitrification, if the atmosphere is not sufficiently oxidizing to oxidize the iron to Fe(III), Fe(II) can be trapped and protected from re-oxidation during cooling by the vitreous matrix. This is what was observed in Italian *sigillata* from the late 1st century BCE.

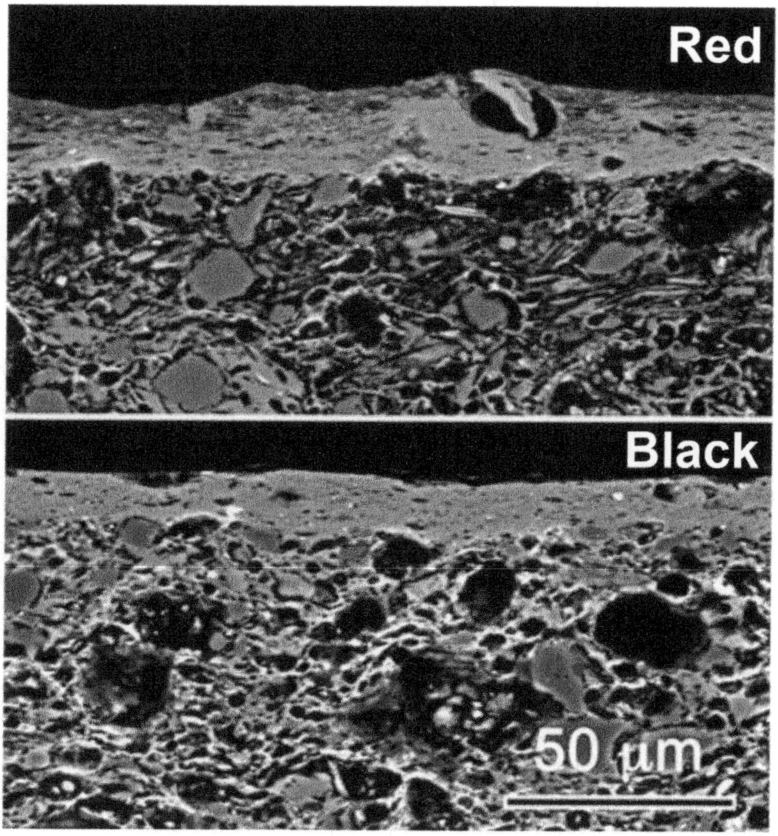

Figure 5.6. *SEM images corresponding to the red and black sides of the pre-sigillata in Figure 5.5(b)*

The spatial resolution in the previous examples, on the order of a few micrometers, does not allow the study of interfaces or heterogeneities in the glaze. For this, it is necessary to work on beamlines equipped with a focusing device to obtain sub-micrometer or even nanometer beams. In the example of Figure 5.7, the study of which was carried out using the ESRF ID21 beamline (European Synchrotron Radiation Facility, Grenoble, France) before the upgrade of the ring, the beam size of the sample was around 0.7 x 0.3 μm^2 (Wang et al. 2016). The differences in composition between the glaze and the pigment zone, richer in calcium, are well highlighted. Anorthite needles ($CaAl_2Si_2O_8$) of the pigmented zone are also visible, as well as the cobalt-rich crystals in the center of the needle clusters. Identification of these cobalt aluminate crystals ($CoAl_2O_4$) was carried out by XANES spectroscopy (Figure 5.7) and by X-ray micro-diffraction.

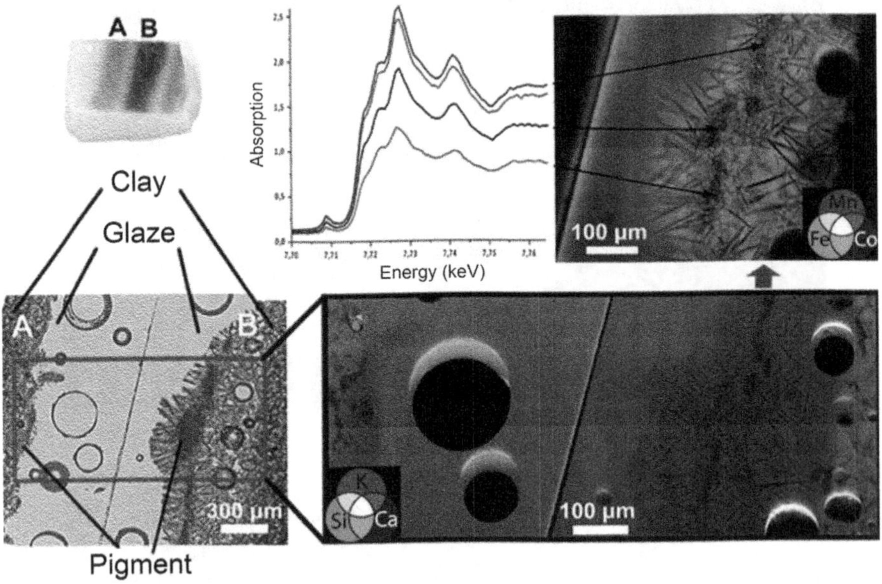

Figure 5.7. *Study of a fragment of porcelain from the Ming dynasty (1368–1644) conducted on the ESRF ID21 beamline with a beam size of 0.7 x 0.3 μm^2 (Wang et al. 2016). Decorative lines A and B were assembled head to tail for simultaneous analysis. The elemental maps were combined to show the differences in body/pigmented zone/glaze compositions (K, Si, Ca) and the distribution of transition elements (Mn, Fe, Co). XANES spectra at the Co K-edge correspond to cobalt aluminate. For a color version of this figure, see www.iste.co.uk/dejoie/synchrotron.zip*

5.3.2. *"Full-field" imaging and spectroscopy*

In this setup, the area to be analyzed is no longer scanned using a micro- or nano-beam but directly radiographed. The spatial resolution is directly that of radiography and depends on the image-forming device. The spectroscopic information at an absorption edge, of the XANES type, is obtained by recording a series of images at different energies around the absorption edge and then by combining these images to create a hypermap, where each pixel is associated with a XANES spectrum. The different steps of the process are explained in Chapter 3. A detailed description of this technique applied to the study of ancient ceramics was also published under the direction of J. Daisy in the book "Synchrotron Radiation" (Sciau and Wang 2019).

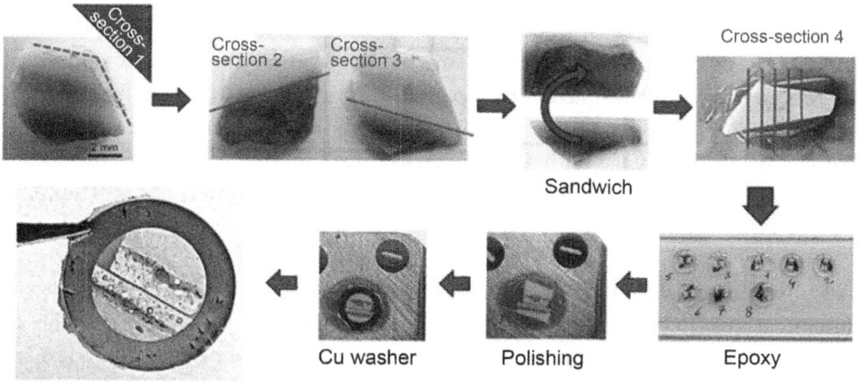

Figure 5.8. *Main steps in sample preparation. Cutting (cross-section 1 to 3) of the two pieces to prepare the "sandwich", cutting (cross-section 4) of slices of approximately 0.5 mm, reinforcement of the slices with an epoxy resin, manual polishing to the desired thickness (30–150mm), and finally attaching a copper washer for easier handling. For a color version of this figure, see www.iste.co.uk/dejoie/synchrotron.zip*

This imaging technique requires working in transmission and for it to be fully effective it is necessary to prepare samples that are suitable for its characteristics and the information sought. In the case of ceramics, the most convenient way to study the different layers is to work on thin cross-sections. These slices, in the examples that follow, were obtained using a process based on the preparation techniques of transmission electron microscopy, the main steps of which are shown in Figure 5.8. The so-called "sandwich" technique, which involves combining two pieces head to tail, was adopted to properly position the area of interest in the center of the sample. This technique, classic in transmission electron microscopy, also preserves sample surfaces and optimizes the analysis.

This imaging technique is particularly well suited to the study of ceramics with iron-rich clay covers (Meirer et al. 2013). In these ceramics, the valence and speciation of iron are, as we have seen for pre-*sigillata*, fundamental parameters for color and the study of firing conditions. In addition, iron is the most absorbent element of the materials constituting these ceramics, generally present in high concentrations, between 5 and 25% in mass oxide equivalent. Under these conditions, it is quite easy to determine section thicknesses that yield good-quality XANES spectra at the K-edge. In cases where the iron concentration varies greatly between the glaze and the clay, it may, however, be interesting to work on two slices with different thicknesses, but in general, a thickness between 40 and 100 μm gives good results. The imaged surface and the resolution are also well adapted to the information sought and allow very reasonable acquisition times. An example of a study relating to a black-glazed Campanian ceramic from the Roman period is presented in Chapter 3. Additionally, to illustrate this type of study, we will focus on a fragment of an Attic vase with a black glaze from the 5th century CE, the study of which was published in the book "synchrotron radiation" cited above (Sciau and Wang 2019).

An absorption calculation from the elemental composition leads to an optimal thickness, at the iron K-edge, of 25 μm and 45 μm for the glaze and the clay respectively. We chose to prioritize the clay by choosing a slice thickness of around 40 μm. The series of X-rays was recorded on the ESRF ID21 beamline following the acquisition strategy and protocol developed for this beamline (Cotte et al. 2017). They were then processed as mentioned in Chapter 3 to build a file associating each pixel of the image with the corresponding XANES spectrum using routines developed on the beamline and now added to the PyMca software package (Solé et al. 2007). Two corrected radiographs from the series, one before and one after the edge, are presented in Figure 5.9. The enlargements show that the spatial resolution obtained with a pixel size of 0.3 x 0.3 μm^2 allows good observation of the details of the clay, while maintaining a sufficiently wide field of view (FOV).

Better spatial resolutions can be obtained on certain beamlines of light but at the expense of the FOV (Table 3.1, Chapter 3). The difference in contrast between the two radiographs comes from the iron and makes it possible to visualize the areas with highest concentrations of this element, such as in the glaze.

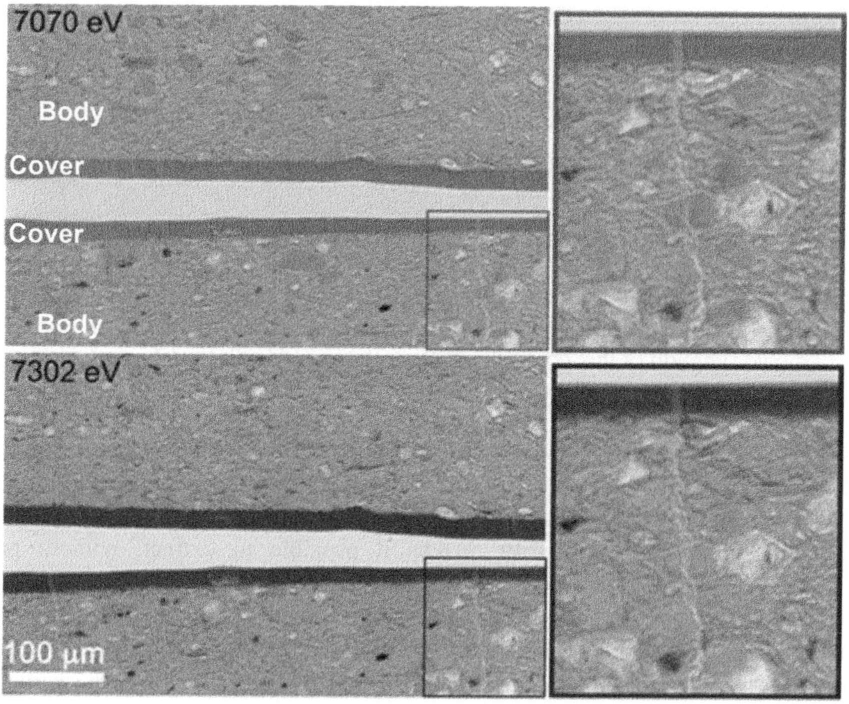

Figure 5.9. *Radiographs of the Attic vase sample before (7,070 eV) and after the iron K-edge (7,302 eV). For a color version of this figure, see www.iste.co.uk/dejoie/synchrotron.zip*

To process the very large number of XANES spectra – several millions – generated by each series of recordings, we used the TXM-Wizard software package (Liu et al. 2012), specially developed for the analysis of this type of data. For each pixel, the extraction of the edge energy position and the jump height makes it possible to visualize the distribution of valence state and concentration (Figure 5.10). With the edge position being correlated with the valence state, this mapping makes it possible to visualize its distribution. As in Figures 5.4 and 5.5, the contrast between the glaze and the clay is very clear. Iron is more oxidized (Fe(III)) in the clay than in the glaze (Fe(II)). The second mapping corresponds to a mapping of the elemental distribution of iron. Glazed ceramics made from iron-rich clays have a more uniform distribution in the glaze than pre-*sigillata* (Figure 5.3) and even *sigillata* (Figure 5.2) ceramics.

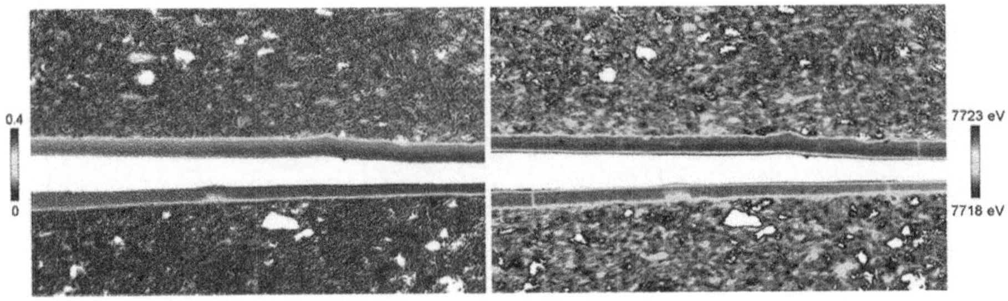

Figure 5.10. *Mappings of the energy position (left) and jump height (right) of the edge, derived from the XANES spectra analysis of each pixel. For a color version of this figure, see www.iste.co.uk/dejoie/synchrotron.zip*

The TXM-Wizard suite exploits, through different types of data processing, the entire XANES spectra and not just the jump position and height. This software includes a statistical analysis of the "clustering" type which, by classifying the spectra based on their similarity, makes it possible to extract, without prior assumptions, characteristic spectra and plot their spatial distribution. A detailed analysis is given in the reference Sciau and Wang (2019) and here we only summarize the main result, which essentially groups the spectra into two categories: a group with an associated spectrum close to that of the maghemite and a second group whose spectra have characteristics of both hercynite and magnetite (see Figure 4 (Sciau and Wang 2019)). Based on this result, all the spectra associated with each pixel of the image were adjusted by the least squares method using the spectra of these three compounds and that of hematite (Figure 5.11). Although the latter was not identified through the statistical analysis, it was nevertheless added in the refinements, confirming its absence in the sample studied. The iron oxide in the clay is mainly maghemite, whereas in the glaze, magnetite and hercynite dominate. The dominant presence of maghemite and the absence of hematite in the clay show that it is indeed a relatively rapid re-oxidation during cooling, which did not allow recrystallization into hematite.

The significant presence of magnetite in the glaze reveals that the temperature during the vitrification of the glaze was not as high as that of the Campanian discussed in Chapter 3. However, we notice that the magnetite/hercynite distribution is not homogeneous in the glaze. The proportion of hercynite is slightly higher on the glaze surface than within, which could indicate that the maximum temperature was not maintained for very long. Unlike Campanian and pre-*sigillata*, the black glazes of Attic vases often have smaller and better vitrified surfaces than within.

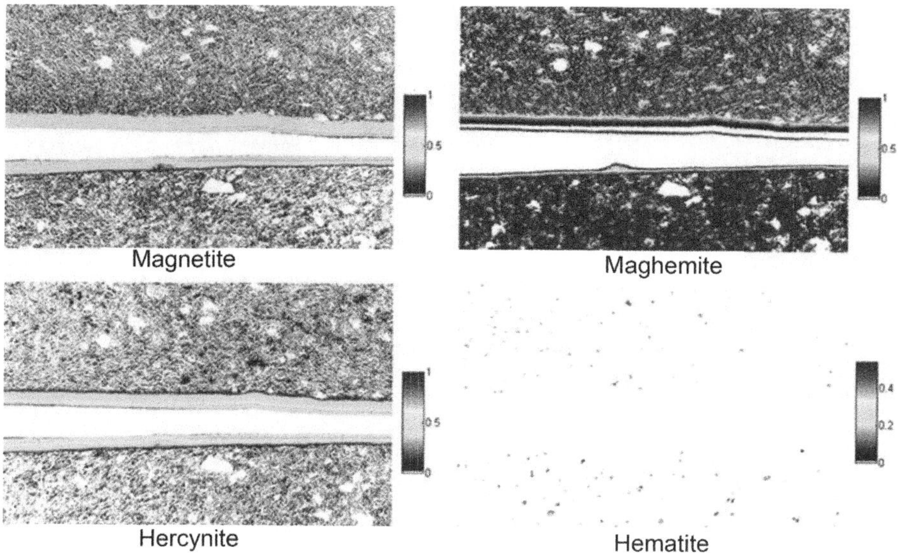

Figure 5.11. *Spatial distributions of the different phases obtained from the refinement of the XANES spectra of each pixel. For a color version of this figure, see www.iste.co.uk/dejoie/synchrotron.zip*

This technique can also prove interesting for imaging the phases of a minor element such as the cobalt-based phases of porcelain with blue and white decoration from the Ming dynasty in China. Although the proportion of cobalt in the blue decoration of the fragment in Figure 4.7 is low (0.3 % CoO), the series of radiographs obtained at the cobalt K-edge, for a 60 mm thick sample, is entirely usable as shown in Figure 5.12. The distribution of cobalt aluminate crystals is perfectly imaged as well as that of Co(II) within the glaze of the pigmented zone. On the other hand, using this technique, it was not possible to study the distribution of cobalt-based phases of blue porcelain from the Yuan dynasty, characterized by high iron concentrations. To compensate for the strong absorption due to iron, it would be necessary to work with thin samples to maintain acceptable transmission, but which unfortunately would not allow for a usable jump at the cobalt edge. Moreover, there is very little cobalt aluminate in these samples and therefore areas of high concentrations of cobalt free of iron. Cobalt-rich crystals are almost exclusively cobalt ferrites ($CoFe_2O_4$), therefore highly concentrated in iron. Only the K-edge of iron could be studied using this technique on these porcelain samples.

Figure 5.12. *Study of the blue pigmented zone of porcelain from the Ming dynasty. Radiography of the sample (left), maps of jump height (center) and cobalt-containing phases (right). For a color version of this figure, see www.iste.co.uk/dejoie/synchrotron.zip*

5.4. Diffraction techniques

5.4.1. *With a monochromatic beam (powder diffraction)*

This is the diffraction technique most commonly used in the analysis of polycrystalline materials such as ceramics. These analyses are primarily carried out using laboratory diffractometers, either directly on the ceramic object or on a piece reduced to powder. As mentioned in the Introduction section, this analysis technique was used very early in the studies of ancient ceramics. The overall mineral composition can be quickly determined and, then with the appearance of diffractometers equipped with micro-sources, more and more information could be obtained about the glazes and decorations. The spatial resolutions accessible on this equipment are, however, not yet sufficient for the study of glaze stratigraphies. The areas analyzed are often too large to be limited solely to the glazed or decorated areas. The diffractograms of the glazes or decorations are superimposed onto those of the clay or adjacent zones, which can strongly limit their use, particularly in terms of quantitative analysis. In the case of *sigillata*, for example, the glazes are generally too thin for this type of analysis (Sciau et al. 2020).

Many synchrotron beamlines allow for micro-diffraction experiments using monochromatic beams in various energy domains. The energy used is not as crucial as for spectroscopy experiments but it can help control the penetration depth or eliminate possible fluorescence issues. The studies we will use to illustrate our examples were carried out on the ALS micro-diffraction beamline (Advanced Light Source, Berkeley, California, USA). These diffraction experiments are, apart from the size of the beam (1–2 μm) and that of the 2D detector (much larger), very similar to those carried out in the laboratory. They were carried out (section 5.3.1) directly on cross-sections prepared for SEM-EDS and Raman spectroscopy analyses. Although this beamline is not optimized for spectroscopic studies, it is nevertheless equipped with an X-ray fluorescence detector used for rapid identification of areas of interest.

The mixed yellow and red glazed *sigillata* (Figure 5.13), commonly referred to as marbled *sigillata*, were a specialty of the Gaulish workshop at La Graufesenque and although representing only a small percentage of pottery productions, they were produced for almost a century and were exported to much of the Roman Empire (Genin 2007). The overlap of the two components, as seen in Figure 5.13, is such that it is very difficult to analyze them separately using a conventional laboratory diffractometer, even equipped with a micro-source.

A crystallographic study was therefore carried out on the ALS micro-diffraction beamline to precisely determine the mineral composition of each component (Leon et al. 2010b). Firstly, the elemental composition of large areas was quickly mapped using the X-ray fluorescence detector to identify the boundaries of the different constituents and allow the recording of characteristic diffraction patterns of the different components (Figure 5.14). These diffraction patterns (Figure 5.14(c)) were collected using a MAR133 CCD camera with a monochromatic beam (8 keV) measuring 2x2 μm^2. The phase analysis was carried out using diffractograms obtained by angular integration of the diffraction rings using the software XMAS developed on this beamline (Figure 5.14(d)).

The patterns of the clay and the red areas of the glaze turned out to be identical to those of "classic" *sigillata*, with a predominantly anorthite in the body and hematite and corundum in the red slip, characteristic of this workshop. However, a new phase was identified from the yellow zone, pseudobrookite, a mixed iron/titanium oxide (Fe_2TiO_5), whose color can vary from yellow to dark brown, depending on its actual composition, as this structure accepts numerous cationic substitutions (Mg, Al, etc.). Once considered as a substitute pigment for chromium-doped compounds used in the coloring of high-temperature ceramics (> 1000°C), this approach is no longer in use due to its cationic exchanges with the supporting ceramic matrix (Dondi et al. 2007). The resulting color after sintering or melting is too strongly dependent on the matrix composition.

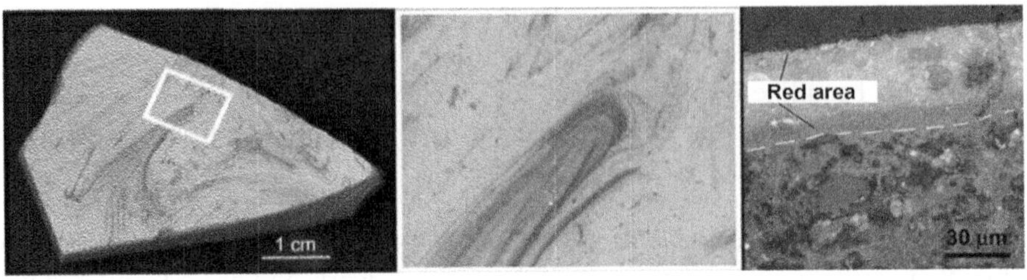

Figure 5.13. *Fragment of marbled La Graufesenque sigillata from the 1st century CE with, on the left, a cross-section view of a predominantly yellow area. For a color version of this figure, see www.iste.co.uk/dejoie/synchrotron.zip*

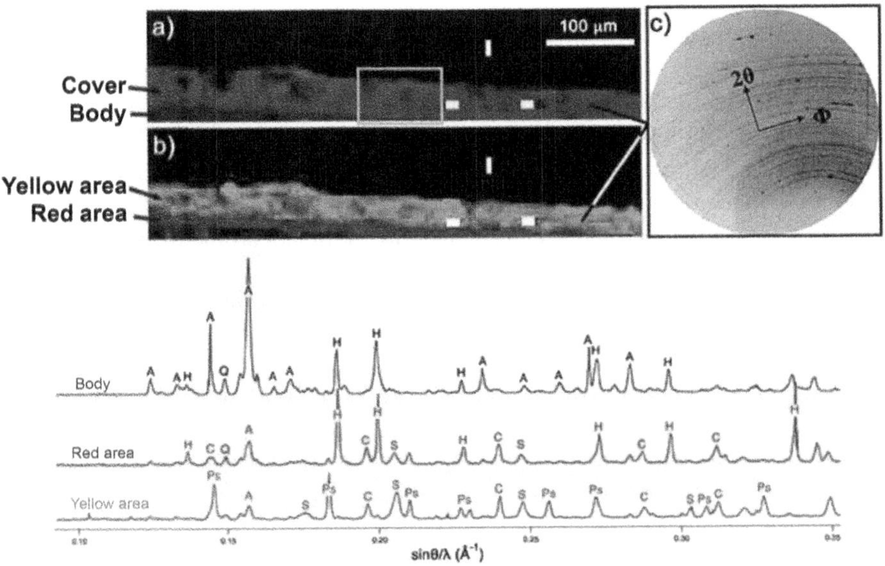

Figure 5.14. *Study of a transverse section of sigillata with marbled glaze (ALS beamline, USA): a) mapping of the K(red)/Ca (blue) distribution highlighting the glaze (rich in K) and the clay (rich in Ca); b) mapping of the Fe(red)/Ti (green) distribution allowing separation the two components of the glaze; c) example of a 2D diffraction pattern taken from the yellow zone (rich in Ti); and d) indexes of the characteristic diffractions of the three components (A: anorthite, H: hematite, Q: quartz, C: corundum, Ps: pseudobrookite and S: spinel). The white marks are Pt deposits to allow the use on different beamlines. For a color version of this figure, see www.iste.co.uk/ dejoie/synchrotron.zip*

After identifying the main crystal phases of each component, a small area, shown in Figure 5.14(a), was selected for a study of their spatial distributions. For this, the diffraction pattern was collected point by point every 2 x 2 μm^2 and then the intensities of the characteristic rings of the main phases were integrated along the azimuthal direction of the diffraction rings, ϕ. The resulting maps are presented in Figure 5.15. These maps confirm that pseudobrookite is only present in the yellow component of the glaze. Spinel and corundum are characteristic phases of both components of the slip. Anorthite is the main crystalline phase in the body, with its high calcium content leading to its formation during the firing of ceramics. Hematite, which is predominant in the red part of the glaze, is almost absent in the yellow one, thus confirming that in this component, iron is mainly associated with titanium.

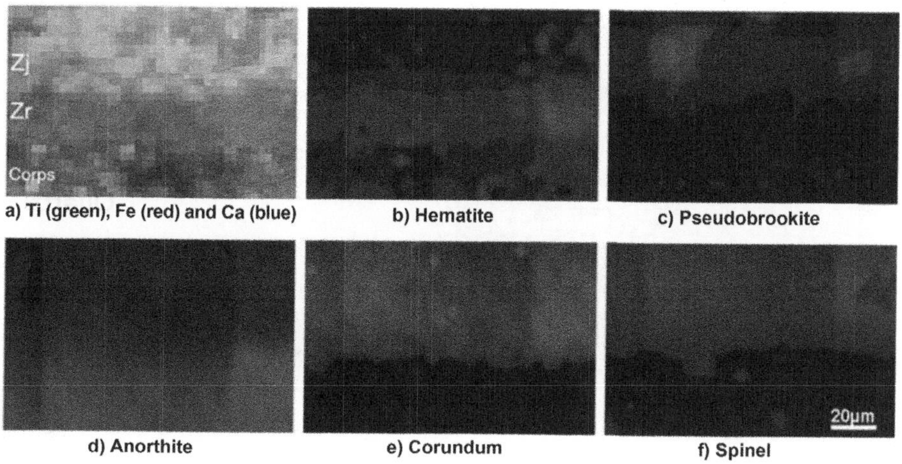

Figure 5.15. *a) 3-color mapping of the elemental composition allowing visualization of the 3 components (clay, red (Zr) and yellow (Zj) components of the glaze) and b) to f) the distribution of the main crystalline phases with anorthitein the clay, spinel and corundum in the glaze, pseudobrookite only in the yellow zone and hematite more strongly concentrated in the red zone of the glaze. For a color version of this figure, see www.iste.co.uk/dejoie/synchrotron.zip*

5.4.2. *With a polychromatic beam (Laue method)*

In the previous example, the crystals present in the glaze analyzed by the microbeam are sufficiently small in size and number to generate a usable powder pattern, i.e. composed of rings of relatively homogeneous intensity along ϕ (Figure 5.16). However, the quartz crystals whose sizes are close to, or in some cases exceed, that of the beam are therefore too few in number and too specifically oriented to form continuous rings. Their diffraction is much more sporadic and the erratic variations between two measurements do not allow any quantitative, or even semi-quantitative, information to be derived. This phenomenon is even more pronounced for the body, where the crystals of many crystalline species are much larger than the area analyzed. The analyzed volume then contains too few crystals to generate a powder diffraction pattern. We find ourselves in single-crystal diffraction conditions, which, in a monochromatic beam, require, for accurate identification, recording of a series of patterns by varying the crystal orientation relative to the incident beam or by varying the energy (wavelength) of the incident beam. This, in the context of the studies presented here, is very difficult, if not impossible, to achieve.

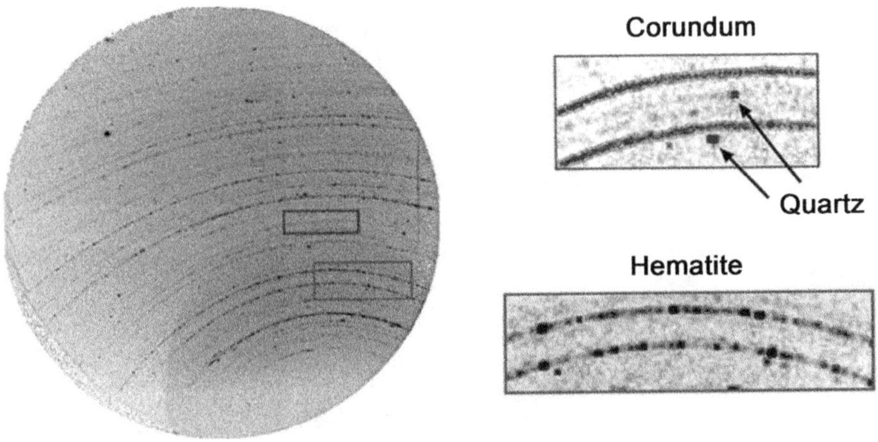

Figure 5.16. *Diffraction pattern of a red zone of glaze collected with a monochromatic beam (8 keV) of 2x2mm (ALS, USA). The size distributions of corundum crystals (nanometric) and hematite crystals (submicrometric) make it possible to obtain continuous and quasi-continuous rings respectively, while that of quartz crystals is more punctual as shown in the insets. For a color version of this figure, see www.iste.co.uk/dejoie/synchrotron.zip*

This obstacle can, however, be easily overcome by using a polychromatic or "white" beam. Instead of adjusting the orientation θ of the crystal to satisfy the diffraction condition ($2d_{hkl}\sin\theta=\lambda$), the wavelength λ is adjusted. Thus, with a polychromatic beam covering a broad wavelength range, we can collect a sufficient number of diffraction spots, regardless of the orientation of the crystal relative to the incident beam, to identify their nature and orientation unambiguously, as shown in

Figure 5.17. The diffraction pattern obtained is called a "Laue pattern", or "Laue diffraction pattern".

The diffraction pattern collected in the body of the ceramic with a polychromatic beam is rich in information and can confirm the presence of large crystals in irradiated volume (Figure 5.17). This pattern corresponds to the contribution of two quartz crystals slightly disoriented relative to each other (Figure 5.18). In contrast, the information is more limited in monochromatic mode due to these two large crystals, whose contribution to the diffraction pattern is reduced to a few punctual spots. However, calcite and hematite can be identified from the partially visible rings.

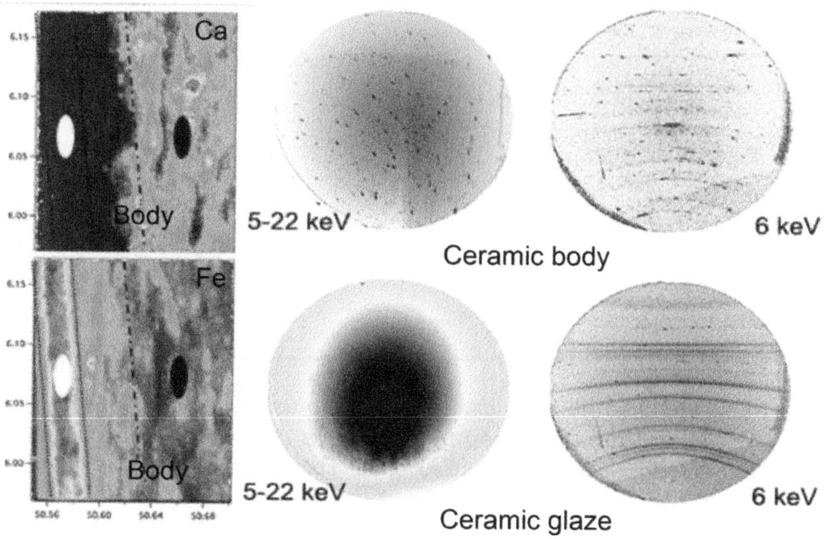

Figure 5.17. *Polychromatic and monochromatic beam diffraction patterns collected in the body (black dot) and glaze (white dot) of pre-sigillata on the ALS microdiffraction beamline (USA). For a color version of this figure, see www.iste.co.uk/dejoie/synchrotron.zip*

The situation is reversed in the glaze, which contains predominantly small crystals (Figure 5.17). The large number of randomly orientated crystals generate, in monochromatic mode, a quasi-theoretical diffraction pattern, with rings whose intensity distributed uniformly along ϕ is proportional to the phase quantity and structure factor. However, in polychromatic mode, the pattern does not contain any usable information, as we can see in Figure 5.17. This pattern can be seen as a superposition of monochromatic patterns generated by each of the energies between 5 and 22 keV or as Laue patterns of the multitude of randomly orientated crystals contained in the irradiated volume. Simultaneously adjusting both orientation and wavelength continuously generates a pattern whose diffraction intensities are so spread out that it becomes unusable.

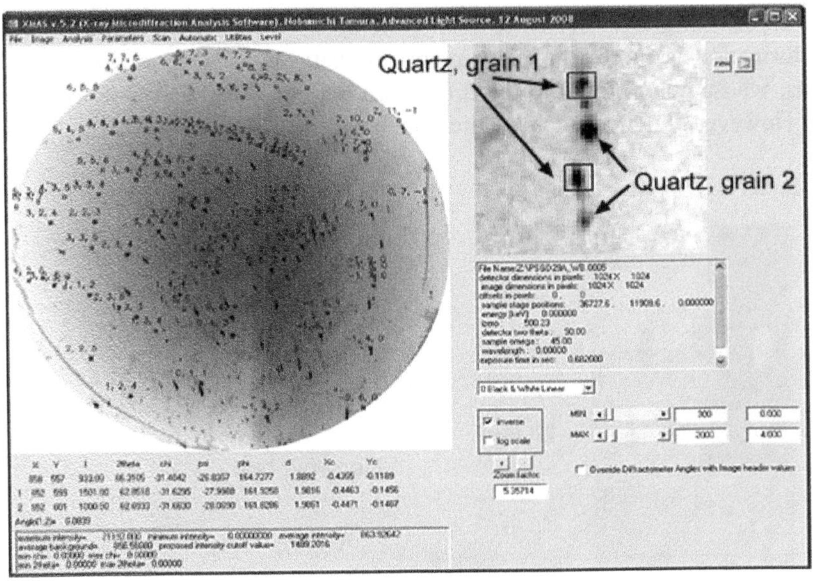

Figure 5.18. *Indexing of the Laue diffraction pattern using the X-ray Microdiffraction Analysis (XMAS) software developed by N. Tamura (see: https://it.lbl.gov/resource/hpc/supported-research-clusters/xmas/). For a color version of this figure, see www.iste.co.uk/dejoie/synchrotron.zip*

As before, it may be interesting to record the diffraction patterns of a defined area step by step. The usable Laue patterns are then indexed using the structures (CIF, Crystallographic Information File) of the different phases present. We can thus obtain phase distribution maps or, as in the example in Figure 5.19, the size distribution and orientation of the crystals for a given phase, in this case, quartz. In this type of representation, each crystallographic orientation is associated with a color, and the fact that no color dominates shows that these crystals do not have a preferential orientation. No quartz crystal was detected in the glaze, as expected, as well as in the upper layers of the body. The first quartz crystals are located a few tens of micrometers from the body/glaze interface. The smoothing of the vessel's surface brought the fine particles of clay minerals to the surface and, conversely, caused the larger grains to migrate inward.

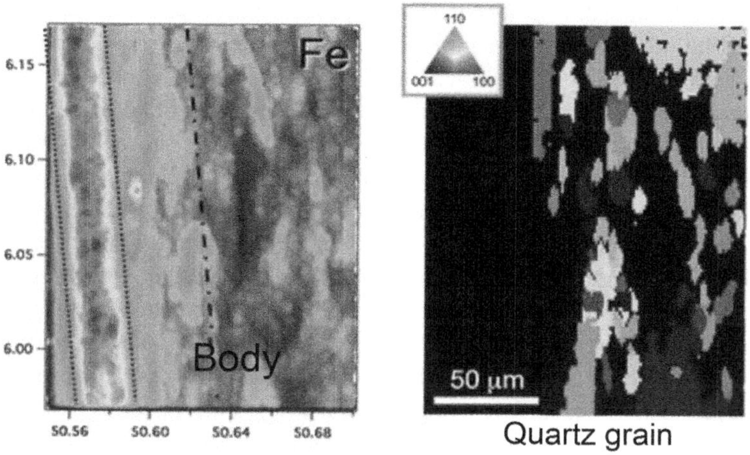

Figure 5.19. *Mapping the distribution of iron and quartz grains obtained after indexing the Laue patterns recorded every 1 x 1 mm² (ALS microdiffraction beamline). For a color version of this figure, see www.iste.co.uk/dejoie/synchrotron.zip*

These two modes of diffraction (monochromatic and polychromatic), addressing populations of crystals of different grain sizes, are complementary, and it is interesting, in certain cases, to combine them to obtain a more complete understanding of the distribution of crystalline species. In this regard, micro-diffraction beamlines such as that of ALS (Berkeley, USA) that allow rapid switching from one mode to another without changing the beam location on the sample, are preferred. If this is not possible and if the research must be carried out on different beamlines, reference points or markers visible on the elementary composition maps should be used. In the absence of easily identifiable natural markers, it may be interesting to add markers, for example, by platinum deposition as for the marbled *sigillata* sample in Figure 5.14 (Sciau et al. 2009; Leon et al. 2010b). Figure 5.20 shows the results of a study combining monochromatic and polychromatic diffraction carried out on the ALS microdiffraction beamline. Additional technical information on this study and examples of studies regarding other types of cultural heritage materials are given in the references (Dejoie et al. 2015, 2018). The *sigillata* crystallites, particularly those produced in the La Graufesenque workshop, are very rarely larger than a few micrometers and analysis with a micrometric monochromatic beam is sufficient to obtain most of the information about the characteristic phases of the body and glaze, as shown in the anorthite and hematite maps in Figure 5.20(b) or those in Figure 5.15. Determining the lattice parameters even made it possible to reveal a systematic difference in the substitution rate of hematite (Fe → Ti/Al) between the body and the glaze, which could be correlated with the manufacturing process (Leon et al. 2015; Sciau et al. 2020). On the other hand, polychromatic beam analysis allows for a more detailed study of quartz crystallites which, without being a characteristic phase of one of the ceramic components, are an important marker of the raw material preparation process. Comparing the quartz maps in Figures 5.19 and 5.20 clearly shows the difference in grain size in the bodies of these two types of Roman ceramics. Polychromatic diffraction also made it possible to highlight the presence of a few large hematite grains in the body as well as pseudobrookite, which is found in too small a quantity, in non-marbled *sigillata*, to be identified from powder patterns under standard conditions, i.e. using a macroscopic beam. These crystals are too few in number and too large in size to be observable by monochromatic micro-diffraction. Polychromatic diffraction is very effective for detecting minor phases, as long as the size of their crystal is, in relation to the beam used, sufficient to generate a usable Laue pattern.

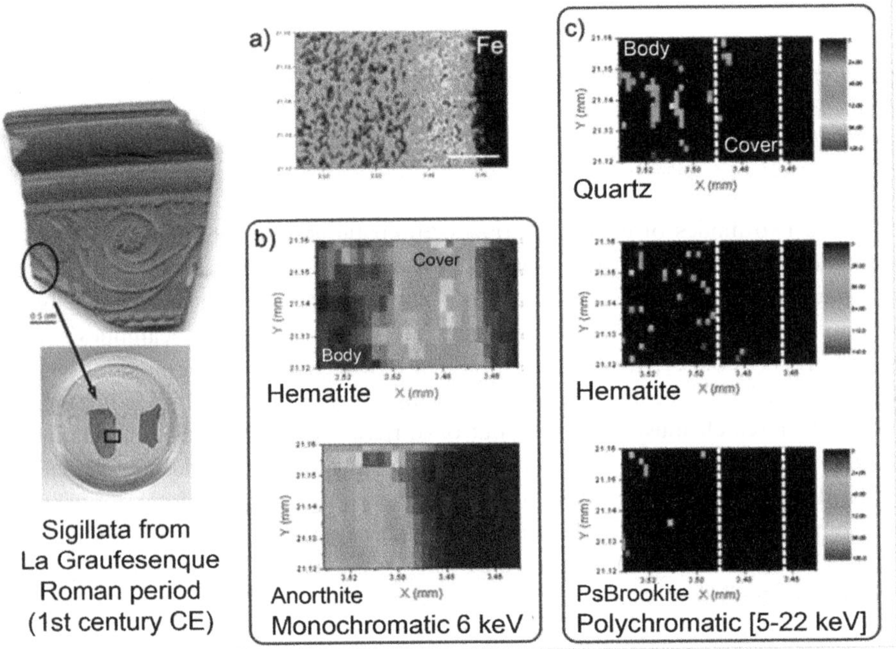

Figure 5.20. *Example of a study coupling monochromatic and polychromatic diffraction (Dejoie et al. 2015). a) Elemental mapping (Fe) of the selected area used for location; b) maps of the distribution of hematite and anorthite obtained in monochromatic; and c) those obtained in polychromatic for quartz, hematite and pseudobrookite. For a color version of this figure, see www.iste.co.uk/dejoie/synchrotron.zip*

5.5. Conclusion

Through these examples, we have attempted to demonstrate that analysis techniques based on synchrotron radiation are particularly well suited to the study of ancient ceramic materials. These techniques make it possible to obtain, in a very reasonable time, structural and chemical information of areas representative of the heterogeneity and complexity of these materials, while maintaining sufficient spatial resolution for precise identification of their various constituents. For example, in the study of marbled *sigillata* (section 5.1.1), it was possible to separate the elemental and mineral compositions of the two components and to precisely identify the nature of the yellow pigment, in this case, pseudobrookite. Similar experiments carried out on classic *sigillata*, i.e. those with only red glaze, have even made it possible to highlight a systematic difference in the lattice parameters of the hematite between the body and the glaze, the magnitude of which depended on the production process used. Coupled with other results, this helped highlight a major evolution in the manufacturing processes of the *sigillata* at the beginning of the 1st century CE (Leon et al. 2015).

One of the main advantages of synchrotron spectroscopic techniques is being able to map not only the elemental composition but also the valence states of selected elements. This makes it possible, as shown in section 5.3.1, to obtain information about the manufacturing processes, difficult to attain using other techniques. Mapping the valence states of iron is a powerful means of reconstructing variations in the atmosphere (oxidizing/reducing) during the firing process. The differences in valence state between vitrified and more porous zones make it possible to follow the changes in the atmosphere between the heating and cooling phases. It is also a good way to understand the technical mastery of the processes used. The nature of the phases containing the selected element can be clarified, if necessary, by acquiring point XANES spectra (Figures 5.3 and 5.7). However, if these measurements can be carried out in transmission under the conditions discussed in section 5.3.2, it is more interesting to use the full-field spectrometry technique (FF-XANES), which allows for the rapid acquisition of all XANES spectra from the studied area with very good resolution ranging from tens to hundreds of nanometers depending on the lines used. The resulting maps provide valuable information on the distribution of the different phases containing the selected element. Thus, in the examples in section 5.3.2, we highlighted the almost complete absence of hematite in the body of Attic ceramics (Figure 5.11). Ferric ions (Fe(III)) are mainly in the maghemite crystals, confirming their oxidation during the cooling phase. However, the significant presence of magnetite and the heterogeneity of the magnetite/hercynite distribution in the glaze provide very relevant information on the temperature reached and its maintenance during the firing phase. The cobalt phase mapping in the second example (Figure 5.12) shows that it is nevertheless possible to obtain information, in this case on the structure of the pigmented zone, of elements even in small proportions if the local concentration of the latter is sufficient to obtain usable transmission XANES spectra.

X-ray diffraction is particularly well suited to studying the crystalline part of these materials. The spatial resolutions, around a micrometer, of the microdiffraction beamlines provide direct access to the distributions of the different phases in the bodies and glazes of ceramics (section 5.4). However, due to the very large dispersion of crystallite size, measurements must often be carried out both in monochromatic (powder diffraction) and polychromatic (Laue diffraction) mode.

5.6. References

Bamps, A. (1883). La céramique américaine au point de vue des élements constitutifs de sa pâte et de sa fabrication. *Congrès International des Américanistes*, 5, 274–281.

Cotte, M., Pouyet, E., Salomé, M., Rivard, C., De Nolf, W., Castillo-Michel, H., Fabris, T., Monico, L., Janssens, K., Wang, T. et al. (2017). The ID21 X-ray and infrared microscopy beamline at the ESRF status and recent applications to artistic materials. *Journal of Analytical Atomic Spectrometry*, 32(3), 477–493.

Dejoie, C., Tamura, N., Kunz, M., Goudeau, P., Sciau, P. (2015). Complementary use of monochromatic and white-beam X-ray micro-diffraction for the investigation of ancient materials. *Journal of Applied Crystallography*, 48, 1522–1533.

Dejoie, C., Autran, P.-O., Bordet, P., Fitch, A.N., Martinetto, P., Sciau, P., Tamura, N., Wright, J. (2018). X-ray diffraction and heterogeneous materials: An adaptive crystallography approach. *Comptes Rendus Physique*, 19(7), 553–560.

Dondi, M., Matteucci, F., Cruciani, G., Gasparotto, G., Tobaldi, D.M. (2007). Pseudobrookite ceramic pigments: Crystal structural, optical and technological properties. *Solid State Sciences*, 9(5), 362–369.

Genin, M. (ed.) (2007). *La Graufesenque (Millau, Aveyron), II. Sigillées lisses et autres productions*. Editions de la Fédération Aquitania, Bordeaux.

Jenkins, J.H.B. (1908). The chemical examination of some substances from the Red Hills of Essex. *Proceedings of the Society of Antiquaries of London*, 22, 182–186.

Leon, Y., Lofrumento, C., Zoppi, A., Carles, R., Castellucci, E.M., Sciau, P. (2010a). Micro-Raman investigation of terra sigillata slips: A comparative study of central Italian and southern Gaul productions. *Journal of Raman Spectroscopy*, 41(11), 1550–1555.

Leon, Y., Sciau, P., Goudeau, P., Tamura, N., Webb, S., Mehta, A. (2010b). The nature of marbled Terra Sigillata slips: A combined μXRF and μXRD investigation. *Applied Physics A-Materials Science & Processing*, 99(2), 419–425.

Leon, Y., Sciau, P., Passelac, M., Sanchez, C., Sablayrolles, R., Goudeau, P., Tamura, N. (2015). Evolution of terra sigillata technology from Italy to Gaul through a multi-technique approach. *Journal of Analytical Atomic Spectrometry*, 30(3), 658–665.

Liu, Y., Meirer, F., Williams, P.A., Wang, J., Andrews, J.C., Pianetta, P. (2012). TXM-Wizard: A program for advanced data collection and evaluation in full-field transmission X-ray microscopy. *Journal of Synchrotron Radiation*, 19(2), 281–287.

Maritan, L. (2019). Archaeo-ceramic 2.0: Investigating ancient ceramics using modern technological approaches. *Archaeological and Anthropological Sciences*, 11(10), 5085–5093.

Meirer, F., Liu, Y., Pouyet, E., Fayard, B., Cotte, M., Sanchez, C., Andrews, J.C., Mehta, A., Sciau, P. (2013). Full-field XANES analysis of Roman ceramics to estimate firing conditions – A novel probe to study hierarchical heterogeneous materials. *Journal of Analytical Atomic Spectrometry*, 28(12), 1870–1883.

Oakley, K.P. (1933). The pottery from the Romano-British site on Thundersbarrow Hill. *The Antiquaries Journal*, 13(2), 134–151.

Sciau, P. and Godet, M. (2021). Spectroscopy and diffraction using the electron microscope. In *Spectroscopy, Diffraction and Tomography in Art and Heritage Science*, Adriaens, M. and Dowsett, M. (eds). Elsevier, Amsterdam, 71–102.

Sciau, P. and Wang, T. (2019). Full-field transmission X-ray microspectroscopy (FF-XANES) applied to cultural heritage materials: The case of ancient ceramics. In *Synchrotron Radiation: Useful and Interesting Applications*, Joseph, D. (ed.). IntechOpen, London.

Sciau, P., Salles, P., Roucau, C., Mehta, A., Benassayag, G. (2009). Applications of focused ion beam for preparation of specimens of ancient ceramic for electron microscopy and synchrotron X-ray studies. *Micron*, 40(5–6), 597–604.

Sciau, P., Leon, Y., Goudeau, P., Fakra, S.C., Webb, S., Mehta, A. (2011). Reverse engineering the ancient ceramic technology based on X-ray fluorescence spectromicroscopy. *Journal of Analytical Atomic Spectrometry*, 26(5), 969–976.

Sciau, P., Sanchez, C., Gliozzo, E. (2020). Ceramic technology: How to characterize terra sigillata ware. *Archaeological and Anthropological Sciences*, 12(9), 211.

Solé, V.A., Papillon, E., Cotte, M., Walter, P., Susini, J. (2007). A multiplatform code for the analysis of energy-dispersive X-ray fluorescence spectra. *Spectrochimica Acta Part B: Atomic Spectroscopy*, 62(1), 63–68.

Tite, M.S. (2008). Ceramic production, provenance and use – A review. *Archaeometry*, 50, 216–231.

Wang, T., Zhu, T.Q., Feng, Z.Y., Fayard, B., Pouyet, E., Cotte, M., De Nolf, W., Salome, M., Sciau, P. (2016). Synchrotron radiation-based multi-analytical approach for studying underglaze color: The microstructure of Chinese Qinghua blue decors (Ming dynasty). *Analytica Chimica Acta*, 928, 20–31.

6

Relevant Synchrotron X-rays Techniques to Study Corroded Iron Cultural Heritage Material

Solenn RÉGUER[1], François MIRAMBET[2], Judith MONNIER[3], Delphine NEFF[4], Eddy FOY[4] and Philippe DILLMANN[4]

[1] *Synchrotron Soleil, Saint-Aubin, France*

[2] *Centre de recherche et de restauration des musées de France, Paris, France*

[3] *ICMPE CNRS, Thiais, France*

[4] *LAPA NIMBE/IRAMAT CEA/CNRS, Université Paris Saclay, Gif-sur-Yvette, France*

6.1. Introduction

Metal, especially iron, used for several thousand years in ancient societies for making objects, tools or weapons, was also used to reinforce monuments, at least since antiquity. Found in archaeological context, it must be stored and conserved either onsite, in storage rooms or in exhibition rooms in museums. Unfortunately, iron alloys are not stable and often suffer corrosion processes. Scientific investigations of ancient iron cultural artifacts can help us to understand corrosion processes and improve restoration and conservation approaches. Indeed, metallic heritage objects have to be protected in the long term. However, this protection is only possible if the materials and their evolution under corrosion are well known. Diagnoses describing their alteration state are thus crucial. For this reason, it is necessary to finely determine and predict the alteration processes, their kinetics, for all iron alloys and all exposition medium, whatever atmosphere, soil or marine and also hydraulic binder.

Synchrotron Radiation, Cultural Heritage, Biomineralization,
coordinated by Catherine DEJOIE, Pauline MARTINETTO and Nobumichi TAMURA.

The corrosion processes over a long period of time (often several centuries), lead to the formation of specific iron corrosion systems in different environments (atmosphere, soil and marine; see Figure 6.1). The corrosion layers may reach thicknesses from a few micrometers to several millimeters. Depending on the local conditions in the corrosion system (O_2, pH), the nature, proportion and distribution of the different phases in presence vary. The systems are thus often heterogeneous at different scales, containing amorphous or crystalline phases, and consisting of a mixture of compounds at the scale of the analysis methods (Neff et al. 2005).

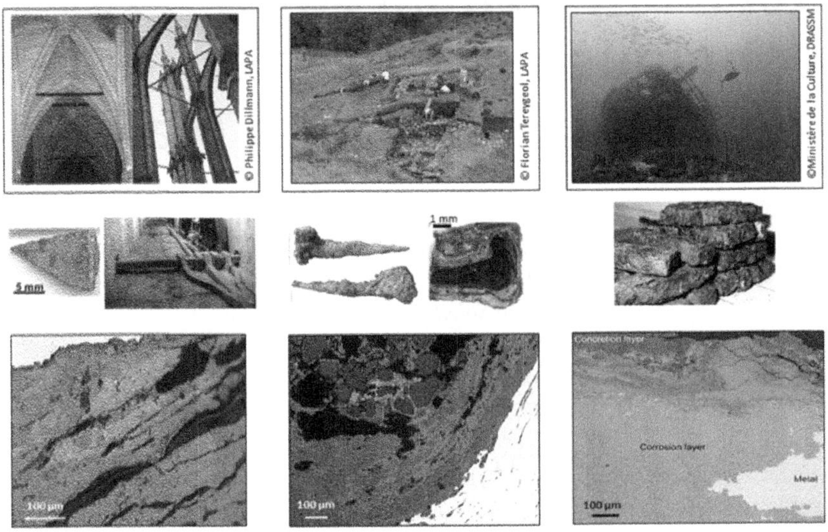

Figure 6.1. *Photographs of some historical and archaeological sites, where the iron is exposed to different corrosion processes (atmospheric, soil and marine); from left to right: medieval cathedral in France and iron bars installed in 1498 in Amiens; French archaeological site (Castel Minier), iron nails from mill of Glinet archaeological site, 12th to 16th century; shipwrecks dated from the Roman period immersed near Les Saintes-Maries-de-la-Mer in the Mediterranean Sea, iron ingots. Micrographies of cross-section of three different samples revealing the corrosion layers morphology and complexity, depending on the original corrosion (atmospheric, soil and marine). For a color version of this figure, see www.iste.co.uk/dejoie/synchrotron.zip*

In order to obtain all the necessary information for an overview of the corrosion system formation and evolution, the physico-chemical properties of the different corrosion products in presence and their distribution at different scale must be determined. Indeed, the phases could be more or less electrochemically reactive

(Monnier et al. 2014). Moreover, during the corrosion processes or protection action, the evolution of the heterogeneous system, which is modified, can lead, for instance, to the transformation from amorphous to crystalline system or the reverse. The role of each corrosion product present in the corrosion layers must be understood as they can control the corrosion kinetic of the whole system. In addition, the reactivity or stability of the system can be influenced by several factors such as the presence of peculiar elements, like P, S or Cl (Réguer et al. 2007a, 2009; Monnier et al. 2011).

Corrosion layers contain different phases, mixed at a micro- to nano-metric scale, implying the use of specific techniques to tackle the heterogeneity of such materials. A specific analytical methodology has been developed to gain insights into ancient iron artifacts, showing the necessity of a multi-scale and multi-techniques approach (Neff et al. 2013). Many modern analytical techniques are employed, and among them, the potential of synchrotron-based techniques was demonstrated, with fruitful results. In fact, many available laboratory techniques allow characterizing the iron-corroded artifacts as initial, necessary and sometimes sufficient steps to obtain useful and complementary information, such as micro-Raman spectroscopy, TEM (Transmission Electron Microscopy) or ToF-SIMS (Time Of Flight – Secondary Ion Mass Spectrometry). Some specific questions may, however, remain and need to be answered. For these issues, synchrotron-based techniques are best suited, as illustrated in this chapter.

The studied objects are usually a set of archaeological and historical iron artifacts from different sites: marine or terrestrial archaeological sites or historical buildings. The samples are generally prepared as cross-sections taken from disposable artifacts (like bars, nails, etc.) to study the entire corrosion system, including the metal core, the corrosion products and the external environment (concretion for the marine sites, adherent soil from land sites). However, studies may also focus on simplified systems known as "models", such as artificially aged metal coupons under controlled conditions, or products derived from chemical syntheses made in the laboratory or commercial products considered as references. The studies on these two latter kinds of samples are necessary to get a permanent confrontation between a natural complex and heterogeneous system and a simplified model.

For ancient samples, a high sensitivity and high lateral resolution is usually required, while studies based on reference samples meet a wider scale, which imply an analytical approach combining micro beam (1–10 µm or less) and macro beam (200 µm–1 mm). The synchrotron radiation-based techniques allow us to obtain all the physico-chemical information and the chemical and structural distribution into the analyzed artifact from millimetric to nanometric scale. For this reason, they have become popular tools to study the ancient iron corrosion artifacts.

The characteristic, capabilities and use of synchrotron experiments for archaeo-metallurgy studies were well expressed in a previous paper (Young 2012). Some advantages were cited and demonstrated, as the non-destructive aspect, fast measurements, versatility with tunable energy, beam size adjustment, sensitivity, possibility of multi-elemental analysis and multiple phases' determination. This chapter aims to give an overview of the studies dedicated to the use of synchrotron techniques (mainly X-ray-based techniques) to study alteration processes and protection solutions of ancient iron artifacts. The studies especially focused on the fine identification (interest of combination of techniques) of the corrosion products and the possibility to follow processes during time (interest of fast measurements). Three main techniques described hereafter are used:

– X-ray fluorescence spectroscopy (XRF) is extensively and successfully used on many ancient artifacts to determine their composition with the advantage of high elemental sensitivity, low detection limit, small beam size availability and fast collection time allowing imaging (elemental maps) acquisitions.

– X-ray powder diffraction (XRD) reveals crucial information, especially with a focused beam at the micrometer scale that allows determining exactly the structure and locating several crystalline products in the corrosion layers. In addition, the quality of the data enables us to describe the structure (using refinement), the degree of crystallinity, the grain size, the texture and any information that can help to understand the behavior of each phase. In addition, it is possible to make some phases quantification based on the powder XRD diagrams.

– X-ray absorption spectroscopy (XAS), almost exclusively found in synchrotron facilities, is a well-suited technique for the determination of phases in presence in the complex corrosion system, as this technique can be equally applied on amorphous or crystalline material and can use macro- to nano-probe. Indeed, the X-ray Absorption Near Edge Structure (XANES) and Scanning Transmission X-ray Microscopy (STXM) are excellent to reflect the environment of a given absorbing atom, especially at Fe K or L-edges; this allows us, toward the extraction of pre-edge and edge characteristics, to access the oxidation state, coordination numbers, site symmetry and structural distortion. In addition, the Extended X-ray Absorption Fine Structure (EXAFS) domain provides direct information on the identity of the first and more distant surrounding atoms and their specific bound distances (Teo and Joy 1981) to fingerprint the local structure of the probed atoms. The phases in presence in the corrosion layers of corroded iron material always contain different oxidation degree for iron (Fe^{2+}, Fe^{3+} or mixed) but also different site coordination: octahedral (VI) or tetrahedral (IV). The key details are mainly given by Fe K-edge investigations, but some information is also provided by other elements such as Cl, S or P, which implies the investigation of other edges. The precise study of X-ray absorption spectra of different references phases enables establishing the relevant criteria to distinguish the phases in order to interpret the spectra obtained from the complex ancient system.

Let us give an overview of these synchrotron-based techniques toward the scientific questions of ancient iron artifacts' alteration and conservation.

6.2. Iron corrosion diagnosis in historical monuments

Iron has been extensively used all around the world for the construction of buildings and also to strengthen their architecture, as for the Khmer temples at Angkor and in Europe for the medieval Gothic cathedrals. For example, in the Amiens cathedral, an iron reinforcement chain was installed in the triforium in 1497 to stop the bending of the pillars of the crossing between the transept and the nave. In this case, materials were exposed to an indoor atmospheric environment. Another example is the impressive iron bars between the flying buttresses of the Beauvais cathedral, put in the monument during its building in the 13th century (Leroy et al. 2015).

In such conditions, part of the metal undergoes corrosion processes due to a cycle alternating humidity and drying on its surface. During the humidification stage, oxidation of the metal substrate is caused by the reduction of some compounds present in the corrosion layers. However, it has been shown that the different compounds have different electrochemical reactivities. To predict the behavior and evolutionary aspect of the object's corrosion, it is essential to identify unambiguously the nature of the corrosion products involved.

6.2.1. *Identification of the corrosion products*

In the field of iron atmospheric corrosion, samples coming from the Amiens cathedral, but also from several other historical buildings have been studied. The atmospheric corrosion systems formed were described (Dillmann et al. 2004; Hoerle et al. 2004) using a combination of many analytical tools, including XRF, XRD and XAS measurements. For instance, corrosion resulting from the degradation of 1,600 years old iron from Indian monuments was examined using XRD and semi-quantitative analyses were carried out to propose a profile of corrosion layer system (Balasubramaniam et al. 2004).

Data treatment methodology of XAS measurements applied on iron archaeological artifacts is exported from various scientific communities (geochemistry, earth science, etc.), showing the same analytical approach even though different systems and objectives are involved. A specific procedure was developed for the XANES of Fe K-edge spectra study of iron mineral (oxides and oxyhydroxides) and glasses: the features have been investigated as a function of the oxidation state and coordination geometry (Wilke et al. 2001; Farges et al. 2004; Bokhoven and Lamberti 2016). In addition, iron to iron nearest distances revealed by EXAFS (Schwertmann and Cornell 2000) depends mainly on the nature of octahedral linkages (face, corner, edge). The protocol can be favorably applied on corrosion products, due to many phases in common.

The interpretation of XAS data at the Fe K-edge of samples from ancient buildings was crucial in the fine characterization of the corrosion layers as it revealed the oxidation degree of iron and its distribution in the corrosion layers, but also detected variations in the atom environment with octahedral or tetrahedral geometries of oxygen around iron. The XANES and EXAFS signals make it possible to confirm which phase is present and distinguish three groups of compounds, depending on the oscillation shapes: oxides (such as maghemite), crystalline oxyhydroxides (such as goethite) and poorly crystalline oxyhydroxides (such as ferrihydrite) as described hereafter. Previous studies used these observations on EXAFS signal and from its Fourier Transform to study the hydrolysis processes and formation of ferric gels, leading to akaganeite, goethite or hematite formation (Bottero et al. 1994; Singh and Grafe 2006). These results show the possibility to recognize the different iron oxyhydroxides (see Figure 6.2):

– Maghemite (γ-Fe_2O_3) and magnetite (Fe_3O_4) possess quite the same oscillation due to their similar local structure. However, a high-quality signal thanks to a high signal/noise ratio allows us to see the difference. In addition, magnetite shows an edge position lower in energy, which allows their discrimination. Indeed, magnetite contains both di and trivalent Fe. The trivalent ions occupy both tetrahedral and octahedral sites, while in maghemite, the tetrahedral and octahedral sites are occupied by only trivalent ions.

– The three oxyhydroxides: goethite α-FeOOH, akaganeite β-FeOOH, and lepidocrocite γ-FeOOH. These compounds are constituted of double chains of octahedral formed by edge sharing (see Schwertmann and Cornell (2000), for illustration and details), implying some difficulties to distinguish them, but the arrangement of the double chains differs.

– Ferrihydrite and feroxyhyte δ-FeOOH, which show featureless EXAFS spectra revealing a weak local order (Drits et al. 1993; Rancourt and Meunier 2008), due to the poor crystallinity of these phases.

Such possible distinction is important as maghemite is considered less reactive than ferrihydrite and feroxyhyte, implying they have different roles in the corrosion mechanism. For samples from ancient monuments, the main problem was the identification of the nature of light marblings which crosses the corrosion products layer.

Thanks to μXAS, acquisitions were performed at different locations, corresponding to the light marbling on four samples of an historical building, to determine whether magnetite (the "conductive" phase) is present or not in the corrosion systems. The pre-edge shape, the edge position and the post-edge oscillations on the spectra obtained on ancient samples were compared to the ones obtained on reference powders. All of the spectra from ancient samples led to the conclusion that no magnetite was present in those light marblings (Monnier et al. 2010b).

6.2.2. *Specific chemical elements from the elaboration processes*

Iron and steel, manufactured before the 19th century, using processes different from those used by modern metallurgy, may have different structures and chemical compositions than those of contemporary materials.

For example, they may contain significant amounts of phosphorus from the ore, and this element can play a beneficial role in atmospheric corrosion behavior: this could explain the exceptional corrosion resistance of the Delhi column (Balasubramaniam 2000), made from wrought iron and dated from the Gupta era (4th–5th centuries).

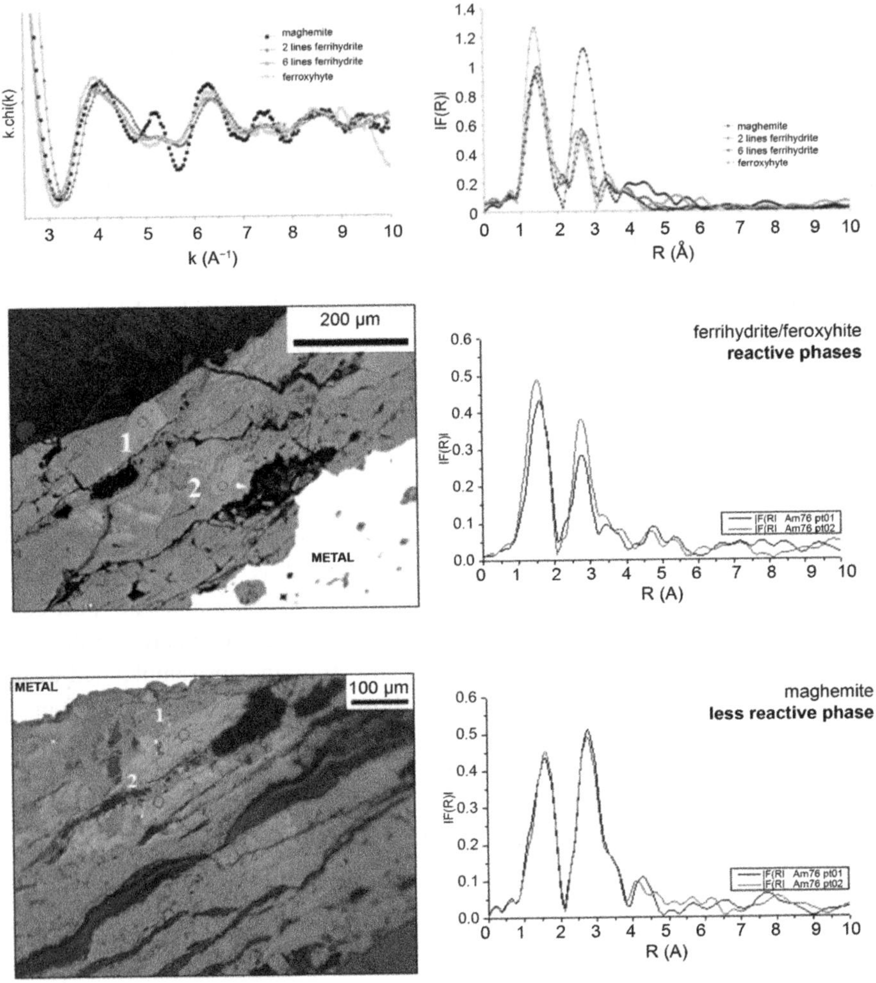

Figure 6.2. *Phases in presence in the case of atmospheric corrosion of iron. EXAFS and radial distribution functions (FT) of reference powders, using macroscopic beam at the Fe K edge (FAME, ESRF). Micrographies from Amiens cathedral samples and points of interest for EXAFS acquisitions at the Fe K edge using a 3 µm beam size (LUCIA, SOLEIL) and FT suggesting the presence of maghemite or ferrihydrite/ feroxyhyte phase in corrosion product layers (see Monnier et al. 2010a). For a color version of this figure, see www.iste.co.uk/dejoie/synchrotron.zip*

The role of phosphorus was investigated on samples from historical Indian monuments such as the Deogarh temple (5th century), and also from Amiens cathedral, where heterogeneous distributions of phosphorus using XRF maps were revealed (Monnier et al. 2010a). Even though present in the entire corrosion system with heterogeneous distribution, P seems more concentrated at the external part of the corrosion system. In addition, P could be present inside the slag inclusion of the metal part.

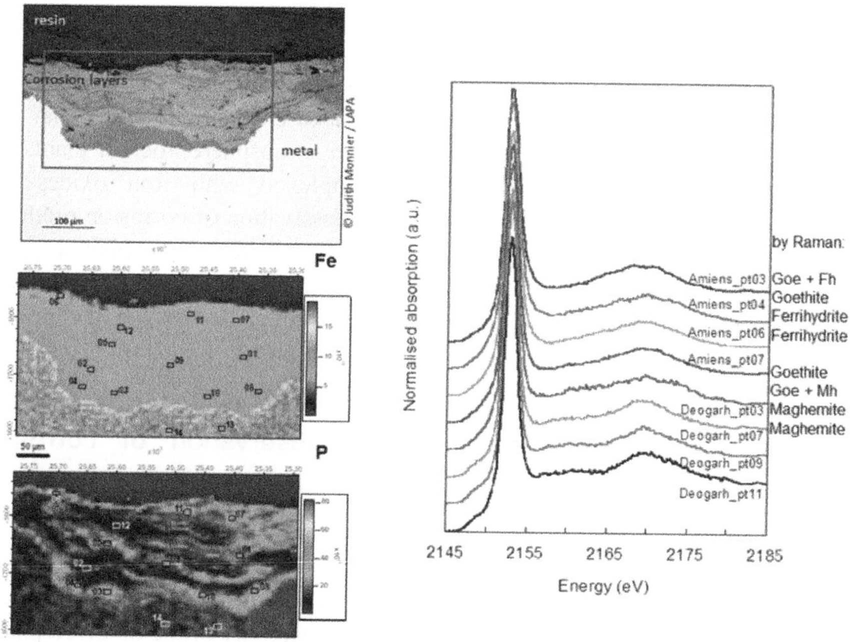

Figure 6.3. *Micrography of Deogarh (India) sample, where the main phases constituting the corrosion system are goethite and maghemite (large marblings); XRF elemental mapping showing the P and Fe distribution and acquired at 2.5 keV using a 3 µm beam size (LUCIA, SOLEIL). Micro XANES spectra at the P K-edge on samples from two monuments (Amiens cathedral and Deogarh temple) with the localization of the analyzed points on the optical microphotograph (see Monnier et al. 2011). For a color version of this figure, see www.iste.co.uk/dejoie/synchrotron.zip*

XANES measurements at the P K-edge specifically examine the P speciation in order to evaluate the effect of the P-containing corrosion products in the corrosion layer and their possible role in the alloy stability (see Figure 6.3). Some studies report results obtained at the P K-edge for mineral studies, especially in the frame of soil depollution (Rose 1997; Vantelon et al. 2007). As many phases may be the same, a similar approach can be applied on corrosion products. In addition, XANES spectra are compared with synthesized references (Monnier et al. 2011). In the metallic substrate, P is supposed to be present in solid solution and in the form of non-metallic inclusions as iron and calcium phosphates, typical of metals of these periods.

On the contrary, two types of phases are identified in the corrosion layer. XANES spectra are in good agreement with the structure of aluminum phosphates and/or calcium phosphates, which are characteristic of exogenous sources that may have been deposited during the long exposure, from atmosphere, rock or plant. P is also present in the form of phosphates complexed with iron oxides and oxyhydroxides, which can be at the origin of the passivation of corrosion products; but no iron phosphate is identified. The iron oxide complexes may explain a certain stabilization of the corrosion system, decreasing their solubility or their interaction with other species and therefore of the structures concerned within the historic monuments.

6.3. In situ experiments for a direct observation of corrosion processes

6.3.1. *How to reveal atmospheric corrosion processes?*

In order to better understand the process occurring in atmospheric corrosion mechanisms (Stratmann 1990), the electrochemical reactivity of the involved phases and transport phenomena are critical aspects. A corrosion mechanism must be understood and monitored. In that way, the products formed during the reduction of electrochemically reactive phases were determined. Such experimentations were realized using in situ measurements in an electrochemical cell designed to have less than 500 μm of thickness of electrolyte above the surface sample. In that specific case, combination of XANES, EXAFS and XRD measurements were performed in transmission or in reflexion mode using a macro beam in order to characterize highly reactive corrosion products associated with the atmospheric corrosion of Fe.

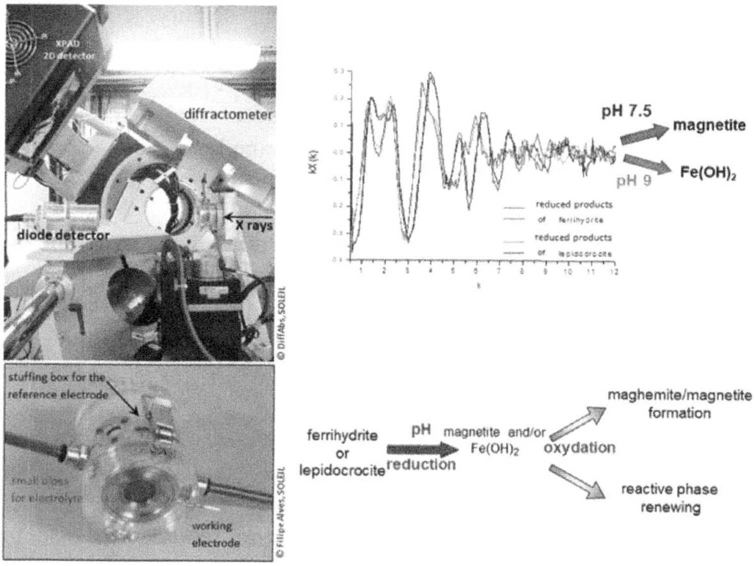

Figure 6.4. *Photographs of the set-up (top left) in transmission mode for XRD and XAS (DiffAbs, SOLEIL) and of the electrochemical cell (bottom left); EXAFS signals (top right) of the reduced products at the end of the in situ reduction processes at different pH; schematic diagram of the phase evolution in the in situ cell (bottom right) (see Monnier et al. 2014). For a color version of this figure, see www.iste.co.uk/dejoie/synchrotron.zip*

The electrochemical reactivity of lepidocrocite and two-line ferrihydrite powders (Monnier et al. 2014) were visible toward the shift of the energy position of the XANES spectra acquired: decrease/increase of the oxidation state during the reduction/re-oxidation process (see Figure 6.4). The two compounds transform into magnetite at lower pH (pH = 7.5), while $Fe(OH)_2$ is formed at higher pH (pH = 9). With the magnetite, the cycle will be stopped, while for the second compound it can continue. Thus, the determination of the compound allows the determination of the state of a specific system: stable or reactive.

6.3.2. *Iron corrosion in cement matrices of ancient building*

The earliest forms of metal reinforcing bars observed in cement and hydraulic binders matrices date from antiquity. Since, it has evolved but has never ceased to be employed in the building of monuments as in the Popes Palace in Avignon or the dungeon of the Vincennes Castle (both 14th century), or latter the Pantheon church in Paris. Finally, since the beginning of the 20th century, concrete has been reinforced with low alloy steel. Damages due to corrosion of iron reinforcement in cement and hydraulic binders are critical. Indeed, corrosion processes imply the formation of corrosion products which can generate crack and binder spalling caused by high tensile stresses generated in the medium. These processes can eventually lead, at the end, to important damage for the monuments.

The initial more or less alkaline conditions depending on the hydraulic binders or concrete environment allow the formation of a passive layer on iron, avoiding further corrosion. However, a decrease of the pH of the solution can lead to the destabilization of the passive layer due to sulfate, carbonate or chlorine species, and thus, let corrosion occurs (Tuuti 1982).

To reveal what is happening during this initiation of the corrosion, in situ monitoring of initial corrosion processes was performed (Grousset et al. 2015). Specific cells composed of mortar and iron bars can simulate the corrosion processes. They were thus designed so that the penetration of a chlorinated solution on the binder/iron interface is realized by capillarity. XRD maps of a phase's distribution within the corrosion layer allows monitoring of the corrosion process.

These experiments allowed (i) locating the zones of the corrosion, thus showing a porous network influence; (ii) following the iron diffusion through the binder "far and fast"; and (iii) determining the first crystalline corrosion products formed: chloride green rust which evolves thereafter into akaganeite.

6.4. Archaeological objects corrosion and stabilization

Corrosion processes in soil are quite different from atmospheric processes previously described as the O_2 and presence of other elements from water and soil can influence the corrosion processes.

6.4.1. *Corrosion in anoxic soils*

Due to the diversity of phases in presence at a very small scale (down to nanometric), spatially precise methods can be used. In that way, the metal/corrosion layer interface was studied using STXM because the physic-chemical processes, even at a nanometric scale, are involved in the global corrosion mechanism at the macroscopic level. The STXM acquisitions demonstrate that in addition to a layer of carbonates, a nanometric layer made of maghemite and magnetite play an interface barrier of few 10 to 50 nm (Leon et al. 2014).

6.4.2. *Specific Cl-containing corrosion products*

In the case of iron corrosion in soil or in a marine environment, chlorides constitute an aggravating factor of instability for the conservation of objects. In the presence of oxygen, such an element causes the dramatic acceleration of corrosion processes and a rapid degradation of the objects in very short periods of time. Cl-containing corrosion products were thus finely studied.

For instance, XANES spectra were acquired at Fe K-edge on ferrous artifacts from four different archeological sites dating from the 12th to the 16th century (Réguer et al. 2007a and 2007b) to determine the Fe oxidation degree and to correlate it with Cl quantification to identify the corrosion products.

The value of Fe K-edge position of the XANES spectrum can be directly used as simple but crucial information. Indeed, the edge position is close to Fe(II) in the region with a high amount of chlorine (16–20% mass) associated with hydroxychloride, while it is shifted to the higher energy, Fe(III) in the region with a lower Cl content of about 4–12% mass and associated with akaganeite. In addition, the pre-edge characteristics evaluation of the XANES spectra acquired on the same samples allowed determining the Fe environment by comparing with the graphs given in literature (Wilke et al. 2001). It showed the presence of Fe(III) and Fe(II) oxidation states correlated to VI-Fe coordinance (octahedral), confirming the presence of akaganeite FeOOH and ferrous hydroxychloride $Fe_2(OH)_3Cl$ respectively.

Such heterogeneous distribution of oxidation degree and description of the distribution of these two Cl-containing phases inside the corrosion system was done; this gave not only key information for any research on corrosion processes involving Cl ions but also significant information about the dechlorination mechanisms (Réguer et al. 2006, 2007a and 2007b).

Cl K-edge XANES and EXAFS investigations were found to be a crucial complement to Fe K-edge measurements for the Cl-containing corrosion products analysis. Later studies (Réguer et al. 2015; Simon et al. 2018) revealed the capacity of this technique to distinguish the Cl various forms toward the characteristics of the spectra. Indeed, in archaeological samples, chloride ions could be located inside the pores and cracks as "free" chlorides or adsorbed at the surface of the grains of the various phases, or they may be trapped inside the structure of iron oxyhydroxides as akaganeite and ferrous hydroxychloride $Fe_2(OH)_3Cl$. The availability of this chlorine during corrosion processes is an indication of the stability of the objects.

In a marine environment case, the characterization of iron bar corrosion products, discovered during archaeological excavations of a Gallo-Roman wreck (1st century BCE) near the coast of Saintes-Maries-de-la-Mer, revealed the two Cl-containing products with a main presence of hydroxychloride $Fe_2(OH)_3Cl$ (Rémazeilles et al. 2009).

To complete the ancient sample characterization, the main phases were separately studied (see Figure 6.5). The ferrous hydroxychloride was carefully studied by a combination of XRD and XAS at Fe and Cl K-edge to precisely determine the crystal structure (Réguer et al. 2015). The study of the inherent ability of chlorine in the akaganeite structure was performed by conducting successive washings of the pure phase (Réguer et al. 2009). The akaganeite powders containing various Cl content were analyzed by XAS at Fe and Cl K-edge, in addition to XRD measurements. The spectra acquired at Cl K-edge revealed a correlation between the amplitude of the EXAFS oscillations and the Cl content adsorbed at the surface of the compound, which could not be observed at the Fe K-edge. The fine structural analysis by this combination of techniques using synchrotron radiation showed that most of the chlorine atoms that could be released, either during the corrosion process or during dechlorination treatment, came mainly from the desorption process of Cl initially adsorbed on the grains than Cl from the structure.

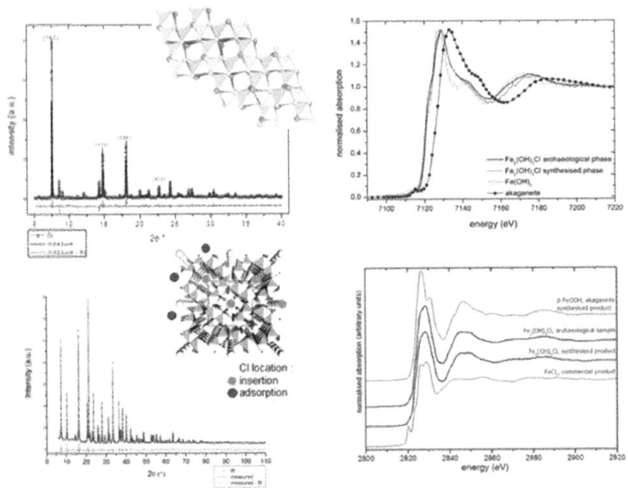

Figure 6.5. *Crystallographic structure of β-FeOOH akaganeite and β-$Fe_2(OH)_3Cl$ ferrous hydroxychloride (see Ståhl et al. 2003; Réguer et al. 2009). XRD patterns of the phases (D2AM, ESRF and DiffAbs, SOLEIL) and XANES spectra at Fe and Cl K edges (LUCIA and DiffAbs, SOLEIL) (Réguer et al. 2015). For a color version of this figure, see www.iste.co.uk/dejoie/synchrotron.zip*

6.4.3. *Dechlorination treatment revealed*

Restoration and conservation strategies involve determining the mechanisms taking place during the stabilization and protection treatments, evaluating their efficiency, optimizing them or developing new protocols.

The evolution of the corrosion products of archaeological objects containing chloride was monitored during the stabilization process (Kergourlay et al. 2018). A sample from a cross-sectional iron bar was placed in a chemical cell containing a sodium-based solution, a dechlorination treatment traditionally used in restoration workshops. The rapid acquisition of XRD maps revealed the distribution of the different corrosion products on a micrometer scale and their transformation during processing. Indeed, it highlighted the disappearance of some products (ferrous hydroxychloride) in favor of more stable ones (magnetite) and the presence of transition compounds (such as "green rust", a turquoise blue corrosion product) observable only during the process, because it transformed into other products at the end of the process. Cl K-edge XANES measurements are in this case a key approach to decipher the corrosion patterns involving Cl ions and to understand by in situ observation the dechlorination processes. (Réguer et al. 2020). This methodology is crucial to define the real treatment processes of ancient iron objects and to optimize them by controlling parameters such as temperature or oxygen content of the treatment solution.

6.5. Protection of ancient iron artifacts

On a very large part of the metallic elements in heritage sciences, it is necessary to preserve the patina, the layers of paint or even the corrosion products because they can contain elements likely to reveal information on the technique and/or the history of the object. The best way to protect the iron with cultural heritage relevance from corrosion degradation is to store all artifacts in a very controlled environment (stable temperature, low relative humidity variations, free of pollutants), which is quite difficult in many cases. In order to protect metals from ageing, protection methods using coatings or corrosion inhibitors are developed to limit exchanges or maybe isolate the metal and the corrosive environment. These methods have to correspond to the deontological request of conservation (aesthetic aspect, removable). In addition, the thickness of the coating must be sufficiently thick to be an effective barrier for corrosion processes and sufficiently thin to be optically transparent.

The well-known treatments used are resin or waxes or more recently polyphenolic molecules (Veys-Renaux et al. 2018) and carboxylate solution.

This latter protection leads to the formation of a layer of several micrometers (Hollner et al. 2010). The protection mechanisms that occurred must be determined by observing the interface between the corroded metal and the protection methods used.

In order to best meet the conservation expectations of the corroded iron objects, new effective protection treatments were studied. These friendly corrosion inhibitors are based on sodium carboxylates NaC_n (formula $CH_3(CH_2)_n$-2COONa, with 7<n<10) extracted from vegetable oils (sunflower, etc.). In situ experiments were conducted by XAS and XRD in an electrochemical cell, to monitor the formation of the protective passivation coating from the first moments of immersion of iron in the treatment solution (Mirambet et al. 2010). These experiments showed that the inhibitory properties of the solution are related to precipitation of a metal soap thin film (<1 μm) made of iron carboxylate. The study of the local structure of this latter phase showed that the environment around the iron is the same, whatever the chain length (n), with a +III oxidation state of iron atoms in an octahedral environment of oxygen surrounding the iron. This protection layer stops diffusion phenomena and reduces the reactivity of certain corrosion products, thus stopping corrosion processes. Furthermore, these experiments confirmed that the conversion process is relatively fast, which is very important for applications on iron ancient objects.

6.6. Conclusion and outlook

The data acquired using synchrotron radiation, obtained through the analysis of ancient objects or in situ experiments, are of primary importance for the detailed characterization of the altered system, including metal, corrosion products and local interface, but also to follow and determine the alteration or the protection mechanism of metal elements in heritage sciences. The study of these complex materials at different scales is crucial to reveal, improve, develop and approve long-term conservation and protection protocols.

This review showed how it is beneficial to use X-ray synchrotron-based techniques, even though they often have to be used in combination with a panel of complementary techniques due to the complexity of the studied system (Neff et al. 2013).

Promising applications of synchrotron radiation techniques are mainly on two bases:

– Fast acquisitions

The use of fast measurement synchrotron radiation-based techniques is fundamental to monitor phase evolution in specific conditions, using quick EXAFS, or fast and fly XRD and XRF acquisitions. The fast acquisitions also allow an increase in the size of the analyzed zones within a sample and thus a better representation of the measurement, for example, using flyscan mode for mapping (Hamon et al. 2020). It is also an opportunity to analyze a series of samples from a corpus, which is essential for studies on heterogeneous materials before making any hypothesis on alteration and protection mechanisms.

– Imaging techniques

Corrosion studies of ancient objects should focus on the transport phenomena taking place in the corrosion layers and their surrounding environment. This question of species transport highlights the importance of porosity and its permeability characteristics on the object's potential reactivity to corrosion processes continuity. In that case, full-field acquisitions or even more deeply, X-ray tomography makes it possible to visualize the internal structure of objects for porosity description and process understanding (Jacot-Guillarmod et al. 2015). Modestly used up to now, both 2D and 3D imaging techniques have to be developed for this topic, to show more rapidly and with higher resolution the internal features of the artifacts. The full-field and X-ray tomography methods promise great opportunities in the field of iron alteration and conservation, but limitation concerns the difficulty of sample preparation and the possible use of the vacuum chamber.

New or not yet currently used methods are also promising for the field as 2D to 3D X-ray Raman multispectral imaging (Georgiou et al. 2019) and X-ray PDF (Cersoy et al. 2014).

6.7. Acknowledgments

Much of the work was carried out by an enthusiastic team: Emmanuel Rocca, Florian Kergourlay, Mandana Saheb, Sophie Grousset, Marine Bayle, Florence Mercier, Céline Rémazeilles, Claire Gervais, Florian Tereygeol, Stephanie Leroy, and so many other valuable colleagues!

The whole methodological approach and the scientific results, including the development and realization of synchrotron experiments would not have been possible without the support of beamline scientists of the involved following beamlines: DiffAbs, LUCIA, HERMES at SOLEIL (France), ID21, FAME, D2AM, ID24 and ID19 at ESRF (France – Eu facility) and SM (10ID-1) at CLS (Canada).

We appreciated the availability, helpful advise and the necessary expertise to have a successful beamtime of Gemma Guillera, Olivier Proux, Jean-Francois Berrar, Delphine Vantelon, Denis Testemale, Marine Cotte, Jean-Louis Hazemann, Jean Susini, Eric Dooryhee, Jean-Louis Hodeau, Pierre Lagarde and so many other valuable people!

The whole study was supported by the FSP, ANR ARCOR, PNRC OdefA and PNRC CARBOTAN (French Ministry of Culture) projects, and also ESRF, SLS, CLS and SOLEIL facilities.

6.8. References

Balasubramaniam, R. (2000). On the corrosion resistance of the Delhi iron pillar. *Corrosion Science*, 42(12), 2103–2129.

Balasubramaniam, R., Saxena, A., Anantharaman, T.R., Réguer, S., Dillmann, P. (2004). A marvel of medieval Indian metallurgy: Thanjavur's forge-welded iron cannon. *Journal of Materials*, 56, 17–23.

Bokhoven J.A. and Lamberti C. (2016). *X-Ray Absorption and X-Ray Emission Spectroscopy: Theory and Applications*. John Wiley & Sons, New York.

Bottero, J.-Y., Manceau, A., Villieras, F., Tchoubar, D. (1994). Structure and mechanisms of formation of iron oxide hydroxide (chloride) polymers. *Langmuir*, 10(1), 316–319.

Cersoy, S., Martinetto, P., Bordet, P., Hodeau, J.-L., Walter, P., Van Elslande, E. (2014) Analysis of archaeological samples using XRPD-PDF and Raman spectroscopy. *Acta Crystallographica Section A: Foundations and Advances*, 70, c863–c863.

Dillmann, P., Mazaudier, F., Hoerlé, S. (2004). Advances in understanding atmospheric corrosion of iron. I. Rust characterisation of ancient ferrous artefacts exposed to indoor atmospheric corrosion. *Corrosion Science*, 46(6), 1401–1429.

Drits, V.A., Sakharov, B.A., Salyn, A.L., Manceau, A. (1993). Structural model for ferrihydrite. *Clay Minerals*, 28(2), 185–207.

Farges, F., Lefrère, Y., Rossano, S., Berthereau, A., Calas, G., Brown, G. (2004). The effect of redox state on the local structural environment of iron in silicate glasses: A combined XAFS spectroscopy, molecular dynamics and bond valence study. *Journal of Non-Crystalline Solids*, 344(3), 176–188.

Georgiou, R., Gueriau, P., Sahle, C.J., Bernard, S., Mirone, A., Garrouste, R., Bergmann, U., Rueff, J.P., Bertrand, L. (2019). Carbon speciation in organic fossils using 2D to 3D X-ray Raman multispectral imaging. *Science Advances*, 5(8), 5019.

Grousset, S., Kergourlay, F., Neff, D., Foy, E., Gallias, J.L., Réguer, S., Noumowe, A., Dillmann, P. (2015). In situ monitoring of corrosion processes by coupled μXRF/μXRD mapping to understand degradation mechanisms of reinforcing bars in hydraulic binders from historical monuments. *Journal of Analytical Atomic Spectrometry*, 30(3), 721–729.

Hamon, C., Réguer, S., Bellot-Gurlet, L., Pailler, Y., Brisotto, V. (2020). Tracking the first bronze metallurgists of Western Europe: Combined use-wear analysis and X-ray fluorescence synchrotron spectroscopy of a stone toolkit from Ploneour-Lanvern (Brittany). *Archaeological and Anthropological Sciences*, 12(1), 14.

Hoerlé, S., Mazaudier, F., Dillmann, P., Santarini, G. (2004). Advances in understanding atmospheric corrosion of iron. II. Mechanistic modelling of wet–dry cycles. *Corrosion Science*, 46(6), 1431–1465.

Hollner, S., Mirambet, F., Rocca, E., Réguer, S. (2010). Evaluation of new non-toxic corrosion inhibitors for the conservation of iron artefacts. *Corrosion Engineering, Science and Technology*, 45(5), 362–366.

Jacot-Guillarmod, M., Rozenbaum, O., L'Hostis, V., Dillmann, P., Neff, D., Gervais, C. (2015). Degradation mechanisms of reinforcing iron rebars in monuments: The role of multiscale porosity in the formation of corrosion products investigated by X-ray tomography. *Journal of Analytical Atomic Spectrometry*, 30(3), 580–587.

Kergourlay, F., Guilminot, E., Neff, D., Rémazeilles, C., Réguer, S., Refait, P., Mirambet, F., Foy, E., Dillmann, P. (2010). Influence of corrosion products nature on dechlorination treatment: The case of wrought iron archaeological ingots stored two years in air before NaOH treatment. *Corrosion Engineering Science and Technology*, 45(5), 407–413.

Kergourlay, F., Réguer, S., Neff, D., Foy, E., Picca, F.-E., Saheb, M., Hustache, S., Mirambet, F., Dillmann, P. (2018). Stabilisation treatment of cultural heritage artefacts: In situ monitoring of marine iron objects dechlorination in alkali solutions. *Corrosion Science*, 132, 21–34.

Leon, Y., Saheb, M., Drouet, E., Neff, D., Foy, E., Leroy, E., Dynes, J.J., Dillmann, P. (2014). Interfacial layer on archaeological mild steel corroded in carbonated anoxic environments studied with coupled micro and nano probes. *Corrosion Science*, 88, 23–35.

Leroy, S., L'Héritier, M., Delqué-Kolic, E., Dumoulin, J.-P., Moreau, C., Dillmann, P. (2015). Consolidation or initial design? Radiocarbon dating of ancient iron alloys sheds light on the reinforcements of French Gothic Cathedrals. *Journal of Archaeological Science*, 53, 190–201.

Mirambet, F., Réguer, S., Rocca, E., Hollner, S., Testemale, D. (2010). A complementary set of electrochemical and X-ray synchrotron techniques to determine the passivation mechanism of iron treated in a new corrosion inhibitor solution specifically developed for the preservation of metallic artefacts. *Applied Physics A: Materials Science & Processing*, 99, 341–349.

Monnier, J., Legrand, L., Bellot-Gurlet, L., Foy, E., Réguer, S., Rocca, E., Dillmann, P., Neff, D., Mirambet, F., Perrin, S. et al. (2008). Study of archaeological artefacts to refine the model of iron long-term indoor atmospheric corrosion. *Journal of Nuclear Materials*, 379(1–3), 105–111.

Monnier, J., Réguer, S., Vantelon, D., Dillmann, P., Neff, D., Guillot, I. (2010a). X-ray absorption study on medieval corrosion layers for the understanding of very long term indoor atmospheric iron corrosion. *Applied Physics A*, 99(2), 399–406.

Monnier, J., Neff, D., Réguer, S., Dillmann, P., Bellot-Gurlet, L., Leroy, E., Foy, E., Legrand, L., Guillot, I. (2010b). A corrosion study on the ferrous medieval reinforcement of the Amiens cathedral. Phase characterisation and localisation by various microprobes. *Corrosion Science*, 52(3), 695–710.

Monnier, J., Vantelon, D., Réguer, S., Dillmann, P. (2011). X-ray absorption spectroscopy study of the various forms of phosphorus in ancient iron samples. *Journal of Analytical Atomic Spectrometry*, 26(5), 885–891.

Monnier, J., Réguer, S., Foy, E., Testemale, D., Mirambet, F., Saheb, M., Dillmann, P., Guillot, I. (2014). XAS and XRD in situ characterisation of reduction and reoxidation processes of iron corrosion products involved in atmospheric corrosion mechanisms. *Corrosion Science*, 78, 293–303.

Neff, D., Dillmann, P., Bellot-Gurlet, L., Beranger, G. (2005). Corrosion of iron archaeological artefacts in soil: Characterization of the corrosion system. *Corrosion Science*, 47(2), 515–535.

Neff, D., Dillmann, P., Réguer, S. (2013). Analytical techniques for the study of corrosion of metallic heritage artefacts: From micrometer to nanometer scales. In *Corrosion and Conservation of Cultural Heritage Metallic Artefacts*, Dillmann, P., Watkinson, D., Angelini, E., Adriaens, A. (eds). Woodhead Publishing, Oxford.

Rancourt, D.G. and Meunier, J.-F. (2008). Constraints on structural models of ferrihydrite as a nanocrystalline material. *American. Mineralogist*, 93(8–9), 1412–1417.

Réguer S., Dillmann P., Mirambet F., Susini J., Lagarde P. (2006). Investigation on Cl corrosion products of iron archaeological artefacts using micro focused synchrotron X-ray absorption spectroscopy. *Applied Physics A*, 83(2), 189–193. doi: 10.1007/s00339-006-3506-3.

Réguer, S., Dillmann, P., Mirambet, F. (2007a). Buried iron archaeological artefacts: Corrosion mechanisms related to the presence of Cl-containing phases. *Corrosion Science*, 49(6), 2726–2744.

Réguer, S., Dillmann, P., Mirambet, F., Susini, J. (2007b). Contribution of the local and structural characterisation for studies of the corrosion mechanisms related to the presence of chlorine on the archaeological ferrous artefacts. In *Corrosion of Metallic Heritage Artefacts. Investigation, Conservation and Prediction of Long Term Behaviour*, Dillmann, P., Beranger, G., Piccardo, P., Matthiesen, H. (eds). Woodhead Publishing, Cambridge.

Réguer, S., Mirambet, F., Dooryhee, E., Hodeau, J.-L., Dillmann, P., Lagarde, P. (2009). Structural evidence for the desalinisation process of akaganeite in the preservation of iron archaeological objects, using synchrotron X-ray powder diffraction and absorption spectroscopy. *Corrosion Sciences*, 51, 2795–2802.

Réguer, S., Mirambet, F., Rémazeilles, C., Vantelon, D., Kergourlay, F., Neff, D., Dillmann, P. (2015) Iron corrosion in archaeological context: Structural refinement of the ferrous hydroxychloride β-$Fe_2(OH)_3Cl$. *Corrosion Science*, 100, 589–598.

Réguer, S., Kergourlay, F., Foy, E., Neff, D., Vantelon, D., Cotte, M., Mirambet, F., Dillmann, P. (2020). XANES at the Cl K-edge as a relevant technique to reveal the iron archaeological artefact dechlorination treatments. *Journal of Analytical Atomic Spectrometry*, 35, 2358-2368. doi: 10.1039/D0JA00122H.

Rémazeilles, C., Neff, D., Kergourlay, F., Foy, E., Conforto, E., Guilminot, E., Réguer, S., Refait, P., Dillmann, P. (2009). Mechanisms of long-term anaerobic corrosion of iron archaeological artefacts in seawater. *Corrosion Science*, 51(12), 2932–2941.

Rose, J., Flank, A.-M., Masion, A., Bottero, J.-Y., Elmerich, P. (1997). Nucleation and growth mechanisms of Fe oxyhydroxide in the presence of PO_4 Ions. 2. P K-edge EXAFS study. *Langmuir*, 13(6), 1827–1834.

Schwertmann, U. and Cornell, R.M. (2000). *Iron Oxydes in the Laboratory*. Wiley-VCH, Weinheim.

Simon, H., Cibin, G., Robbins, P., Day, S., Tang, S., Freestone, I., Schofield, E. (2018). A synchrotron-based study of the Mary Rose iron cannonballs. *Angewandte Chemie International Edition*, 57, 7390–7395.

Singh, B. and Grafe, M. (2006). *Synchrotron-Based Techniques in Soils And Sediments*. Elsevier, Amsterdam.

Ståhl K., Nielsen K., Jiang J., Lebech B., Hanson J.C., Norby P., van Lanschot J. (2003). On the akaganéite crystal structure, phase transformations and possible role in post-excavational corrosion of iron artifacts. *Corrosion Science*, 45(11), 2563–2575. doi: 10.1016/S0010-938X(03)00078-7.

Stratmann, M. (1990). The atmospheric corrosion of iron and steel. *Metallurgica I Odlewnictwo*, 16, 46–52.

Teo, B.K. and Joy, D.C. (1981). *EXAFS Spectroscopy*. Springer, New York.

Tuuti, K. (1982). *Corrosion of Steel in Conrete*. Swedish Cement and Concrete Research Institute, Stockholm.

Vantelon, D., Hofmann, A., Hanselmann, K., Flank, A.-M. (2007). Speciation of phosphate in iron rich mineral deposit in a high mountain lake environment, a study at the micron scale at the LUCIA beamline. *AIP Conference Proceedings*, 882, 232–234.

Veys-Renaux, D., Réguer, S., Bellot-Gurlet, L., Mirambet, F., Rocca, E. (2018). Conversion of steel by polyphenolic model molecules: Corrosion inhibition mechanism by rutin, esculin, esculetol. *Corrosion Science*, 136, 1–8.

Wilke, M., Farges, F., Petit; P.-E., Brown, G., Martin, F. (2001). Oxidation state and coordination of Fe in minerals: An Fe K-XANES spectroscopic study. *American Mineralogist*, 86(5–6), 714–730.

Young, M.L. (2012). Archaeometallurgy using synchrotron radiation: A review. *Reports on Progress in Physics*, 75(3), 036504.

7

Synchrotron X-Ray Microdiffraction Studies of the Mortars of Ancient Roman Concretes

Marie D. JACKSON[1], Heng CHEN[2], Jacob G. PETERSON[3], Cagla MERAL AKGUL[4] and Bryony RICHARDS[5]

[1] *Department of Geology and Geophysics, University of Utah, Salt Lake City, USA*

[2] *Shandong Provincial Key Lab for the Preparation and Measurement of Building Materials, University of Jinan, China*

[3] *Department of Geology and Geologic Engineering, Colorado School of Mines, Golden, USA*

[4] *Department of Civil Engineering, Middle East Technical University, Ankara, Turkey*

[5] *Energy and Geoscience Institute, University of Utah, Salt Lake City, USA*

7.1. Introduction

The mineralogical processes that underlie the extraordinary chemical and mechanical resilience of the concrete elements of late Republican and Imperial age architectural monuments in Rome and massive maritime harbor structures of the Mediterranean region have remained a mystery for 2,000 years (Vitruvius, *de Architectura* 30–27 BCE (Granger 1931, 1934; Gros et al. 1997), Pliny, *Naturalis Historia* 70 CE (Eicholz 1962), Alberti, *de re Aedificatoria*, 1486 (Ramage 1827; Blake 1947; MacDonald 1982; Artioli et al. 2019). Recent investigations of drill core samples of the mortars that bind the ancient concretes with synchrotron

Synchrotron Radiation, Cultural Heritage, Biomineralization,
coordinated by Catherine DEJOIE, Pauline MARTINETTO and Nobumichi TAMURA.

radiation X-ray microdiffraction (μXRD) and X-ray microfluorescence (μXRF) applications have revealed a great deal of information about the cementitious processes that produce the extreme durability of the architectural and maritime structures (Jackson et al. 2013a, 2013b, 2014, 2017). The drill cores sample the concrete deep in the structures so that poorly crystalline and crystalline cementing phases remain relatively free from carbonation and surficial degradation. The cores come from the Markets of Trajan (100–110 CE) in Rome, acquired in collaboration with the archaeologists of the Sovrintendenza Capitolina di Beni Culturali (Jackson et al. 2009, 2014). They also come from marine concrete structures from Mediterranean harbors acquired through the ROMACONS drilling project (Brandon et al. 2014) (Figure 7.1).

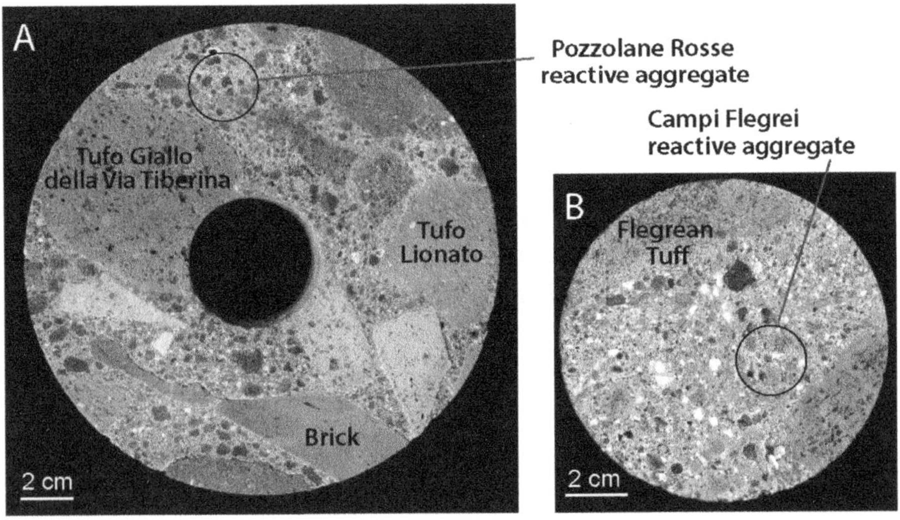

Figure 7.1. *Photographs of drill cores of ancient Roman concrete. A) Markets of Trajan (100–110 CE), Rome. B) Baianus Sinus breakwater (ca. 55 BCE), Bay of Pozzuoli. For a color version of this figure, see www.iste.co.uk/dejoie/synchrotron.zip*

Late Republican and Imperial era Roman concretes are composed of a complex mortar (*materia*) that binds a self-reinforcing conglomeratic framework of decicentimeter-sized coarse aggregate, or *caementa*, described by the Roman architect and engineer, Marcus Vitruvius Pollio, ca. 30 BCE (*de Architectura* 2.4.1–3, 2.6.1–6, 2.5.1–3, 5.12.2–6). The coarse aggregate is composed principally of volcanic tuff, carbonate rock (limestone) and brick, as well as dense lava in some concrete substructures. The raw materials of the mortars are lime (CaO), calcined from limestone in kilns at about 900°C (Adam 1994), and scoriaceous and/or pumiceous tephra quarried from the volcanic districts of central Italy, which extend from north of Rome to the Gulf of Naples (Jackson et al. 2010; Marra et al. 2011, 2016; Jackson 2014). The tephra is composed of volcanic glass, crystal fragments and rock fragments ejected by explosive pyroclastic eruptions; ash is the < 2 mm diameter component. Tuff is the rock that develops from the consolidation of these vitric, crystal and lithic particles. Tephra forms the reactive aggregate of the mortars; tuff forms the conglomeratic coarse aggregate of many of the concretes (Jackson and Marra 2006) (Figure 7.1). Primary volcanic crystals of the tephra, mainly leucite, sanidine and clinopyroxene, as well as authigenic surface coatings produced through alteration in the pyroclastic deposits, mainly opal, poorly crystalline clay mineral and zeolites (De Gennaro et al. 2000; Jackson et al. 2010), form the foundational mineral assemblages of the tephra and tuff aggregates.

The latest Republican and earliest Imperial eras, ca. 55 BCE–14 CE, mark a time of rapid experimentation and technological innovation in mortar and concrete technologies, which led to wide-scale production of highly resilient concrete materials in architectural monuments and marine harbor structures (Van Deman 1912a, 1912b; Blake 1947; Gerding 2002; Oleson et al. 2006; Gotti et al. 2008; Goldsworthy and Min 2009; Jackson et al. 2010, 2011; Jackson and Kosso 2013, Brandon et al. 2014). Vitruvius wrote *de Architectura* for Octavian, later Emperor Augustus, during this transformative period. Roman builders produced complex dimension stone and concrete monuments, such as the Theater of Marcellus (45–11 BCE) (Ciancio Rossetto 2008) and the Tomb of Caecilia Metella (Seymour et al. 2021) in Rome, and the enormous concrete harbor structure at Caesarea Maritimi (23–15 BCE) on the coast of present-day Israel (Boyce et al. 2004) with these refined concretes (Jackson 2017). These building projects employed a precise selection of volcanic aggregate(s), improved methods for lime calcination and hydration, and novel installation protocols that have produced extraordinary chemical and mechanical resilience for two millennia. Those technologies were further refined to create high-performance concretes for the daring and highly complex domed structures of Imperial era monumental architecture, as exemplified by the Markets and Forum of Trajan and the Pantheon (126 CE) in Rome (Lancaster 2005; Marder and Jones 2015; Vitti and Vitti 2019). Meanwhile, marine harbor structures were constructed at Portus Claudius (ca. 50 CE), Portus Neronis (ca. 65 CE) and Portus Traianus (ca. 115 CE), which remain cohesive and intact today (Brandon et al. 2014). In ca. 70CE, Pliny the Elder described these as marine structures that (become) "a single stone mass (*fierem unum lapidem*), impregnable to the waves and every day stronger (*et fortiorem cotidie*)" (*Naturalis Historia* 35.166).

This chapter reviews methods developed with synchrotron μXRF and μXRD applications at beamline 12.3.2 of the Advanced Light Source at the Lawrence

Berkeley National Laboratory to describe the crystalline cementing fabrics of the mortars that bind the ancient concretes (Figures 7.2–7.4). Instructive examples show point analyses with the monochromatic X-ray beam (8–10 KeV) in the cementing matrix and scoriae of the mortar of the substructure of the Markets of Trajan (Figures 7.5 and 7.6), the mineral assemblages in a mapped grid of μXRD points in the pumice aggregate of the vault of the Basilica Ulpia in the Forum of Trajan (Figure 7.7) and maps of the *d*-spacings of mineral cements in fine-scale microenvironments of the mortar of the *Baianus Sinus* breakwater (Figure 7.8). Finally, the integration of monochromatic beam analyses of a fine-grained, crystalline calcium–aluminum–silicate hydrate, Al-tobermorite, with polychromatic beam analyses of a coarse-grained, crystalline calcium chloroaluminate, hydrocalumite, are illustrated, also in the *Baianus Sinus* breakwater (Figure 7.9). When the μXRF and μXRD analyses from micrometer-scale maps of cementing microstructures are integrated with additional microanalyses, they provide profound insights into the processes that produce extraordinary durability and resilience in the ancient concrete structures. These μXRF and μXRD methods have also been applied to investigations of mineral cements in basaltic tephra in drill cores from Surtsey volcano, Iceland, a young oceanic island and geologic analog of Roman marine concrete (Jackson et al. 2019, 2024). The integrated results form the foundation of experimental Roman prototypes for environmentally friendly concrete infrastructure in structural and marine applications with projected service lives of hundreds of years.

7.2. Standard powder X-ray diffraction studies of Roman mortars

Visual inspection of a hand sample of Roman mortar reveals a highly heterogeneous fabric that contains numerous cementing microenvironments – in the tephra aggregate and its interfacial zones, in the cementing matrix, in relict lime clasts and in relict voids (Figure 7.1). These microenvironments contain a complex spatial (and temporal) distribution of crystalline phases that record the evolving interaction of interstitial fluids with the reactive tephra aggregate and mineral cements over the 2,000-year life span of the ancient concrete structures (Jackson et al. 2012, 2013a, 2017, 2018) (Figure 7.2).

A standard X-ray diffraction (XRD) analysis of an ancient Roman mortar produces a complicated result, since the geologic and cementitious mineral assemblages – primary volcanic crystals, authigenic mineral surface coatings, pozzolanic and post-pozzolanic cementing phases – are ground together to produce a fine powder. Even when the cementing matrix is carefully sieved from the aggregate particles traces of primary volcanic crystals are detected in the powder diffraction results (Chiari et al. 1992, 1996; Stanislao et al. 2011; Vola et al. 2011; Jackson 2014; Rispoli et al. 2019a, 2019b, 2020).

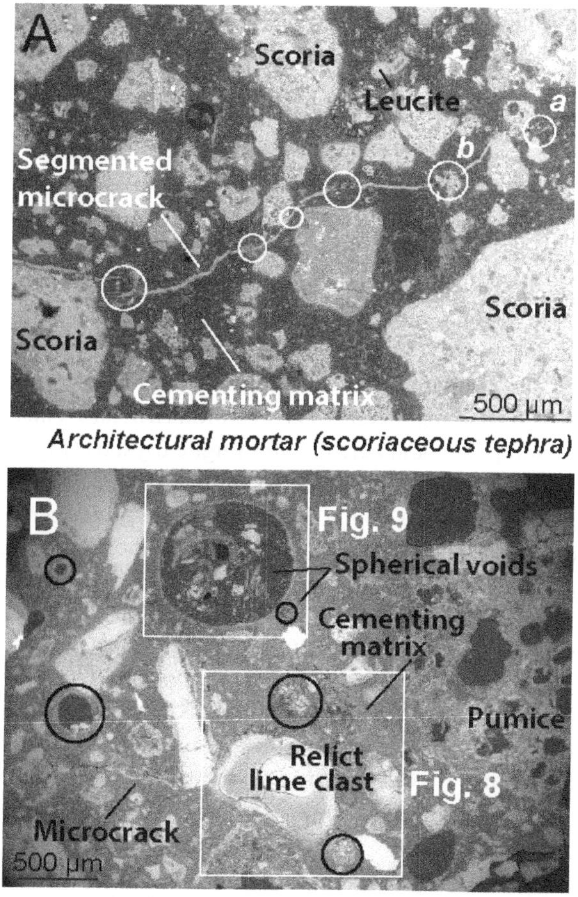

Marine mortar (pumiceous tephra)

Figure 7.2. *SEM-BSE images of typical Roman mortar fabrics. (A) Markets of Trajan substructure. A cohesive cementing matrix with platey strätlingite crystals (Figure 7.5) binds Pozzolane Rosse tephra from the Alban Hills volcanic district. Circles show steps in the segments of a microcrack that has been repaired with mineral cements and terminates at microscoria (a, b). (B) Baianus Sinus breakwater. A complex cementing matrix binds relict lime clasts and pumiceous tephra from the Campi Flegrei volcanic district (Figure 7.8). Circles show spherical voids that concentrate calcium sulfoaluminate and chloroaluminate crystals (Figure 7.9). A microcrack that has been repaired with mineral cements traverses the cementing matrix*

For powder XRD analyses of the mortars of the Theater of Marcellus concretes, for example, the cementing matrix was carefully disaggregated from the coarse sand-sized tephra aggregate and sieved (Figure 7.3(a)). Even so, analyses of the < 0.6 mm fraction mainly record primary volcanic crystal fragments in microscoria, quarried from the mid-Pleistocene Pozzolane Rosse pyroclastic flow from the nearby Alban Hills volcanic district (Jackson et al. 2010; Marra et al. 2011). These are clinopyroxene (diopside) and leucite, a potassic silicate ($KAlSi_2O_6$). Analcime ($NaAl(Si_2O_6)\bullet H_2O$) is a zeolite produced through the replacement of leucite in the pyroclastic deposit. The powder XRD analyses also record the carbonation of the poorly crystalline calcium–aluminum–silicate–hydrate (C-A-S-H) binding phase of the mortar to calcite and vaterite. Only the powder XRD analyses of the mortar of the humid substructure of the theater, inundated by Tiber River fluids innumerable times over the past two millennia, detect traces of strätlingite mineral cement. This is a highly stable calcium–aluminum–silica–hydrate mineral that gives good compressive strength to modern cement-based concretes containing supplemental cementitious materials (MacDowell 1991; Okoronkwo and Glasser 2016). SEM images show common platey and acicular strätlingite crystals (Figure 7.3(b)), yet the *d*-spacing intensity is weak in the powdered mortar sample, relative to calcite and the volcanic crystal fragments. Despite the importance of authigenic silicate mineral cements to the chemical and mechanical resilience of the imperial era architectural and marine concretes (Jackson et al. 2014, 2017), these phases are only weakly detectable in most powder XRD analyses of ancient Roman mortars (e.g. Vola et al. 2011; De Luca et al. 2015; Rispoli et al. 2019a, 2019b, 2020).

Although many archaeometric studies include XRD studies of powdered mortar samples, these analyses give little precise information about the spatial distribution, reaction mechanisms and time-lapse progression of cementing processes. Instead, the components of the mortar aggregates – primary crystal fragments in the case of volcanic rock or detrital minerals in the case of sedimentary rocks – dominate the diffraction patterns, as in the Theater of Marcellus mortar (Figure 7.3). Crushed powders of bulk mortar specimens thus provide useful guidelines for establishing the lithological provenance of volcanic aggregates and, potentially, the long-term carbonation of the cementing systems (Chiari et al. 1992, 1996; Stanislao et al. 2011; Vola et al. 2011; De Luca et al. 2015; Columbu et al. 2018; Rispoli et al. 2019a, 2019b, 2020; Secco 2019). The investigations of the crystalline phases that record the evolving post-pozzolanic processes that produce silicate mineral cements – strätlingite ($Ca_2Al(AlSi)O_2(OH)_{10}\cdot 2.25H_2O$ (Rinaldi et al. 1990), Al-tobermorite ($[Ca_4(Si_{5.5}Al_{0.5}O_{17}H_2)]Ca_{0.2}\cdot Na_{0.1}\cdot 4H_2O$) (Taylor 1992) and various zeolites (Stanislao et al. 2011; Vola et al. 2011; Jackson et al. 2017; Rispoli et al. 2019a, 2020) – benefit from in situ microanalyses, since powdering the mortar destroys the spatial and temporal relationships among these phases. The diverse microenvironments that produce these cementing phases can also be investigated with scanning electron microscopy (SEM) and their chemical variations identified with Energy Dispersive Spectroscopy (EDS) and Raman Spectroscopy; however, the micrometer-scale variations in the cementing mineral assemblages and their crystallographic parameters can best be described with X-ray microdiffraction.

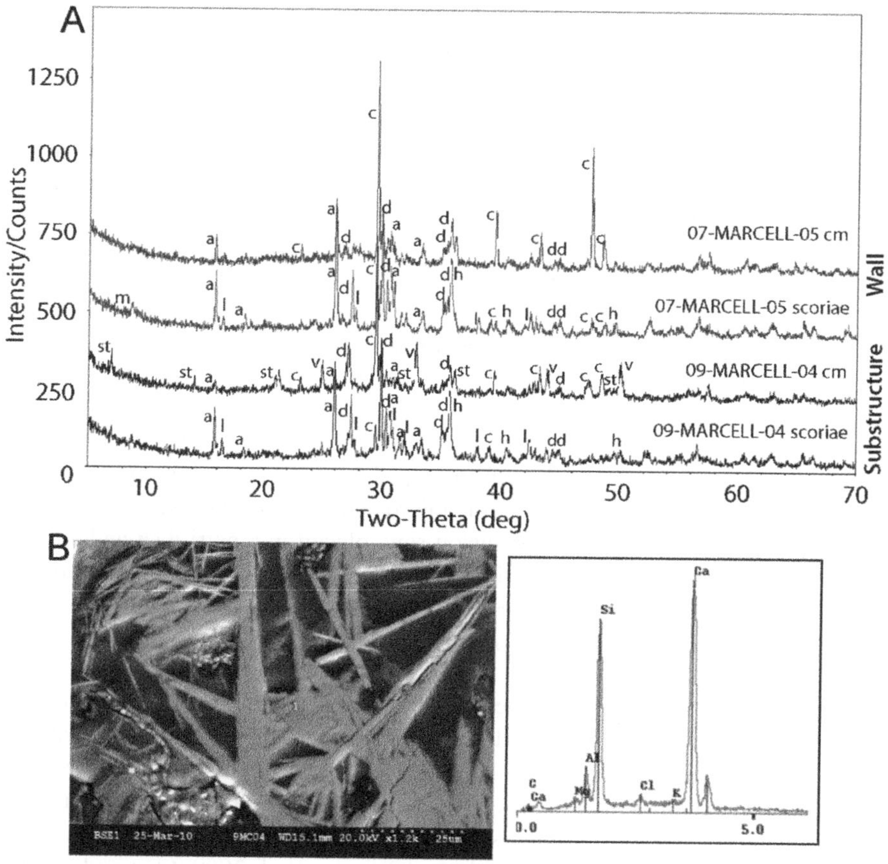

Figure 7.3. *Theater of Marcellus (44–11 BCE), Rome, standard X-ray diffraction results of powdered mortar samples. (A) X-ray diffractograms of the cementitious binding matrix (< 0.6 mm fraction) and scoriae from the Pozzolane Rosse pyroclastic flow (l, leucite; d, diopside; a, analcime; v, vaterite; c, calcite; h, hematite; st, strätlingite). (B) SEM–EDS image and analysis of strätlingite microstructures in the mortar of the concrete substructure (after Jackson et al. 2011). For a color version of this figure, see www.iste.co.uk/dejoie/synchrotron.zip*

7.3. Synchrotron μXRD and μXRF studies of ancient Roman mortars

Investigations with synchrotron radiation, the electromagnetic energy emitted by charged particles when they are accelerated radially, or perpendicular, to their velocity, have provided a comprehensive platform for gaining insights into the nano- and micrometer-scale characteristics of the heterogeneous fabrics of the mortars that bind Roman conglomeratic concretes. The electrons move through a strong magnetic field to produce brilliant beams of light that are transmitted through components of the mortar fabrics to produce diffraction analyses of poorly crystalline and crystalline phases in diverse cementitious microenvironments (e.g. Jackson et al. 2013a, 2013b, 2014, 2017).

The application of synchrotron μXRD and μXRF to Roman cementitious materials has required new methods for sample preparation, the adjustment of beamline settings to optimize the acquisition of crystallographic data, and procedures for refining data analysis through beamline software and other software programs. The sample preparation begins with a polished thin section of a mortar sample at standard 0.3 mm thickness. Superglue™ adhesive is used to glue the thin mortar slice to a glass slide.

Petrographic studies are made in plane and cross polarized light to identify and record instructive microstructures for further analysis. SEM imaging and analysis is also employed. The area selected for microanalyses is then sliced from the glass-mounted thin section with a precision saw, placed in a petri dish and covered with nitromethane until the glue is dissolved and the mortar slice begins to detach from the glass slide. A length of X-ray transparent adhesive tape (which does not produce a diffraction pattern) is then pressed onto the mortar slice.

At beamline 12.3.2 of the Advanced Light Source, the mortar slice mounted on tape is attached to an aluminum holder and loaded in transmission mode, with the detector generally placed at 39° with respect to the incident beam (Figure 7.4). Submicron-sized crystals are investigated with powder microdiffraction using a monochromatic beam and larger single crystals are investigated with Laue microdiffraction using a polychromatic beam (Tamura et al. 2009; Stan and Tamura 2018). The monochromatic X-ray beam of 8 or 10 keV is focused to 2 × 5 μm spot size and a DECTRIS Pilatus 1M area detector placed at about 150-mm records Debye–Scherrer rings from crystalline phases.

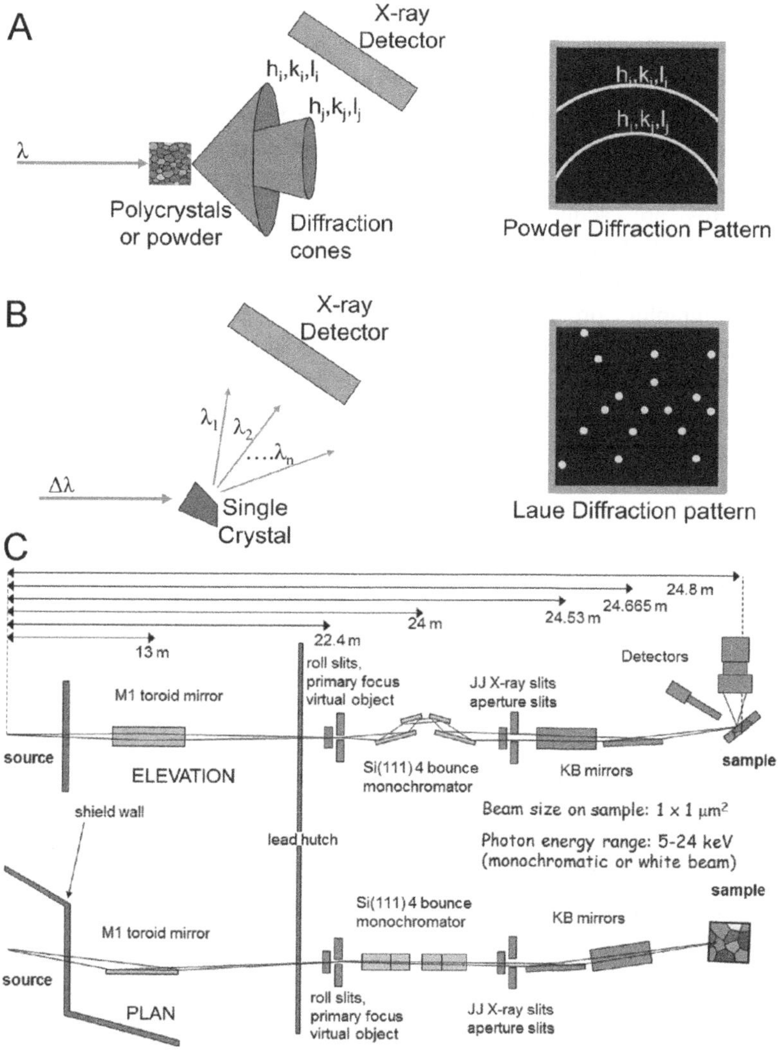

of

Figure 7.4. *Beamline 12.3.2 at the Advanced Light Source. (A). A polychromatic (white) beam is used for polycrystalline samples with grain size larger than the beam size (2–3 μm) and a monochromatic beam is used for polycrystalline assemblages with grain size smaller than the beam size. (B). Elevation and plan view of the 12.3.2 microdiffraction beamline (after Tamura et al. 2009). For a color version of this figure, see www.iste.co.uk/dejoie/synchrotron.zip*

The experimental geometry is calibrated using α-Al2O3 powder. Diffraction plots are produced with *d*-spacing reflections that can be integrated radially for 2θ 3–54° mainly over a 76° arch segment (χ) around the cone of diffraction (Figures 7.4(a), 7.6(c), 7.9(d)). The settings cover the large interlayer *d*-spacings of layered cementitious hydrates and nanocrystalline clay minerals, as well as the smaller *d*-spacings of zeolite mineral cements. Single point analyses on instructive microstructures are acquired in some instances (Figures 7.5 and 7.6) or point-to-point μXRD maps with 3–20 μm step size create a grid of analyses that reveal the fine-scale distribution of minerals (Figures 7.7–7.9). The mineral *d*-spacings and crystallographic preferred orientations, where present, are referenced to qualitative calcium (Ca), iron (Fe), titanium (Ti) or potassium (K) μXRF scans, where numbered points correspond to a specific analysis that can be referenced spatially in the map grid or data set. Single-crystal Laue microdiffraction scans may also be collected on crystals > 5 μm with the sample loaded in transmission to the X-ray beam (Figures 7.4(b) and 7.9(f)). The data are processed using the XMAS Laue pattern indexing software (Tamura 2014). Further analyses use additional plotting programs and X-ray diffraction software.

From these micrometer-scale maps, the spatial distribution, crystallographic characteristics and chemical microenvironments of mineral cements in relict clasts of hydrated lime, the cementing matrix, pumiceous and scoriaceous tephra aggregates and their interfacial zones can be deciphered. Instructive examples are provided from the mortars of concretes from the Markets and Forum of Trajan in Rome (Figures 7.5–7.7) and the mortar of the *Baianus Sinus* breakwater concrete in the Bay of Pozzuoli. Successive investigations of the same microstructure can then be achieved with various diffraction and spectroscopic methods (Jackson et al. 2013a, 2013b, 2014, 2017; Seymour et al. 2021). These methods have been applied to other ceramic materials in cultural heritage as well as pyroclastic rocks (Dejoie et al. 2014, 2015; Leon et al. 2015; Jackson et al. 2019, 2024).

7.4. Architectural mortars – Markets and Forum of Trajan

A central discovery derived from synchrotron μXRD and μXRF investigations of the Roman architectural concretes is that the mortars have highly heterogeneous fabrics composed of reactive tephra aggregate and crystalline cementing phases that develop in diverse microenvironments (Jackson et al. 2014). Conventional portland cement concretes represent a more-or-less binary system of cement paste and inert sand and gravel aggregates (e.g. Taylor 1997, pp 351–355; Landis and Bolander 2009). By contrast, the principal binding component of Roman mortar is a cementing matrix (Figure 7.5) composed of partially reacted tephra particles 0.1–1 mm in diameter, calcium–aluminum–silicate–hydrate (C-A-S-H) binding phase and various crystalline cementing minerals. In the wall and substructure concretes of the Markets of Trajan (Vitti 2007), this reactive aggregate is mainly scoriaceous tephra from the Pozzolane Rosse pyroclastic flow (Jackson et al. 2009, 2014). Alteration in the pyroclastic flow deposit produced reactive clay mineral (halloysite) and zeolite (phillipsite and chabazite) surface coatings; relicts of these coatings are preserved in Pozzolane Rosse scoriae in wall and substructure mortars (Jackson et al. 2010). XRD analyses of powders of the cementing matrix of the Markets of Trajan wall mortars detect a trace of strätlingite, but the patterns are dominated by volcanic crystal fragments and calcite, similar to those of the Theater of Marcellus (Figure 7.2) (Jackson et al. 2009, Figure 3).

7.4.1. *Cementing matrix*

Synchrotron μXRD and μXRF investigations of a typical mortar sample from a drill core of the substructure of the Markets of Trajan reveal that the mineralogical fabric of the cementing matrix is dominated by strätlingite (Figure 7.5(a), (b)). The green, yellow and orange areas of the potassium μXRF map show Pozzolane Rosse scoriae with highly potassic groundmass and leucite crystals; the green, yellow and orange in the calcium μXRF map indicate diopside crystals and cementing hydrates, principally the poorly crystalline C-A-S-H binding phase and strätlingite crystals. The monochromatic beam μXRD spot analyses detect strätlingite in diverse microstructures (Figure 7.5(c)): in partially reacted microscoria (site #15), at the contacts, or interfacial zones, of larger scoriae within the cementing matrix (sites #10, #22, #23) and in the vesicles of the larger clasts of scoriaceous aggregate (sites #1, #5, #6).

Arc-bending fracture tests and mechanical analyses of reproductions of this mortar demonstrate that the acicular and blade-like crystals of strätlingite add toughness in the same way that fiber additives reinforce the paste of portland cement concretes (Samuelli Ferretti 1995; Brune et al. 2013; Jackson et al. 2014). The strätlingite crystals grow in situ in the Roman architectural mortar, however, reinforcing both the cementing matrix and the reactive interfacial zones of scoriae; they record the chemical evolution of cementitious processes over two millennia. The SEM overview of the cementing fabric of the mortar (Figure 7.2(a)) shows a microcrack adjacent to the micrometer-scale μXRD analyses of Figure 7.5. The microcrack is composed of multiple segments that traverse the strätlingite-enriched cementing matrix. It terminates in a zone of Pozzolane Rosse microscoriae, which apparently has produced an obstacle to further propagation (Figure 7.2(a), (b)). The microcrack walls appear to be effectively re-bonded with mineral cements, including calcite.

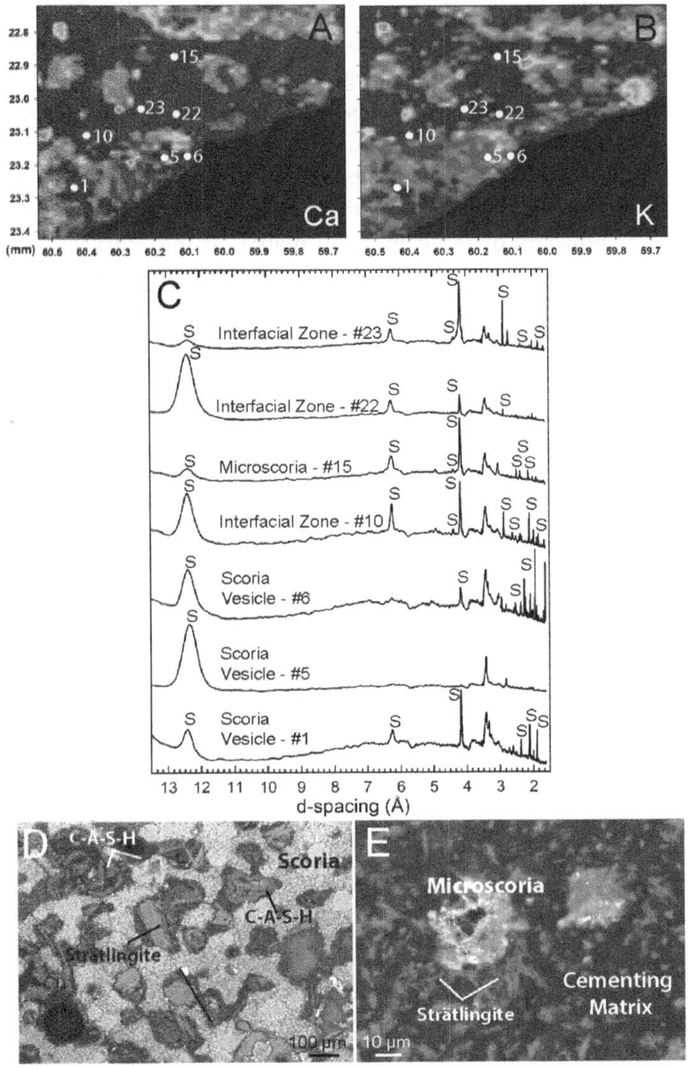

Figure 7.5. *Markets of Trajan (100–110 CE), Rome, strätlingite mineral cements, drill core from the concrete substructure south of via Biberati (Vitti 2007). (A, B) Calcium and potassium μXRF maps of the cementing matrix (blue) surrounding Pozzolane Rosse scoriae (green) show sites of μXRD analyses (monochromatic beam, 8 KeV). (C) Strätlingite d-spacings (S) in the cementitious binding matrix, interfacial zones and scoria vesicles. (D) SEM-BSE images of strätlingite crystals in scoria vesicles near sites #5, #6. (E) Strätlingite crystals in microscoria near sites #22, #23. For a color version of this figure, see www.iste.co.uk/dejoie/synchrotron.zip*

7.4.2. *Vesicular scoria*

Synchrotron μXRF and μXRD investigation of a large Pozzolane Rosse scoria, adjacent to the cementing matrix mapped in Figure 7.5, identifies crystalline cementing hydrates in an internal vesicle (Figure 7.6). The calcium μXRF map shows open vesicles (blue), the scoria groundmass (green) and cementing hydrates (yellow and orange); the potassium μXRF map shows the potassic groundmass (green) and slight potassium enrichment in the area of cementing hydrates. The μXRD spot analyses (sites #1, #2, #3) detect strätlingite as well as Al-tobermorite, an unusual layered mineral that is a crystalline analog for the cross-linked layered structure of poorly crystalline calcium–aluminum–silicate–hydrate (C-A-S-H) binding phase in environmentally friendly cement-based concretes containing supplemental cementitious materials (Myers et al. 2015). Al-tobermorite is the principal crystalline cementing phase in ancient Roman marine concretes (Jackson et al. 2013a, 2013b, 2017); however, it has not been previously observed in the architectural concretes of the Roman monuments. The 11.31 Å interlayer spacing indicates aluminum content of 0.5–1.0 Al/(Al+Si) (Barnes and Scheetz 1991; Jackson et al. 2013a) and the Debye–Scherrer diffraction plot detects a crystallographic preferred orientation. SEM images of nearby vesicles show a zone of altered scoria groundmass (Figure 7.6(e)), a rim of C-A-S-H along the vesicle surface (Figure 7.6(f)) and platey strätlingite (S) (Figure 7.6(e), (f)). Al-tobermorite (T) occurs as fine accumulations or as fans of acicular crystals. The strätlingite and Al-tobermorite appear to postdate the production of the C-A-S-H binding phase, which is initially produced pozzolanically through reactions with hydrated lime (Jackson et al. 2014). The presence of the crystals in the humid environment of the substructure suggests that vesicles in the reactive groundmass of the scoria may concentrate cations, as compared with fluid flow in interconnected pores of the cementitious matrix, leading to high localized pH and alkalinity that produces low-temperature Al-tobermorite crystallization.

7.4.3. *Pumiceous aggregate*

The vast vaulted concrete ceilings of the imperial age monuments in Rome contain lightweight caementa, principally conglomeratic clasts of scoriae and/or pumiceous tuff (Lancaster et al. 2011). The mortars of the vaulted concrete structures of the Great Hall in the Markets of Trajan and the Basilica Ulpia in the Forum of Trajan also contain pumice as fine aggregate in a mix with Pozzolane Rosse scoriaceous tephra (Jackson et al. 2009; Bianchi et al. 2011; Marra et al. 2013). The proportion of pumice added to the aggregate mix, about 30–40 vol %, decreased the unit weight of the mortar by 5–20 weight %, as compared to the unit weight of the mortars of the wall and substructure concretes. It is thought that Romans used the pumice to reduce the weight of the vaulted ceiling and, perhaps, to develop other high-performance characteristics such as increased interfacial cohesion with the cementing matrix (Jackson et al. 2009).

Synchrotron μXRF and μXRD investigation of a pumice clast in the mortar of a fragment of the collapsed Basilica Ulpia vaulted ceiling (Bianchi et al. 2011, Figure 3, crollo no. 13) provides insights into the alteration of the volcanic glass in the pumiceous aggregate and its role in maintaining the cohesion of the concrete vaulted ceiling of the monument (Figure 7.7). Here, a typical small pumice, 0.1 by 0.2 mm in size, and its contact with the cementing matrix have been mapped with a grid of μXRD analyses (Figure 7.7(a), (b)) and calcium and potassium μXRF analyses (Figure 7.7(c), (d)) at 10 μm step size. The powder diffraction pattern acquired at each site, focused to 2 × 5 μm spot size, has been evaluated to identify the phase assemblage and mineralogic characteristics. The cementing matrix is shown by vaterite, small areas of aragonite and calcite; these appear to record carbonation of the binding phases. The fabric of an adjacent tuff particle has coarse zeolite mineral cements that are too large to decipher from the Debye–Scherrer rings, but may represent analcime and phillipsite that formed in the pyroclastic flow deposit. The pumice clast, however, is dominated by very-fine grained potassic phillipsite with some alteration to calcite (the green squares with yellow circles) (Figures 7.7(a)). The phillipsite seems to derive, in part, from the dissolution of sanidine crystal fragments (mapped in the lower right corner of Figure 7.7(a)). The addition of the pumiceous aggregate seems to have produced post-pozzolanic reactions that generated authigenic phillipsite mineral cements in the mortar of the vault concrete.

7.5. Marine harbor mortar – *Baianus Sinus* Bay of Pozzuoli

The mortars of the Roman marine concretes have highly heterogeneous fabrics that are composed of reactive tephra aggregate and crystalline cementing phases in diverse microenvironments, but the aggregate of the eleven harbors drilled by the ROMACONS project is mainly derived from pumiceous deposits of the Campi Flegrei and Somma–Vesuvius volcanic districts surrounding the Gulf of Naples (Figures 7.1(b) and 7.2(b)) (Jackson et al. 2012, 2013a; Brandon et al. 2014; Marra et al. 2016). The principal components of the mortar are a complex cementing matrix that surrounds pumiceous tephra, relict lime clasts and spherical voids that include various cementing hydrates (Figure 7.2(b)). These microstructures are ubiquitous in the drill cores of the harbors drilled by ROMACONS, suggesting that Romans developed a standardized practice for mixing and installing the marine concretes (Brandon et al. 2014). A central discovery derived from synchrotron μXRD and μXRF investigations of the marine mortars is that cementing processes occur first through a pozzolanic phase through reactions with $Ca(OH)_2$ and then through a post-pozzolanic phase after $Ca(OH)_2$ has been fully consumed (Jackson et al. 2017).

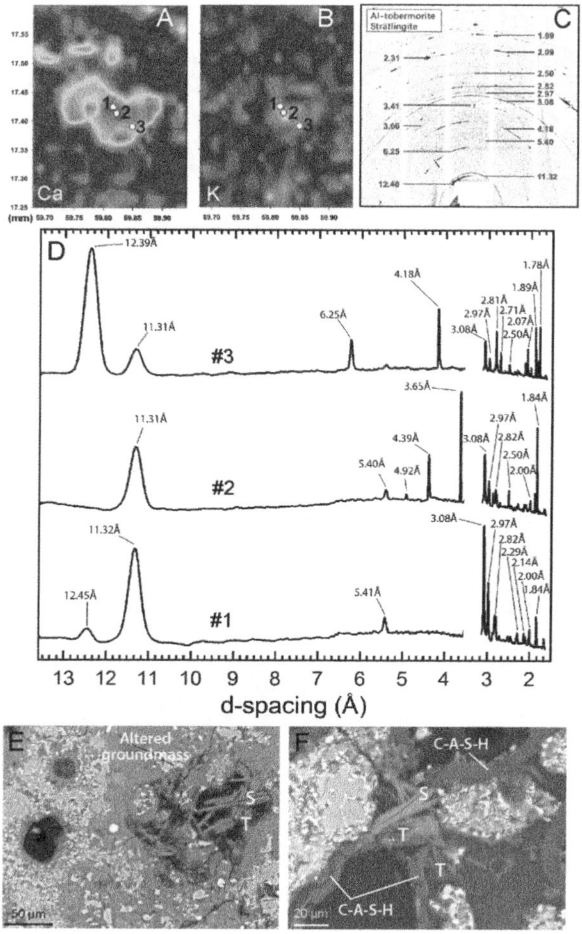

Figure 7.6. *Markets of Trajan, Rome, scoria vesicle, drill core from the concrete substructure south of via Biberati (Vitti 2007). (A, B) Calcium and potassium μXRF maps of the Pozzolane Rosse scoria vesicle and ground mass showing sites of μXRD analyses (monochromatic beam, 8 KeV). (C) Debye–Scherrer diffraction plot showing strätlingite and Al-tobermorite reflections, site #1. (D) Strätlingite and Al-tobermorite d-spacings at sites #1, #2, #3. (E, F) SEM-BSE images of strätlingite and Al-tobermorite in vesicles near sites #1, #2, #3. For a color version of this figure, see www.iste.co.uk/dejoie/synchrotron.zip*

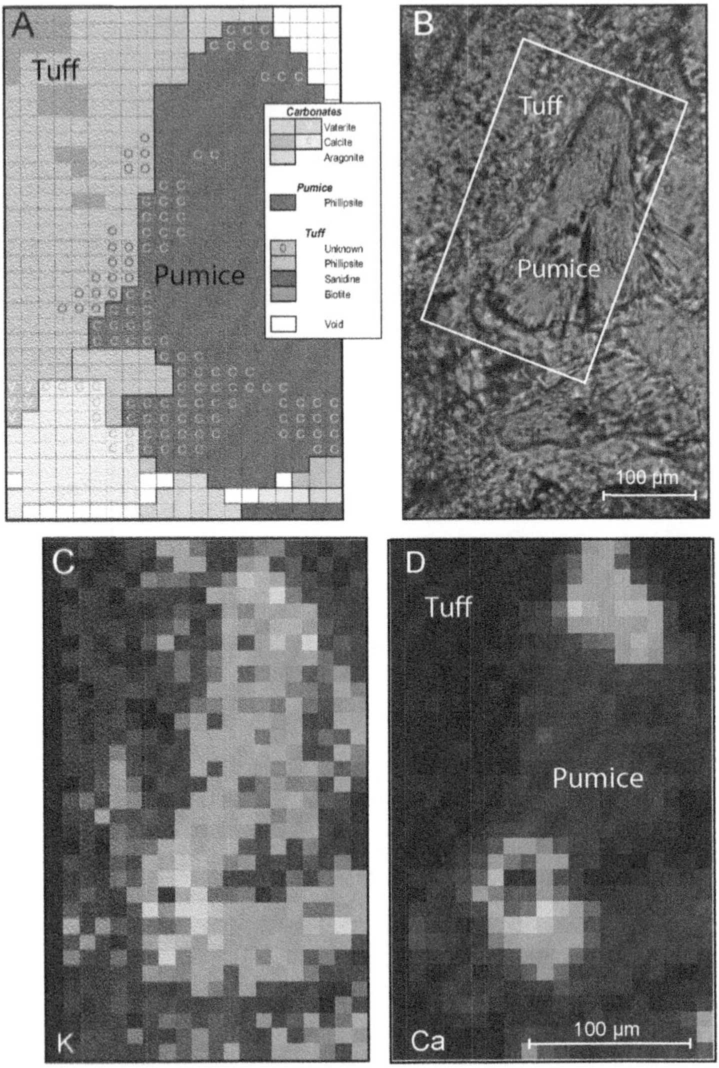

Figure 7.7. *Forum of Trajan (dedicated 112 CE), Rome, pumice aggregate, fallen block of the vaulted ceiling of the Basilica Ulpia, crollo no. 13 (Bianchi et al. 2011). (A) Grid of μXRD analyses, 10 μm step size, through a pumice clast and its interfacial zone with the cementing matrix. (B) Petrographic image of the pumice and cementing matrix, plane polarized light, rectangle shows region of μXRD and μXRF maps. (C, D) potassium and calcium μXRF maps corresponding to the grid of μXRD analyses. For a color version of this figure, see www.iste.co.uk/dejoie/synchrotron.zip*

7.5.1. *Relict lime clast and cementing matrix*

The calcium μXRF map of Figure 7.8 illustrates some of the diverse cementing environments of the mortar of the *Baianus Sinus* breakwater, a $6 \times 6 \times 10$ m^3 concrete block in the Bay of Pozzuoli near present-day Naples (Jackson et al. 2013a; Brandon et al. 2014). Point analyses along a traverse from the perimeter of a relict lime clast to its center (sites #13 – #18) show *d*-spacings for Al-tobermorite with 11.22 Å interlayer spacing in the calcium-enriched interior of the clast; a broad reflection centered at 4.65 Å indicates C-A-S-H in the interfacial zone with the cementing matrix (site #11). This microstructure seems to represent early pozzolanic dissolution of the perimeter of a hydrated lime clast, production of C-A-S-H binding phase and subsequent crystallization of Al-tobermorite in the interior; it is a ubiquitous component of the mortars of the concrete harbor structures (Jackson et al. 2012, 2013a, 2018; Jackson 2014).

A very different mineral assemblage occurs nearby, in relict pores within the cementing matrix that are shown by calcium enrichment in the coupled μXRD and μXRF maps (Figure 7.8(a), (b)). The principal phases mapped in the 550×100 μm grid of μXRD analyses at 10 μm step size across the relict voids are hydrous, layered calcium aluminate crystals: ettringite (a calcium sulfoaluminate, $Ca_2Al(OH)_{6.5}Cl_{0.5}{\bullet}3(H_2O)$), with interlayer *d*-spacings of 9.56 and 7.85 Å, respectively (Figure 7.8(b), (c)). The 11.23 Å *d*-spacing of Al-tobermorite and 3.28 Å *d*-spacing of vaterite are also mapped, but these crystals occur mainly in the cementing matrix. The most pronounced accumulations of ettringite and hydrocalumite occur in the large pore with calcium enrichment (sites a, b) (site #22). Other accumulations detected by point analyses at sites #21, #22 occur nearby (Figure 7.8(d)). These phases seem to record the immediate capture of SO_4^{2-} and Cl^- anions from seawater by the pozzolanic $Ca(OH)^2$ system. The hydrocalumite typically forms large plates, up to 50–60 μm (Jackson 2014, Figure 7.3(c)), and the ettringite commonly acquires a rough, mosaic fabric with traces of silica, sulfur, sodium and potassium in its bulk composition (Jackson et al. 2012, Figures 10 and 11). Remarkably, hydrocalumite or ettringite are rarely detected in the cementing matrix by μXRD or SEM-EDS studies of the marine mortars and instead occur systematically in pores and micropores (Jackson et al. 2012, 2013b; Jackson 2014).

By contrast, the principal crystalline phase in the cementing matrix is Al-tobermorite, shown by the distribution of the 11.23 Å *d*-spacing in the grid of μXRD analyses (Figure 7.8(b)), the point analyses at sites #25 and #26, and the point analyses at sites #7 and #29 (Figure 7.8(d)). The Al-tobermorite that crystallizes in relict lime clasts during the period of pozzolanic reaction has substantially more calcium relative to the more siliceous post-pozzolanic Al-tobermorite that crystallizes in the cementing matrix and pumiceous aggregate over the long history of the marine concrete structures (Jackson et al. 2017). The paucity of vaterite (and calcite) in the cementing matrix suggests that carbonation of the calcium–silicate–hydrate minerals was minimal during the > 2,000 years of submersion in the seawater environment. The platey and acicular Al-tobermorite crystals seem to reinforce the C-A-S-H binding fabric in the marine mortars in the same way that stratlingite reinforces the fine-scale fabric of the architectural mortar system. Variations of these processes have recently been explored by MacFarlane et al. (2021).

7.5.2. *Relict spherical void*

Some of the most intriguing microstructures of the marine mortars are nearly perfectly spherical relict voids in the cementing matrix, which contain large plates of hydrocalumite crystals surrounded by a rim of acicular Al-tobermorite crystals (Figure 7.2(b)). The μXRF map of Figure 7.9(a) and (b) shows calcium enrichment in the center of one of these spherical structures, about 0.8 mm in diameter. The μXRD spot analyses with the monochromatic beam (8 KeV) detect the *d*-spacings of the fine Al-tobermorite crystals at the rim of the void (sites #11, #8, #7) with regular, well-defined Debye–Scherrer diffraction patterns (Figure 7.9(d), site #11). They indicate an interlayer spacing of about 11.16–11.24 Å (Figure 7.9(f)) and the distribution of this *d*-spacing is mapped in a transect across the spherical structure (Figure 7.9(b)).

The monochromatic beam diffraction of the larger hydrocalumite crystals, however, produces a random pattern of spots that is indecipherable (Figure 7.9(e), site #1); there are not sufficient randomly oriented crystals in the beam to replicate the diffraction pattern of a finely powdered sample. Laue diffraction with the polychromatic beam of the single crystal at site #1 describes the hydrocalumite structure (Figure 7.9(g)) and confirms that the large crystal diffracts as a single crystal under the micron-size beam. The spatial distribution of the Laue reflection intensity in the radial traverse (Figure 7.9(b), right figure) indicates that hydrocalumite crystals occur mainly in the center of the relict void. They apparently record the sequestration of Cl^- from seawater in the hydrocalumite crystal structure.

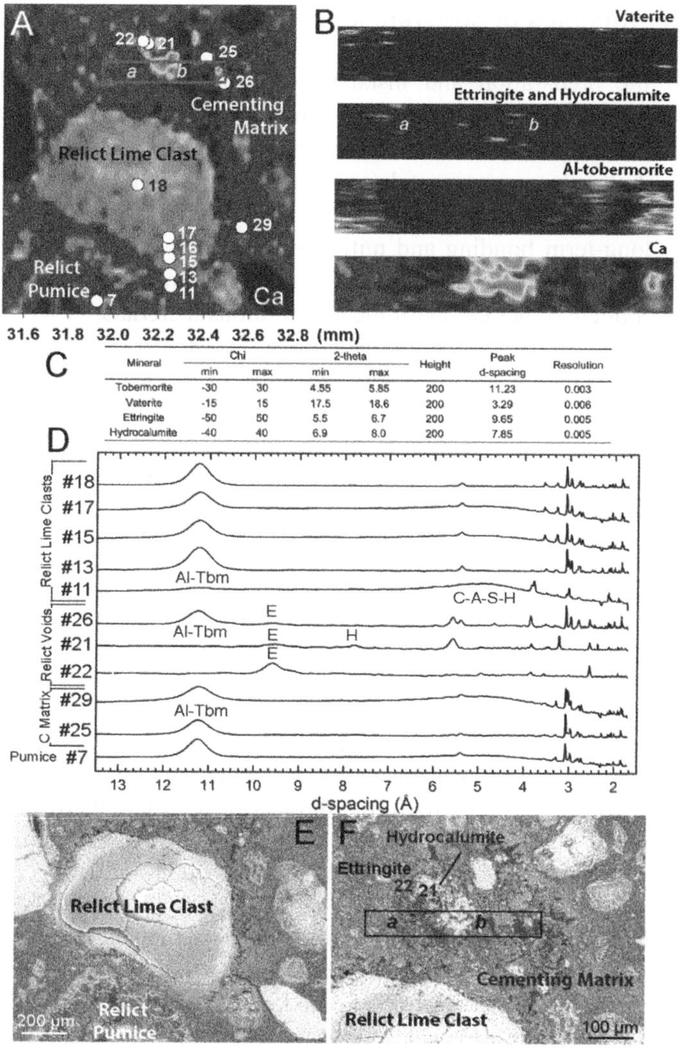

Mineral	Chi		2-theta		Height	Peak d-spacing	Resolution
	min	max	min	max			
Tobermorite	-30	30	4.55	5.85	200	11.23	0.003
Vaterite	-15	15	17.5	18.6	200	3.29	0.006
Ettringite	-50	50	5.5	6.7	200	9.65	0.005
Hydrocalumite	-40	40	6.9	8.0	200	7.85	0.005

Figure 7.8. *Baianus Sinus breakwater (ca. 55 BCE), Bay of Pozzuoli, 1.3 m below surface (Brandon et al. 2014). (A) µXRF calcium map of a relict lime clast and relict pores in the cementing matrix. (B) µXRD maps (monochromatic beam, 8 KeV) showing the distribution of d-spacings with the Ca µXRF map for reference. (C) Parameters employed in creating the d-spacing maps of (B). (D) Mineral phases detected through µXRD. (E, F) SEM-BSE images of the microstructures mapped with µXRF and µXRD. For a color version of this figure, see www.iste.co.uk/dejoie/synchrotron.zip*

7.6. Roman principles of concrete longevity

Although the preparation and installation of the architectural and maritime concretes followed different procedures and used different reactive volcanic aggregates, Roman builders seem to have intentionally developed parallel cementing processes in both the monumental and marine structures. The outcomes of synchrotron μXRD and μXRF investigations of the concretes to date have described the nature of long-term bonding and mitigation of decay in the ancient cementing fabrics, so that Roman builders' expertise and methods can be more fully understood and, perhaps, applied to modern maritime cementitious materials.

The principal cementitious microenvironments in the ancient mortars that are described here with μXRD include: (1) the cementing matrix of the architectural mortars that include partially reacted tephra particles and strätlingite mineral cements (Figure 7.5); (2) the internal vesicles of the reactive scoria aggregate in the architectural mortars that crystallize strätlingite and Al-tobermorite at cool ambient temperatures (Figure 7.6); (3) the reactive pumiceous aggregate of the vaulted ceiling mortars that crystallize phillipsite, also at ambient temperatures (Figure 7.7); (4) the relict lime clasts in the marine mortars that are derived from the hydrated lime that catalyzed pozzolanic reactions and produce C-A-S-H binding phase and Al-tobermorite (Figure 7.8); and (5) the spherical pores in the cementing matrix of the marine mortars that have been filled first with calcium–aluminate mineral cements and apparently evolve over time to produce Al-tobermorite (Figure 7.9). The micrometer-scale μXRD maps of these microenvironments illustrate the pervasive production of strätlingite in the mortars of the Markets of Trajan and the associated potential for increased fracture toughness in the architectural concretes (Jackson et al. 2014).

Furthermore, they record the first evidence for low-temperature crystallization of Al-tobermorite in the architectural wall concretes of the imperial age monuments in Rome. They give new perspectives to studies of the material characteristics of C-A-S-H and Al-tobermorite in the relict lime clasts and cementing matrix of the Baianus Sinus mortar as well as the sequestration of SO_4^{2-} and Cl^- in relict spherical voids of the cementing matrix of the marine mortars. Also, they add to the growing body of evidence for mechanical and chemical resilience that is derived from a great diversity of microstructural environments, which produce wide-ranging mineral cements in micrometer-scale domains of the massive conglomeratic concrete structures.

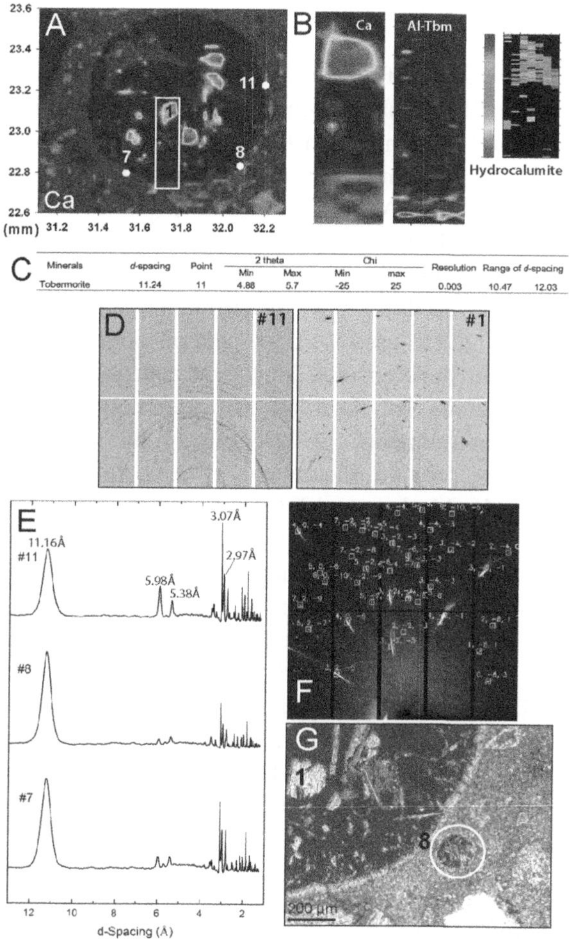

Minerals	*d*-spacing	Point	2 theta		Chi		Resolution	Range of *d*-spacing	
			Min	Max	Min	max			
Tobermorite	11.24	11	4.88	5.7	-25	25	0.003	10.47	12.03

Figure 7.9. *Baianus Sinus breakwater, Bay of Pozzuoli, spherical microstructures, drill core sample about 1.3 m below surface. (A) µXRF calcium map of a relict spherical pore in the cementing matrix of the mortar. (B) µXRF calcium map, µXRD map of the 11.2 Å d-spacing of fine Al-tobermorite (monochromatic beam, 8 KeV) and coarse hydrocalumite (polychromatic beam) at the same scale in the relict void. (C) Parameters employed in creating the Al-tobermorite d-spacing map. (D) Debye–Scherrer diffraction plots (monochromatic beam) showing the d-spacings of fine crystals of Al-tobermorite (site #11) as compared to the irregular spots of coarse crystals in the center of the spherical void (site #1). (E) Al-tobermorite d-spacings of sites #7, #8, #11. (F) Laue pattern (polychromatic beam) of coarse crystals of hydrocalumite. (G) SEM-BSE image showing a smaller spherical relict void adjacent to the larger spherical microstructure. For a color version of this figure, see www.iste.co.uk/dejoie/synchrotron.zip*

7.7. Conclusion

Synchrotron X-ray microdiffraction analyses of ancient Roman mortars have provided an exceptionally valuable window into the cementing fabrics of the concretes of architectural monuments in Rome and marine harbor structures in the Mediterranean region, yielding many discoveries into the exceptional resilience and longevity of these materials. At the Advanced Light Source beamline 12.3.2, coupled μXRD and μXRF micrometer-scale maps are created for cementing microstructures in thin (30 μm) slices of the mortars. A monochromatic X-ray beam in transmission mode is used for powder diffraction of fine grained (< 1 μm) crystals; polychromatic light is used for Laue patterns on larger single crystals. The cementing processes that have ensured long-term cohesion of the ancient Roman concretes for two millennia are recorded by wide-ranging crystalline phases in highly diverse microenvironments. Descriptions of the μXRF and μXRD data acquired at the beamline provide illustrations of cementing microstructures in the concrete structures of the Markets and Forum of Trajan in Rome and the *Baianus Sinus* breakwater in the Bay of Pozzuoli. Instructive examples of μXRD and μXRF results for these materials describe reaction pathways that generate strätlingite in the architectural mortars and Al-tobermorite in the marine harbor mortars as well as the first Al-tobermorite to be identified in the architectural mortars; they also describe the sequestration of ettringite and hydrocalumite in the spherical microstructures in the marine mortar. These observations provide the foundation for future studies that will further investigate the reactive components of the ancient concretes and how the evolving compositions of interstitial fluids in these microstructural environments determine the types of cementitious phases that develop and their subsequent reaction progress over time.

7.8. Acknowledgments

This research used beamline 12.3.2, which is a resource of the Advanced Light Source, a U.S. DOE Office of Science User Facility under contract no. DE-AC02-05CH11231. DOE ARPA-e award DE-AR0001139 also supported this research. We thank the archaeologists of the Sovrintendenza Capitolina ai Beni Culturali in Rome for invaluable research collaborations over the years. Marie Jackson is a principal in Caementa LLC.

7.9. References

Adam, J.-P. (1994). *Roman Building: Materials and Techniques*. Indiana University Press, Bloomington.

Alberti, L.B. (1988). *On the Art of Building in Ten Books*. Translated by J. Rykwert, N. Leach, R. Tavernor. The MIT Press, Cambridge.

Artioli, G., Secco, M., Addis, A. (2019). The Vitruvian legacy: Mortars and binders before and after the Roman world. *European Mineralogy Union Notes in Mineralogy*, 20(4), 151–202.

Barnes, M.W. and Scheetz, B.E. (1991). The chemistry of Al-tobermorite and its coexisting phases at 175°C. In *Specialty Cements with Advanced Properties*, Scheetz, B.E., Landers, A.G., Oder, I., Jennings, H. (eds). Materials Research Society, Warrendale.

Bianchi, E., Meneghini R., Jackson, M.D., Brune, P., Marra, F. (2011). Archaeological, structural, and compositional observations of the concrete architecture of the Basilica Ulpia and Trajan's Forum. *Comm. Hum. Litt.*, 128, 72–93.

Blake, M.E. (1947). *Ancient Roman Construction in Italy from the Prehistoric Period to Augustus*. Carnegie Institute, Washington.

Boyce, J.I., Reinhardt, E.G., Raban, A., Pozza, M.R. (2004). Marine magnetic survey of a submerged Roman Harbour, Caesarea Maritima, Israel. *The International Journal of Nautical Archaeology*, 33(1), 122–136.

Brandon, C., Hohlfelder, R.L., Jackson, M.D., Oleson, J.P. (2014). *Building for Eternity: The History and Technology of Roman Concrete Engineering in the Sea*. Oxbow Books, Oxford.

Brune, P., Ingraffea, A.R., Jackson, M.D., Perucchio, R. (2013). The fracture toughness of an Imperial Roman mortar. *Engineering Fracture Mechanics*, 102, 65–76.

Callebat, L., Gros, P., Jacquemard, C. (1999). *Vitruve, de l'architecture*. Les Belles Lettres, Paris.

Chiari, G., Santarelli, M.L., Torracca, G. (1992). Caratterizzazione delle malte antiche mediante l'analisi di campioni non frazionati. *Materiali e Structure*, 2, 111–137.

Chiari, G., Cimitan, L., Della Ventura, G., Filetici, M.G., Santarelli, M.L., Torraca, G. (1996). Le malte pozzolaniche del mausoleo di Sant'Elena e le pozzolane di Torpignattara. *Materiali e Structure*, 6, 1–36.

Ciancio Rossetto, P. (2008). Teatro di Marcello. Novità sulla struttura e organizzazione dell'area post-scaenam, *Studi Romani*, LVI(1–4), 3–23.

Columbu, S., Liscia, C., Sitziaa, F., Lorenzettib, G., Lezzerinic, M., Pagnottab, S., Raneric, S., Legnaiolib, S., Palleschib, V., Gallellod, G. et al. (2018). Mineralogical, petrographic and physical–mechanical study of Roman construction materials from the Maritime Theatre of Hadrian's Villa (Rome, Italy). *Measurement*, 127, 264–276.

De Gennaro, M., Cappelletti, P., Langella, A., Perrotta, A., Scarpati, C. (2000). Genesis of zeolites in the Neapolitan Yellow Tuff: Geological, volcanological and mineralogical evidence. *Contributions to Mineralogy and Petrology*, 139, 17–35.

De Luca, R., Miriello, D., Pecci, A., Domínguez-Bella, S., Bernal-Casasola, D., Cottica, D., Bloise, A., Crisci, G.M. (2015). Archaeometric study of Mortars from the Garum Shop at Pompeii, Campania, Italy. *Geoarchaeology*, 30, 330–351.

Dejoie, C., Sciau, P., Li, W., Noé, L., Mehta, A., Chen, K., Luo, H., Kunz, M., Tamura, N., Liu, Z. (2014). Learning from the past: Rare ε-Fe_2O_3 in the ancient black-glazed Jian (Tenmoku) wares. *Scientific Reports*, 4(1), 1–9.

Dejoie, C., Tamura, N., Kunz, M., Goudeau, P., Sciau, P. (2015). Complementary use of monochromatic and white-beam X-ray micro-diffraction for the investigation of ancient materials. *Journal of Applied Crystallography*, 48(5), 1522–1533.

Eicholz, D.E. (1962). *Pliny, Natural History, Volume X: Books 36–37*. Harvard University Press, Cambridge.

Gerding, H. (2002). The tomb of Caecilia Metella: Tumulus, tropaeum, and thymele. Doctoral Thesis, Lund University, Lund.

Goldsworthy, H. and Min, Z. (2009). Mortar studies towards the replication of Roman concrete. *Archaeometry*, 51, 932–946.

Gotti, E., Oleson, J.P., Bottalico, L., Brandon, C., Cucitore, R., Hohlfelder, R.L. (2008). A comparison of the chemical and engineering characteristics of ancient Roman hydraulic concrete with a modern reproduction of Vitruvian hydraulic concrete. *Archaeometry*, 50(4), 576–590.

Granger, F. (1931). *Vitruvius on Architecture, Volume I: Books 1–5*. Harvard University Press, Cambridge.

Granger, F. (1934). *Vitruvius, on Architecture, Volume II: Books 6–10*. Harvard University Press, Cambridge.

Gros, P., Corso, A., Romano, E. (1997). *Vitruvio, de architectura*. Giulio Einaudi, Turin.

Jackson, M.D. (2014). Seawater concretes and their material characteristics. In *Building for Eternity the History and Technology of Roman Concrete Engineering in the Sea*, Oleson, J.P. (ed.). Oxbow Books, Oxford, 141–187.

Jackson, M.D. (2017). Technological confidence in Roman architectural and maritime concrete production. *Proceedings 5th International Workshop on the Archaeology of Roman Construction: Arquelogía de la Construcción V*, University of Oxford, 11–12 April, 15–28.

Jackson, M.D. and Kosso, C.K. (2013). Scientia in republican era stone and concrete masonry. In *A Companion to the Archaeology of the Roman Republic*, 1st edition, Evans, J.R. (ed.). Blackwell, Oxford, 268–284.

Jackson, M.D. and Marra, F. (2006) Roman stone masonry: Volcanic foundations of the ancient city. *American Journal of Archaeology*. 110, 403–436.

Jackson, M.D., Logan, J.M., Scheetz, B.E., Deocampo, D.M., Cawood, C.G., Marra, F., Vitti, M., Ungaro, L. (2009). Assessment of material characteristics of ancient concretes, Grande Aula, Markets of Trajan, Rome. *Journal of Archaeological Science*, 36, 2481–2492.

Jackson, M.D., Deocampo, D., Marra, F., Scheetz, B.E. (2010). Mid-Pleistocene volcanic ash in ancient Roman concretes. *Geoarchaeology*, 25, 36–74.

Jackson, M.D., Ciancio Rossetto, P., Kosso, C.K., Buonfiglio, M., Marra, F. (2011). Building materials of the Theater of Marcellus, Rome. *Archaeometry*, 4(4), 728–742.

Jackson, M.D., Vola, G., Oleson, J.P., Scheetz, B., Brandon, C., Hohlfelder, R. (2012). Cement microstructures and durability in ancient Roman seawater concretes. In *Historic Mortars: Characterisation, Assessment and Repair*, Válek, J., Hughes, J.J., Groot, C.J.W.P. (eds). Springer, Dordrecht, 49–76.

Jackson, M.D., Chae, S.R., Mulcahy, S.R., Meral, C., Taylor, R., Li, P., Moon, J., Yoon, S., Emwas, A.H., Vola, G. et al. (2013a). Unlocking the secrets of Al-tobermorite in Roman seawater concrete. *American Mineralogist*, 98, 1669–1687.

Jackson, M.D., Moon, J., Gotti, E., Taylor, R., Chae, S.R., Kunz, M., Emwas, A.-H., Meral, C., Guttmann, P., Levitz, P. et al. (2013b). Material and elastic properties of Al-tobermorite in ancient Roman seawater concrete. *Journal of the American Ceramic Society*, 96(8), 2598–2606.

Jackson, M.D., Landis, E.N., Brune, P.F., Vitti, M., Chen, H., Li, Q., Kunz, M., Wenk, H.-R., Monteiro, P.J.M., Ingraffea, A.R. (2014). Mechanical resilience and cementitious processes in imperial Roman architectural mortar. *Proceedings of the National Academy of Sciences*, 111(52), 18484–18489.

Jackson, M.D., Mulcahy, S.R., Chen, H., Li, Y., Li, Q., Cappelletti, P., Wenk, H.-R. (2017). Phillipsite and Al-tobermorite produced by cementitious water-rock reactions in Roman marine concrete. *American Mineralogist*, 102, 1435–1450.

Jackson, M.D., Oleson, J.P., Moon, J., Zhang, Y., Chen, H., Gudmundsson, M.T. (2018). Extreme durability in ancient Roman concretes. *American Ceramic Society Bulletin*, 97(5), 22–28.

Jackson, M.D., Couper, S., Stan, C.V., Ivarsson, M., Czabaj, M.W., Tamura, N., Parkinson, D., Miyagi, L.M., Moore, J.G. (2019). Authigenic mineral texture in submarine 1979 basalt drill core, Surtsey volcano, Iceland. *Geochemistry, Geophysics, Geosystems*, 20(7), 3751–3773.

Jackson, M.D., Heap, M.J., Vola, G., Ardit, M., Rhodes, J.M., Peterson, J.G., Tamura, N., Gudmundsson, M.T. (2024). Material and mechanical properties of young oceanic basalt, Surtsey volcano, Iceland. *Geological Society of America Bulletin*, 136. doi: 10.1130/B37037.1.

Lancaster, L.C. (2005). *Concrete Vaulted Construction in Imperial Rome*. Cambridge University Press, Cambridge.

Lancaster, L.C., Sottili, G., Marra, F., Ventura, G. (2011). Provenancing of lightweight volcanic stones used in Ancient Roman concrete vaulting: Evidence from Rome. *Archaeometry*, 53(4), 707–727.

Landis, E.N. and Bolander J.E. (2009). Explicit representation of physical processes in concrete fracture. *Journal of Physics D: Applied Physics*, 42, 214002.

Leon, Y., Sciau, P., Passelac, M., Sanchez, C., Sablayrolles, R., Goudeau, P., Tamura, N. (2015). Evolution of terra sigillata technology from Italy to Gaul through a multi-technique approach. *Journal of Analytical Atomic Spectrometry*, 30(3), 658–665.

MacDonald, W.L. (1982). *The Architecture of the Roman Empire*, Revised edition. Yale University Press, New Haven.

MacDowell, J.F. (1991). Strätlingite and hydrogarnet from calcium aluminosilicate glass cements. In *Specialty Cements with Advanced Properties*, Scheetz, B.E., Landers, A.G., Odler, I., Jennings, H. (eds). Material Research Society, 159–179.

MacFarlane, J., Vanorio, T., Monteiro, P.J.M. (2021). Multi-scale imaging, strength and permeability measurements: Understanding the durability of Roman marine concrete. *Construction and Building Materials*, 272, 121812.

Marder, T.A. and Jones, M.W. (eds) (2015). *The Pantheon: From Antiquity to the Present*. Cambridge University Press, Cambridge.

Marra, F., Deocampo, D., Jackson, M.D., Ventura, G. (2011). The Alban Hills and Monti Sabatini volcanic products used in ancient Roman masonry (Italy): An integrated stratigraphic, archaeological, environmental and geochemical approach. *Earth Science Reviews*, 108(3–4), 115–136.

Marra, F., D'Ambrosio, E., Sottili, G., Ventura, G. (2013). Geochemical fingerprints of volcanic material: Identification of a pumice trade route from Pompeii to Rome. *Geological Society of America Bulletin*, 125(3–4), 556–577.

Marra, F., Anzidea, M., Benin, A., D'Ambrosio, E., Gaeta, M., Ventura, G., Cavallo, A. (2016). Petro-chemical features and source areas of volcanic aggregates used in ancient Roman maritime concretes. *Journal of Volcanology and Geothermal Research*, 328, 59–69.

Massazza, F. (1998) Pozzolana and pozzolanic cements. In *Lea's Chemistry of Cement and Concrete*, 4th edition, Hewlitt, P. (ed.). Arnold, London, 471–632.

Myers, R.J., L'Hôpital, E., Provis, J.L., Lothenbach, B. (2015). Composition-solubility-structure relationships in calcium (alkali) aluminosilicate hydrate (C-(N,K-)A-S-H). *Dalton Transactions*, 44, 13530–13544.

Okoronkwo, M.U. and Glasser, F.P. (2016). Stability of strätlingite in the CASH system. *Materials and Structures*. doi: 10.1617/s11527-015-0789-x.

Oleson, J.P., Brandon, C., Botalico, L., Cucitore, R., Gotti. E., Hohlfelder, R.L. (2006). Reproducing a Roman Maritime structure with Vitruvian pozzolanic concrete. *Journal of Roman Archaeology*, 19, 129–152.

Ramage, C.T. (1827). On the material which the Romans employed in the buildings. *Edinburgh Philological Journal*, April–September, 85–90.

Rinaldi, R., Sacerdoti, M., Passaglia, E. (1990). Stratlingite: Crystal structure, chemistry, and a reexamination of its polytype vertumnite. *European Journal of Mineralogy*, 2, 841–849.

Rispoli, C., De Bonis, A., Guarino, V., Graziano, S.F., Di Benedetto, C., Esopsito, R., Morra, V., Cappelletti, P. (2019a). The ancient pozzolanic mortars of the thermal complex of Baia (Campi Flegrei, Italy). *Journal of Cultural Heritage* 40, 143–154.

Rispoli, C., Graziano, S.F., Di Benedetto, C., De Bonis, A., Guarino, V., Esposito, R., Morra, V., Cappelletti, P. (2019b). New insights of historical mortars beyond Pompei: The example of Villa del Pezzolo, Sorrento Peninsula. *Minerals*, 9, 575. doi: 10.3390/min9100575.

Rispoli, C., De Bonis, A., Esposito, R., Graziano, S.F., Langella, A., Mercurio, M., Morra, V., Cappelletti, P. (2020). Unveiling the secrets of Roman craftsmanship: Mortars from Piscina Mirabilis (Campi Flegrei, Italy). *Archaeological and Anthropological Sciences*, 12, 1. doi: 10.1007/s12520-019-00964-8.

Samuelli Ferretti, A. (1995). Proposte per lo studio teorico-sperimentale della statica dei monumenti in opus caementicium. *Materiali e Strutture* 7, 63–83.

Secco, M., Peviato, C., Addis, A., Zago, G., Kamsteeg, A., Dilaria, S., Canovaro, C., Artioli, G., Bonetto, J. (2019). Mineralogical clustering of the structural mortars from the Sarno Baths, Pompeii: A tool to interpret construction techniques and relative chronologies. *Journal of Cultural Heritage*, 40, 265–273.

Seymour, L.M., Tamura, N., Jackson, M.D., Masic, A. (2021) Reactive binder and aggregate interfacial zones in the mortar of Tomb of Caecilia Metella concrete, 1C BCE, Rome. *Journal of the American Ceramic Society*, 105(2), 1503–1518.

Stan, C.V. and Tamura, N. (2018). Synchrotron X-ray microdiffraction and fluorescence imaging of mineral and rock samples. *Journal of Visualized Experiments*, 136. doi: 10.3791/57874.

Stanislao, C., Rispoli, C., Vola, G., Cappelletti, P., Morra, V., de'Gennaro, M. (2011). Contribution to the knowledge of ancient Roman seawater concretes: Phlegrean pozzolan adopted in the construction of the harbor at Soli-Pompeiopolis (Mersin, Turkey). *Periodico di Mineralogia*, 80(3), 471–488.

Tamura, N. (2014). XMAS: A versatile tool for analyzing synchrotron microdiffraction data. In *Strain and Dislocation Gradients from Diffraction*, Ice, G. and Barabash, B. (eds). World Scientific, Singapore, 125–155.

Tamura, N., Kunz, M., Chen, K., Celestre, R.S., MacDowell, A.A., Warwick, T. (2009). A superbend X-ray microdiffraction beamline at the advanced light source. *Materials Science and Engineering A*, 524(1–2), 28–32.

Taylor, H.F.W. (1992) Tobermorite, jennite, and cement gel. *Zeitschrift fiir Kristallographie*, 202, 418–450.

Taylor, H.F.W. (1997). *Cement Chemistry*. Thomas Telford, London.

Van Deman, E.B. (1912a). Methods of determining the date of Roman concrete monuments (first paper). *American Journal of Archaeology*, 16, 230–251.

Van Deman, E.B. (1912b). Methods of determining the date of Roman concrete monuments (second paper). *American Journal of Archaeology*, 16, 387–432.

Vitti, M. (2007). Mercati di Traiano. In *Il Museo dei Fori Imperiali nei Mercati di Traiano*, Ungaro, L. (ed.). Electra, Milan, 5–19.

Vitti, M. and Vitti, P. (2019). Experimental innovations in vaulting construction under Trajan and Hadrian. In *The Trajanic Building Programme at Rome and Its Ports*, Keay, S. and Triantafillou, C. (eds). Highfield, Oxford.

Vola, G., Gotti, E., Brandon, C., Oleson, J.P., Hohlfelder, R.L. (2011). Chemical, mineralogical and petrographic characterization of ancient Roman hydraulic concrete cores from Santa Liberata, Grosetto, Italy and Cesarea, Palestina, Israel. *Periodico di Mineralogia*, 80(2), 99–120.

8

Biomineralization in Sea Urchin Spines: A View on Amorphous Calcium Carbonate Occurrence, Stabilization and Crystallization

Marie ALBÉRIC[1] and Ronald SEIDEL[2]

[1] *Laboratoire chimie de la matière condensée, Sorbonne University, Paris, France*

[2] *Center for Molecular and Cellular Bioengineering, Technische Universitat Dresden, Germany*

8.1. Introduction

"Biomineralization refers to the processes by which living organisms form minerals" (Lowenstam and Weiner 1989). At the interface between geology, biochemistry, biology, chemistry, crystallography and materials science, the field of biomineralization emerged in the 1930s, along with the advances of many analytical techniques such as optical and electronic microscopies, biochemistry methods based on macromolecules extraction, protein separation and chromatography as well as X-ray diffraction (Lowenstam and Weiner 1989). These different techniques revealed the complex ultrastructures and the nanocomposite nature of biominerals as well as the important role of organic matrices and macromolecules in controlling crystal growth (Lowenstam and Weiner 1989).

In recent decades, synchrotron radiation (SR) sources have been increasingly used for the characterization of biominerals as they provide high spectral, angular and spatial resolution. The broad spectrum of SR techniques, with different

Synchrotron Radiation, Cultural Heritage, Biomineralization,
coordinated by Catherine DEJOIE, Pauline MARTINETTO and Nobumichi TAMURA.

wavelengths (from infrared to Gamma rays) at high fluxes, is truly beneficial to our understanding of the composition, the ultrastructure and the architecture of mineralized tissues at various scales – from the atomic level for elemental and crystallographic studies and the submicrometer scale for spatially resolved analyses when characterizing the organic–inorganic interactions, to the macroscale, allowing us, for example, to link the architecture of different tissue types with their mechanical and ecological functions.

One of the first SR experiments on biominerals was performed by Aizenberg et al. (1995). They studied the crystal texture of calcitic sponge spicules with high-resolution X-ray diffraction (HR-XRD) at the X7B beamline of the National Synchrotron Light Source at Brookhaven National Laboratory (New York, USA). Five years later, another milestone in the field of biomineral research was the study by Zizak et al. (2000). They took advantage of the high flux of synchrotron radiation allowing small beam size and low beam divergence to perform spatially resolved scanning Small and Wide Angle X-ray Scattering (SAXS and WAXS) experiments at the SAXS beamline of ELETTRA (Trieste, Italy). Thanks to this setup exhibiting a beam diameter of 20 μm, they were able to measure the size and the orientation of nanometer size mineral particles at the bone–cartilage interface. With the emergence of nanobeam sizes, scanning SAXS was then achieved at the ID13 beamline of ESRF (Grenoble, France) with a spatial resolution of 50 nm, which allowed very fine mapping of bone tissue, including osteons, canaliculi and even single collagen microfibrils (Wagermaier et al. 2013).

Apart from X-ray diffraction and scattering methods, many more SR-techniques are nowadays commonly used in the field of biomineralization such as X-ray absorption spectroscopies, spectroscopy combined with electron imaging, X-ray spectromicroscopy and X-ray computed tomography (Beniash et al. 2014). The constant evolution of synchrotron beamlines, in particular the development of nanosize beams and the coupling of different techniques, opens up a wide range of possibilities to explore the nano- and micro-heterogeneities of biominerals. For example, recent studies combined nano-X-ray fluorescence tomography, nano-SAXS and nanoholotomography to reveal the nanometer- to micrometer-scale organization of the Mg-rich layered pattern in the arms of calcitic brittle stars (Polishchuk et al. 2017; Seknazi et al. 2019). In addition, SR-experiments – where data can be acquired in a few milliseconds – allow performing time-resolved (or in situ) measurements and thus, following the emergence of nano- and micro-structures during their formation and eventually crystallization (Stawski and Benning 2013).

An exhaustive review of the use of synchrotron radiation techniques in the field of biomineralization would require and deserve a complete book. Instead, here, we present how a set of complementary SR-techniques was decisive for the specific study of sea urchin spines and in vitro amorphous calcium carbonate (ACC) systems. The results of the studies presented in this chapter contribute to a better understanding of the mechanisms that drive biomineralization in adult sea urchins, with a focus on the occurrence, the stabilization and the crystallization of ACC phases.

8.2. Biomineralization in sea urchins

Adult sea urchins exhibit multiple skeletal elements: (1) the test (body shell) that is composed of individual plates, called test plates; (2) the spines; (3) the pedicellariae; and (4) the teeth. The teeth are self-sharpening and among the stiffest biomineralized tissues studied (Amini and Miserez 2013). The test encloses the internal organs of the sea urchins and thus protects them from the environment. The spines fulfill multiple functions, including locomotion and defense. They function as sacrificial appendages for protecting the test (Dubois and Ameye 2001) by absorbing energy when they break (Strathmann 1981). Therefore, spines not only break quite easily but also regenerate rapidly between 0.11 and 0.18 mm/day, depending on the species (Heatfield 1971; Gorzelak et al. 2011, 2017).

The skeletal part of the test plates and the spines exhibit a characteristic lightweight mineralized microstructure, the so-called "stereom", which is generally defined by a porous, three-dimensional arrangement of mineralized trabeculae also called "struts" (Figure 8.1). The stereom microstructure is characteristic for all echinoderm skeletons, including sea urchins (Echinoidea), brittle stars (Ophiuroidea), sea stars (Asteroidea), sea cucumbers (Holothuroidea) and sea lilies (Crinoidea). During the lifetime of the animals, the stereom is covered and filled with living tissues containing different cell types. In sea urchins, the cells responsible for the skeletogenesis are called sclerocytes (Pilkington, 1969, Markel and Roser 1983) or alternatively calcoblasts (Heatfield and Travis 1975).

Different types of stereom have been described and classified for sea urchin test plates (e.g. imperforate (no pores), microperforate, irregular perforate, galleried, labyrinthic and laminar) that vary in terms of geometry and dimensions of the struts and pores (Smith 1980). Figure 8.1 depicts representative stereom microstructures of the spines and test plates of *Paracentrotus lividus* sea urchins. The main difference in terms of microstructure between the *P. lividus* spines and the test plates is the presence of dense outer wedges in the spines, which consist of "imperforate" stereom, also called septa (Figure 8.1(f)).

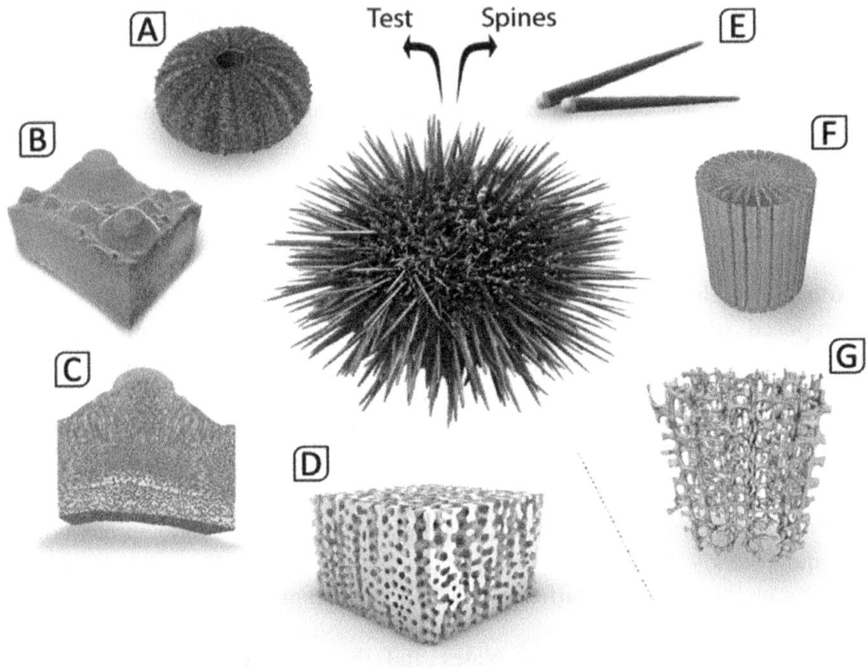

Figure 8.1. *A) Test of sea urchin Paracentrotus lividus; B) 3D rendering of a cropped volume of a test plate acquired by microcomputed tomography at the Anatomix beamline (SOLEIL)(Weitkamp et al. 2017); C) virtual section through the 3D volume; D) exemplary stereom architecture; E) spines of P. lividus; F) 3D rendering of a volume acquired by microcomputed tomography at the Tomcat beamline (SLS); and G) 3D volume of the inner stereom of the spine. For a color version of this figure, see www.iste.co.uk/dejoie/synchrotron.zip*

Early studies on biomineralization in sea urchins mainly focused on their embryonic stages where calcitic triradiate spicules are formed (Benson et al. 1986; Brown et al. 1995; Beniash et al. 1997, 1999). An amorphous calcium carbonate (ACC) phase was first evidenced during the growth of *P. lividus* sea urchin spicules by X-ray diffraction, infrared spectroscopy and transmission electron microscopy (Beniash et al. 1997, 1999). It was therefore assumed that "ACC transforms into calcite during the growth of sea urchin larval spicule" (Beniash et al. 1997). About a decade later, X-ray absorption near-edge spectroscopy collected at the Calcium (Ca) K-edge allowed monitoring of the transition from ACC to calcite of *Strongylocentrotus purpuratus* and *P. lividus* spicules at different developmental embryonic stages by measuring the short-range order (SRO) around the Ca atom (Politi et al. 2006). Moreover, two ACC phases, called ACC type I and II, having different Ca $L_{2,3}$-edge spectral signatures were identified by X-ray photoemission electron spectromicroscopy (X-PEEM) of *S. purpuratus* spicules (Politi et al. 2008; Gong et al. 2012). By comparison of the Ca $L_{2,3}$-edge spectra with synthetic analogues, biogenic ACC Type I was shown to be hydrated and ACC type II, anhydrous (Politi et al. 2008; Gong et al. 2012; Albéric et al. 2019). These findings revealed that the development of sea urchin embryonic spicules involves a two-step reaction of the transformation of hydrated ACC into anhydrous ACC and, subsequently into calcite (Politi et al. 2008; Gong et al. 2012).

The spines of adult sea urchin, in particular the stereom, also form via the deposition of ACC precursors as shown by infrared spectroscopy and electron diffraction (Politi et al. 2004). In addition, it was proposed that contrary to the stereom, the septa formation might occur through the direct precipitation of calcite without involving ACC precursors (Moureaux et al. 2010). However, using X-PEEM mapping of freshly regenerated sea urchin spines, the study presented in section 8.3. Albéric et al. (2019) suggests that both stereom and septa form via ACC precursors.

The occurrence of ACC phases in sea urchin biominerals as well as in many other organisms (Addadi et al. 2003) is quite surprising as ACC is a thermodynamically and kinetically unstable phase under ambient conditions. In the last few decades, many research groups have studied the stabilization and the crystallization of ACC phases by considering in vitro systems. They showed that synthetic ACC can be stabilized temporarily in the presence of inorganic additives such as H_2O, Mg^{2+} and PO_4^{3-} and organic molecules such as polyaspartic acid (for reviews, see Addadi et al. (2003) and Fernandez-Martinez et al. (2017)). However, the nature of the stabilization (kinetic or thermodynamic) and the detailed mechanisms behind ACC stabilization and crystallization are not yet fully understood. Using pair distribution function (PDF) analysis of in situ total X-ray scattering measurements acquired during the induced crystallization of ACC, the study presented in section 8.4 (Albéric et al. 2018a) sheds light into this topic.

The skeletal part of the spines and test plates of *P. lividus* adult sea urchins are mainly composed of Mg-calcite (3–10 mol% Mg), stable anhydrous ACC (8–10 wt.%) and intra-crystalline organic molecules (1.4–1.8 wt.%) (Albéric et al. 2018b). The stable anhydrous ACC phase is proposed to be a remnant phase of the biomineralization processes, which appears kinetically trapped within the calcite crystals during the two-step reaction ACC-$H_2O \rightarrow$ ACC $\rightarrow$calcite (Albéric et al.

2018b). The presence of remnant ACC in sea urchin biominerals make them unique model systems for studying the biomineralization mechanisms occurring through the crystallization of amorphous precursors. The two studies outlined in section 8.5 (Albéric et al. 2018b, 2022) took advantage of this peculiar property by respectively performing powder HR-XRD and SAXS experiments during the induced crystallization of the stable biogenic ACC phase.

Besides their complex convoluted micromorphologies and nanocomposite nature, the test plates and the spines of sea urchins behave as single crystals of calcite, when observed under polarized light microscopy (Raup 1959) and analyzed with X-ray and electron diffraction techniques (Donnay and Pawson 1969; Tsipursky and Buseck 1993; Larsson and Christy 2008). In *P. lividus* sea urchins, the biogenic calcite *c*-axis is parallel to the long axis of the spines (Donnay and Pawson 1969) and tangent to the surface of the test (Raup 1959). Figure 8.2(a) and (b) display X-ray diffraction patterns of a spine showing diffraction spots (or Bragg spots) characteristic of a single crystal. When reduced to powder (particles ~100 μm), the spines present a X-ray diffraction pattern showing some of the Debye–Scherrer rings, typical of a polycrystal (Figure 8.2(b)), whereas all the Debye–Scherrer rings are observed when the powdered sample is rotated (Figure 8.2(c)) due to a complete average of the different orientations of the particles.

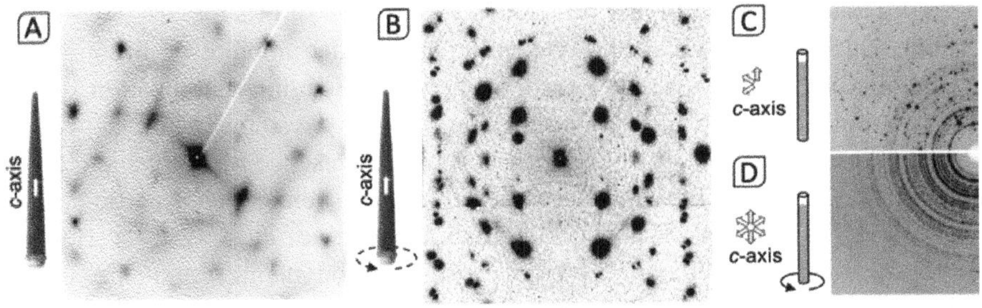

Figure 8.2. *A) X-ray diffraction pattern of an intact spine with its c-axis oriented perpendicular to the beam; B) sum of several X-ray diffraction patterns acquired while rotating from 0 to 180° every 5° an intact spine with its c-axis oriented perpendicular to the beam and X-ray diffraction patterns of a capillary containing spine fine powder, in fixed position C) and when rotated D). The c-axis orientations are indicated by the purple arrows. For a color version of this figure, see www.iste.co.uk/dejoie/synchrotron.zip*

The diffraction spots in Figure 8.2(a) show some diffuse scattering, which reveal the mosaic nature of the biogenic single crystals of the *P. lividus* spines also reported in a recent study (Cölfen et al. 2022). Most of the "real" single crystals as opposed to "ideal" ones are indeed mosaic crystals that consist of slightly misaligned submicrometer-sized crystallites (or mosaic blocks) (Als-Nielsen and McMorrow 2011) but are still largely coherent (Blake et al. 1984). The mosaicity in echinoderm calcite crystals was measured by XRD measurements and electron backscatter diffraction (Blake et al. 1984; Goetz et al. 2010) and was suggested to be due to the presence of defects within the crystals such as intra-crystalline macromolecules. Although not widely accepted, the formation of such crystals was suggested to occur via the highly oriented aggregation of nanocrystallites into a mesocrystal intermediate that further leads to the final single-crystalline structure by fusion (Cölfen and Antonietti 2005). It was especially suggested for the formation of sea urchin spines that were thus speculated to have a mesocrystalline structure (Seto et al. 2012; Cölfen et al. 2022), although some authors rather proposed a secondary nucleation crystallization mechanism through ACC nanoparticles (Weiner and Addadi 2011). Finally, crystal twinning was identified in sea urchin biominerals (Larsson and Christy 2008), also likely induced by the presence of intra-crystalline macromolecules (Pokroy et al. 2007).

8.3. Occurrence of ACC precursors in regenerated sea urchin spines

8.3.1. *Why X-ray photoemission electron spectromicroscopy and how?*

X-PEEM was shown to be a valuable technique for the study of biomineralization processes in sea urchins (Politi et al. 2008; Killian et al. 2009; Gong et al. 2012; Albéric et al. 2019), in particular for identifying the spatial distribution of ACC precursor nanoparticles (~40 nm) within freshly grown biominerals. Indeed, by combining X-ray absorption near edge structure (XANES) at the Ca $L_{2,3}$-edge with electron imaging, Ca $L_{2,3}$-edge spectrum mapping can be obtained at a spatial resolution down to 20 × 20 nm (Metzler et al. 2008; Gilbert 2014). X-PEEM experiments performed on embryonic spicules of *S. purpuratus* sea urchins showed the presence of two ACC precursors exhibiting different Ca $L_{2,3}$-edge spectra (Politi et al. 2008), whose spatial distribution was further determined within the freshly grown spicules (Gong et al. 2012). For the later study, flat and smooth samples must be prepared as X-PEEM is a very surface sensitive technique. A detailed protocol of the sample preparation, optimized for the conservation of amorphous phases in freshly grown biominerals, is reported in Gilbert (2014).

In the study presented here (Albéric et al. 2019), freshly regenerated spines after fracture were severed from *S. purpuratus* adult sea urchins. The spines were embedded and polished before analysis following the procedure in Gilbert (2014). X-PEEM experiments were performed on a PEEM--3 microscope at the 11.0.1.1 beamline of the Advanced Light Source (ALS, Lawrence Berkeley National Laboratory, CA, USA) providing high brightness, small beam size, high spectral resolution and high flux (Chopdekar 2018). Stacks of X-ray absorption images were acquired at specific photon energies across the Ca $L_{2,3}$-edge, i.e. ranging from 340 to 360 eV. Therefore, for each image pixel, the full Ca $L_{2,3}$-edge spectrum was captured.

The acquired stacks were first drift-corrected with the PEEM Vision software (Scholl 2014). Second, data analysis was performed using the Igor Pro software and the G.G. macros (GG Macros 2019) that allow the obtaining of Ca component maps displaying the spatial distribution of the different $CaCO_3$ phases. The spectra of each pixel of the Ca component maps consist of a linear combination of the reference spectra of ACC-H_2O, ACC and calcite. Therefore, the Ca component maps are represented as RGB maps where the ACC-H_2O phase is displayed in red, the ACC phase in green and the calcite phase in blue. The three component spectra of the references ACC-H_2O, ACC and calcite are obtained by fitting averaged spectra of the three phases respectively.

8.3.2. *Ca $L_{2,3}$-edge XANES spectra*

The regeneration of sea urchin spines starts with the formation of the inner stereom in the form of microspines as shown in Figure 8.3(a). Calcite, as well as both hydrated and anhydrous ACC, were evidenced in the growing microspines based on their distinct Ca $L_{2,3}$-edge XANES spectra (Figure 8.3(b)).

Ca $L_{2,3}$-edge XANES spectra of calcite presents the characteristic four peaks corresponding to the crystal field split of the L_2 and L_3 lines (De Groot et al. 1994; Ko et al. 2007). Peaks related to the highest energy L_2 white lines are called 1 and 2 and the ones corresponding to the lowest energy L_3 white lines are named 3 and 4. The Ca $L_{2,3}$-edge spectra of both ACC-H_2O and ACC exhibit peaks at position 1 and 3, but not 4. Differences between ACC-H_2O and ACC are observed in peak position 2.

Whereas the Ca $L_{2,3}$-edge spectrum of ACC shows a peak at position 2, the spectrum of ACC-H_2O only presents a shoulder at this position. These variations have been related to varying levels of disorder of the carbonate ions in the two ACC structures in comparison to calcite (Rez and Blackwell 2011).

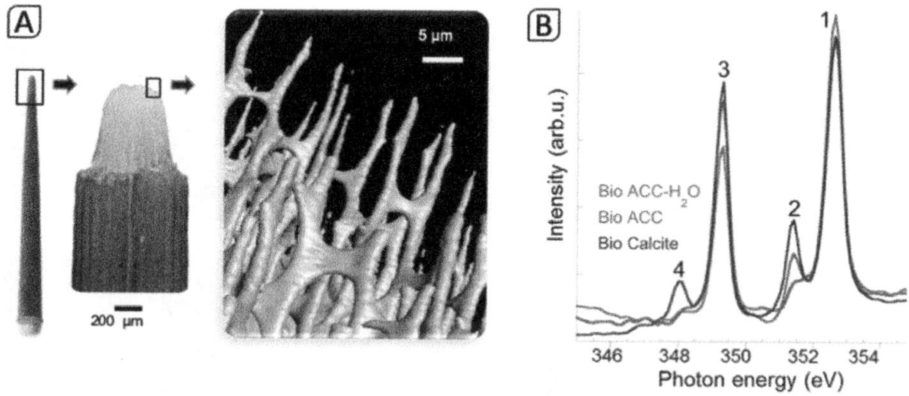

Figure 8.3. *A) Growing tip of a spine composed of microspines at the very tip that are visualized using µCT; and B) Ca $L_{2,3}$-edge XANES spectra of biogenic ACC-H_2O, ACC and calcite. For a color version of this figure, see www.iste.co.uk/dejoie/synchrotron.zip*

The Ca $L_{2,3}$-edge spectra of the three calcium carbonates, identified during the growth of adult sea urchin spines, are similar to those measured in sea urchin embryo spicules (Politi et al. 2008; Gong et al. 2012). It was thus concluded that similar ACC precursors are involved during embryonic and adult biomineralization pathways and that the two-step reaction (ACC-H_2O → ACC → calcite) also occurs during the growth of adult sea urchin spines.

8.3.3. *Calcium component maps*

Figure 8.4 shows the Ca component maps of two regions in a growing spine: (1) a microspine at the tip, where the growth is fast (100 µm/day) (Figure 8.4(a)) and (2) a growing septa at the outer edge, where the growth is slow (10 µm/day) (Gorzelak et al. 2011, 2017) (Figure 8.4(b)). In both regions, calcite dominates over the amorphous precursors. In the slow growing region (Figure 8.4(b)), both ACC and ACC-H_2O phases are detected, whereas in the region of fast growth, only ACC is detected. Assuming the two-step reaction (ACC-H_2O→ACC →calcite), it appears that the transformation is slower during septa formation than during microspine growth. These results relate well to the different growth rates measured in both regions (Gorzelak et al. 2017) and suggest that stereom as well as septa form via the same ACC precursors, however, with different kinetics, possibly due to various ACC stabilities.

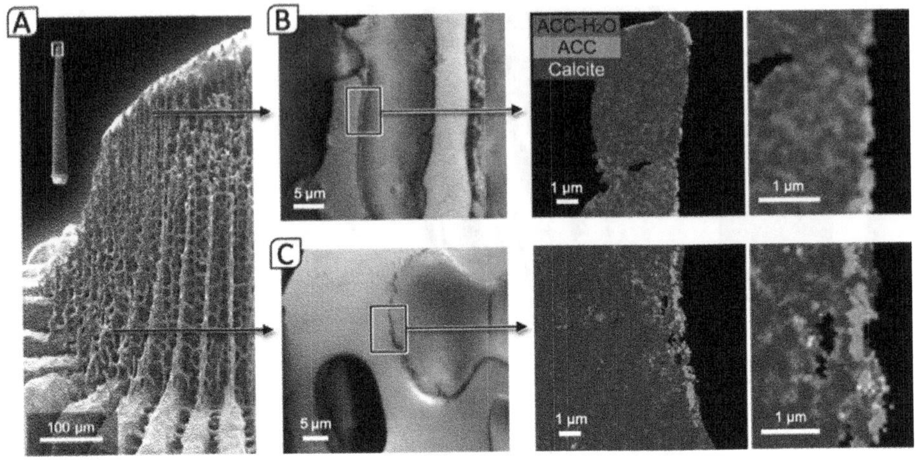

Figure 8.4. *A) SEM micrograph of the tip of regenerated spine; B) Ca component maps of a microspine at the tip of a growing spine; and C) Ca component maps of a forming septa. ACC-H_2O (red), ACC (green) and calcite (blue). For a color version of this figure, see www.iste.co.uk/dejoie/synchrotron.zip*

8.4. ACC structure and stability: role of water and ion impurities

8.4.1. *ACCs structure*

Since the evidence of amorphous calcium carbonate phases in the spicule of soft corals, *Leptogorgia pulchra* and *Leptogorgia rigida* in 1953, biogenic ACC has been found in many other marine and freshwater animals as well as in plants (Weiner et al. 2003). ACC is described as amorphous because it does not diffract X-rays nor electrons and shows isotropy under polarized light, which reflects its lack of long-range order (LRO) (Addadi et al. 2003).

Nevertheless, biogenic ACC presents a short-range order (SRO) up to 4 Å around the Ca atom as revealed by Ca K-edge Extended X-ray Absorption Fine Structure (EXAFS) experiments that provide information about the bond distances and the coordination symmetry around the Ca atom (Taylor et al. 1993; Hasse et al. 2000; Levi-Kalisman et al. 2002; Politi et al. 2006). ACCs from various animals were shown to present different SRO due to their diverse levels of stabilities (Levi-Kalisman et al. 2002), likely related to the presence of impurities such as Mg^{2+} and PO_4^{3-} ions, which are known to stabilize ACC. Therefore, it appears that different polymorphs of ACC exist in both biogenic and synthetic systems (Cartwright et al. 2012; Fernandez-Martinez et al. 2017).

Aiming for a better understanding of the transformation from amorphous precursors to crystalline calcium carbonate polymorphs (e.g. calcite, aragonite and vaterite), the structures of the amorphous and crystalline phases were compared. It was shown that the SRO around the Ca atom of biogenic ACC and that of calcite from the embryonic sea urchin spicules were similar, suggesting that the SRO of the final calcite was encoded in the ACC's SRO (Politi et al. 2006). In addition, ACCs synthetized in the presence of Mg^{2+} ions or polyaspartic acid also present SRO resembling those of calcite or vaterite (Lam et al. 2007; Cartwright et al. 2012). Time-resolved measurements were then performed on synthetic ACC systems to follow in detail the continuous structural rearrangements occurring during the crystallization. In situ SAXS experiments showed the formation of ACC nanoparticles, growing in size until their crystallization (Bolze et al. 2002). Furthermore, in situ Wide-Angle X-ray Scattering (WAXS) analyses allowed the structural study of crystal formation (Bots et al. 2012; Stawski and Benning 2013).

8.4.2. *Why pair distribution function analysis and how?*

In addition to EXAFS, the structure of ACC was also studied by nuclear magnetic resonance (NMR) techniques, infrared and Raman spectroscopies (Fernandez-Martinez et al. 2017). Although very informative, this wide range of analytical techniques only provides information on the SRO up to a few Å and does not allow the analysis of the ACC structure over larger distances (Addadi et al. 2003). Alternatively, atomic pair distribution function (PDF) analysis of total X-ray scattering provides information of the local structure from the SRO up to 40 Å, including the medium-range order (MRO) (Michel et al. 2008; Goodwin et al. 2010; Radha et al. 2012; Reeder et al. 2013; Schmidt et al. 2014; Albéric et al. 2018a; Jensen et al. 2018).

Experimentally, the total X-ray scattering of the sample is measured over a large q-range (up to 30 nm^{-1} for ACCs samples) (Figure 8.5(a)). After integration of the 2D pattern, the total scattering profile S(q) is obtained, reduced and normalized to determine F(q), which is subsequently Fourier-transformed in order to calculate the reduced pair distribution function G(r) (Figure 8.5(b)–(d)). Figure 8.5(d) shows the typical PDF of a synthetic ACC phase with MRO up to 8–10 Å (Michel et al. 2008; Goodwin et al. 2010; Radha et al. 2012, Reeder et al. 2013; Schmidt et al. 2014; Albéric et al. 2018a; Jensen et al. 2018). G(r) shows maxima at specific r(Å) positions corresponding to the average bond distances between pairs of atoms separated in the solid with high probability (Figure 8.5(d)). However, in a PDF analysis, a peak is not atom specific – in contrast with those of XAS – and may correspond to a few atom pairs. Nevertheless, it is possible to deconvolute the contribution of each pair by calculating the partial PDF for the distinct pairs of atoms (Michel et al. 2008). In the case of ACC, the peak at 2.4 Å contains mainly the contributions from Ca-O pairs, the peak at 2.9 Å, the O-O pairs, the peak at 4.1 Å, the Ca-O and Ca-Ca pairs and the peak at 6.2 Å mainly the Ca-Ca pairs (Michel et al. 2008; Saharay et al. 2013; Bushuev et al. 2015; Jensen et al. 2018).

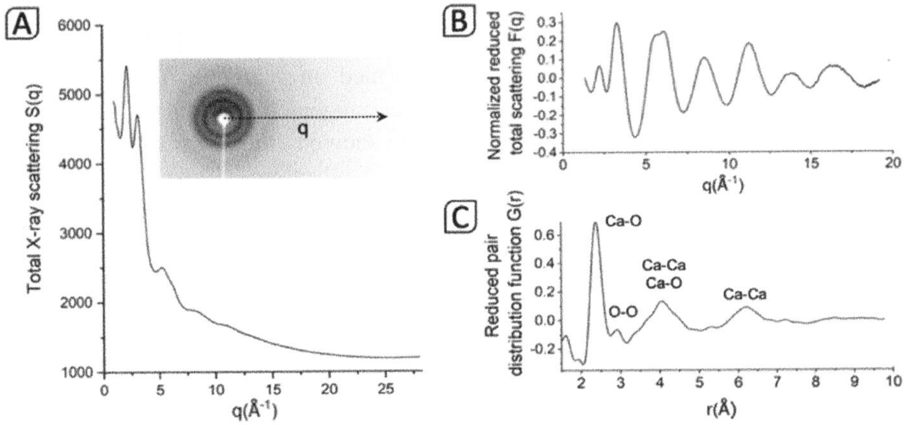

Figure 8.5. *A) WAXS pattern of ACC obtained at ID11 with a large q-range and its respective total X-ray scattering integration profile, S(q); B) normalized reduced total scattering, F(q); and C) reduced pair function, G(r) with the main inter-atomic distance position indicated: Ca-O, O-O and Ca-Ca*

Analogous to time-resolved SAXS//WAXS experiments, in situ total X-ray scattering with PDF analysis is a valuable technique to study the mechanisms behind ACC stabilization and crystallization, allowing us to follow the molecular rearrangement of the SRO and MRO up to the appearance of the LRO of the crystalline calcite.

Therefore, in situ total X-ray scattering was measured at ID11, the materials science beamline of the European Synchrotron Research Facility (ESRF) (Grenoble). In order to further understand the role of Mg^{2+} ions on the stability and structure of ACC, PDF experiments were also performed with ACC samples synthesized in the presence of low amount of Mg^{2+} ions. The powder ACC samples were mounted on a cylindrical sample holder and placed inside a Linkam heating stage (Albéric et al. 2018a). High beam energy of 87 KeV, only available with SR, was used in order to access a large q-range needed for high resolved PDF analysis of the X-ray scattering data. In addition, a highly sensitive 2D detector with fast read out times was used to acquire high temporal resolution (every 60s) with reasonable signal-to-noise ratio needed for PDF analysis. Data were processed using the Fit2D software (Hammersley et al. 1996) and pair distribution functions were calculated using the PDFgetX3 software (Juhas et al. 2013).

8.4.3. *PDFs of ACC and Mg-ACC*

At biologically relevant concentrations of Mg^{2+} (2.4% Mg:Ca), the PDF profile is dominated by the scattering of Ca^{2+} ions and carbonates rather than Mg^{2+} ions, which allows studying the effect of Mg^{2+} ions on the ACC structure. Figure 8.6 shows that the PDFs of hydrated additive free ACC (in red) and hydrated Mg-ACC (in blue) are very similar within the resolution and the sensitivity of the experiment.

These results suggest that the SRO and the MRO of hydrated ACC is not affected by the presence of biological relevant amounts of Mg^{2+} ions. These findings are supported by a previous study that shows changes in the PDF occurring at concentrations starting from 40% Mg:Ca (Radha et al. 2012). However, due to the overlap of several atomic pairs in the PDF, scattering techniques might be less sensitive to small variations in the SRO than spectroscopic methods such as XAS.

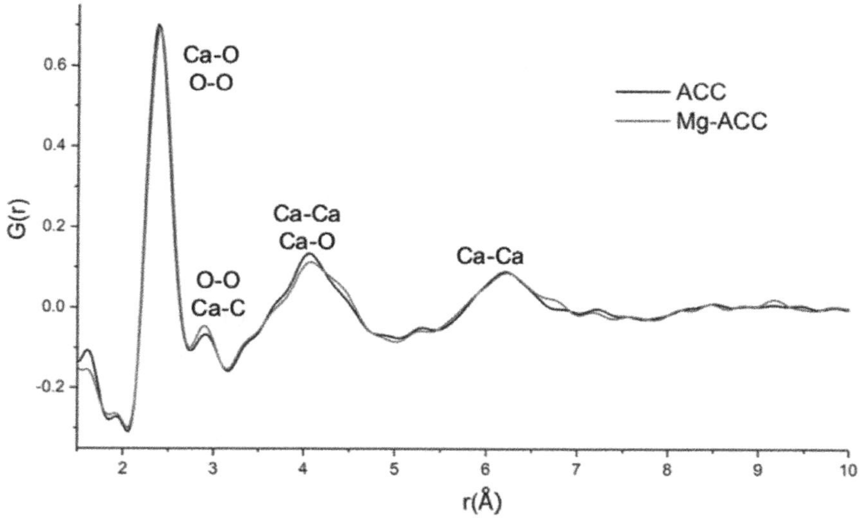

Figure 8.6. *Reduced pair distribution function G(r) of additive free ACC in black and Mg-ACC (2.4% Mg:Ca) in pink, the inter-atomic distances are indicated. For a color version of this figure, see www.iste.co.uk/dejoie/synchrotron.zip*

8.4.4. *In situ PDFs during the crystallization of ACC and Mg-ACC*

The effect of Mg^{2+} ions on the molecular rearrangement of the ACC structure during its crystallization was further studied by in situ PDF analysis of ACC and Mg-ACC upon crystallization induced by heating (from 25 to 400°C). Taking advantage of the highly sensitive 2D detector available at ID11, a PDF could be calculated every 2°C (heating rate of 2°C/min).

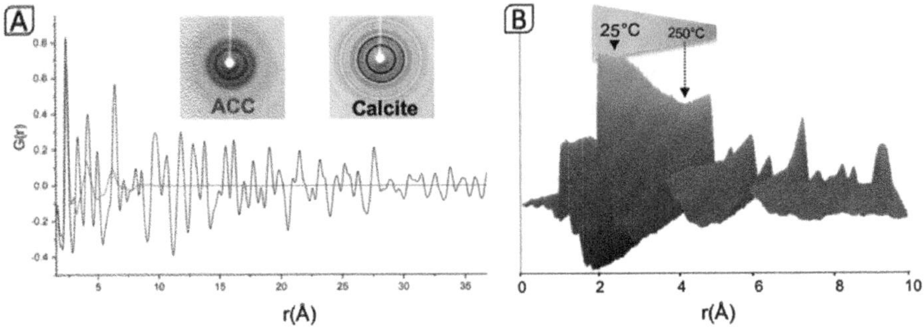

Figure 8.7. *A) PDF of additive free ACC and calcite, the WAXS signal of both phases are displayed; and B) in situ PDF during the crystallization of additive free ACC induced by heating, the arrow indicates the onset of crystallization. For a color version of this figure, see www.iste.co.uk/dejoie/synchrotron.zip*

Figure 8.7(a) compares the PDFs of ACC without additives at room temperature and once it is crystallized into calcite after heating at 200°C. The PDFs of ACC and calcite are clearly different, not only at the LRO but also at the SRO and MRO, suggesting that this specific ACC sample did not encode the calcite SRO.

The PDFs obtained during the transformation of ACC into calcite show a progressive decrease in intensity of the 2.4 Å peak, before the onset of crystallization, indicated by the arrow in Figure 8.7(b). In addition, peak shifts are observed for the 2.4 and 4.1 Å peaks during the dehydration of ACC, a step required for its crystallization (see Figure 8.8). The fact that these changes are progressive suggests that the crystallization occurs via solid-state transformation rather than dissolution/reprecipitation mechanisms (Ihli et al. 2013; Zou et al. 2015).

Differential maps of PDFs as a function of temperature (dPDF/dT) were calculated in order to better visualize the structural changes reflected in peak intensities and shifts during heating and between samples (Figure 8.8). The color scale of the dPDF/dT maps indicates the rate of the intensity changes.

The appearance of adjacent blue and red vertical stripes designates a peak shift. Although SRO and MRO of ACC and Mg-ACC are very similar (Figure 8.6), the local structure evolves differently during dehydration (Figure 8.8). The peak at 4.1 Å, corresponding to the Ca-O and Ca-Ca pairs, shifts to lower distance, r, at higher temperatures for Mg-ACC than for ACC. These observations underline the role of water in stabilizing Mg-ACC, due to its interaction with Mg^{2+} ions.

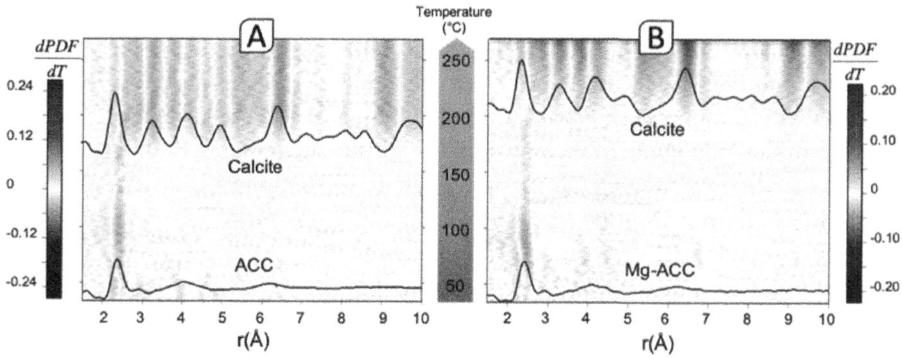

Figure 8.8. *Differential maps (dPDF/dT) of A) additive free ACC and B) Mg-ACC as a function of temperature, ranging from room temperature to 300°C. For a color version of this figure, see www.iste.co.uk/dejoie/synchrotron.zip*

8.5. Induced crystallization of remnant ACC in sea urchin spines

8.5.1. *Why high-resolution X-ray diffraction and how?*

As mentioned in the introduction, high-resolution X-ray powder diffraction (HR-XRD) based on synchrotron radiation was one of the first experiments performed on biominerals at synchrotron sources (Aizenberg et al. 1995). The authors measured anisotropic textural parameters (coherence length and angular spread) of the calcite crystals of sponge spicules. This anisotropy was attributed to the adsorption of intra-crystalline organic molecules on preferential crystallographic calcite planes. The coherence length, which is the average distance between imperfections in specific crystallographic directions, is determined from the width of the diffraction peaks by applying the Scherrer formula. Similar studies performed on sea urchin spines showed that intra-crystalline organic molecules are preferentially adsorbed on planes parallel to the calcite *c*-axis (Aizenberg et al. 1997).

The anisotropy in the crystal textures of biominerals was further studied by the measure of lattice distortions (for a review, see Zolotoyabko (2017)). High angular resolution is a prerequisite for detecting small lattice distortions that are calculated based on lattice parameters obtained from Rietveld refinement analysis of HR-XRD profiles. By this approach, Pokroy et al. (2004) showed that aragonitic shells of *Acanthocardia tuberculate* bivalve exhibit anisotropic lattice distortions with respect to geological aragonite.

The data presented in this chapter (Albéric et al. 2018b) were acquired at ID22, the high-resolution powder diffraction beamline of ESRF (Grenoble). The peaks of the HR-XRD profiles acquired at ID22 are very narrow, with an instrumental contribution to the FWHM of about 0.003° (Fitch 2004). The samples were ground to fine powders, mounted and measured in spinning glass capillaries to reduce the preferred-orientation effect of the powdered single crystals (Figure 8.9).

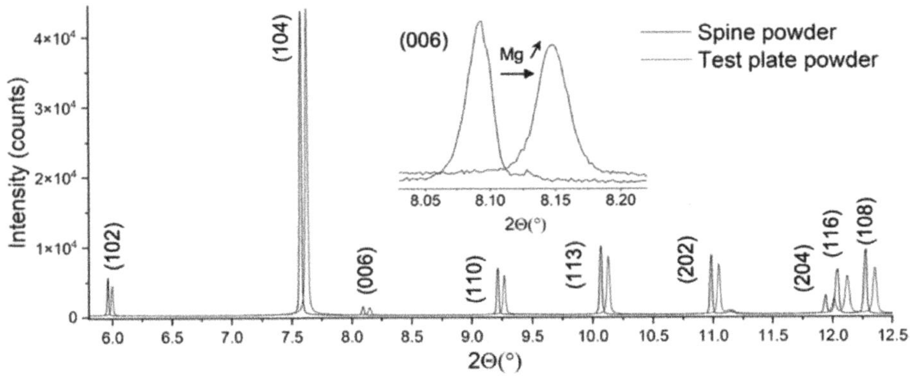

Figure 8.9. *Part of the HR-XRD profiles of sea urchin spines (in purple) and test plate (in green) powders measured at ID22 (ESRF) with λ=0.399952 Å, the XRD reflections are indicated. For a color version of this figure, see www.iste.co.uk/dejoie/synchrotron.zip*

Although Rietveld refinements are needed to accurately determine lattice parameters, lattice distortions can be first appreciated when considering the peak positions of HR-XRD profiles. Figure 8.9 shows HR-XRD profiles for powdered sea urchin spines and test plates. All peaks of the test plates diffractogram are shifted to higher 2Θ positions in comparison to the spine profile and therefore, to lower *d*-spacing ($\sin \Theta = \lambda/2d$, with λ being the wavelength of the synchrotron radiation). Sea urchin spines incorporate 3mol% of Mg, whereas test plates incorporate 10 mol%. The presence of higher amount of Mg in the test plates than in the spines likely explains the peak shift to lower *d*-spacing for the former with respect to the latter. Indeed, the incorporation of Mg^{2+} ions, smaller than Ca^{2+} ions within the $CaCO_3$ lattice, reduces the values of the lattice parameters.

The important effect of Mg ions on lattice distortions must therefore be taken into consideration when aiming at calculating the lattice distortions induced by the incorporation of organic molecules. An equation for this correction was established by Pokroy et al. (2006).

The Rietveld refinement method is a powerful technique that consists of fitting an experimental HR-XRD profile, which includes the structural and instrumental parameters, with a calculated XRD profile (Rietveld 1969). Rietveld refinement of the data presented in this chapter was performed with the FULLPROF software package (Rodriguez Carvajal 1993) in the calcite hexagonal settings. Calcite belongs to the hexagonal crystal family, the trigonal crystal system and has a hexagonal unit cell with a=b=4.990 Å, c=17.061 Å, $\alpha = \beta = 90°$ and $\gamma = 120°$ (Effenberger et al. 1981).

The higher lattice parameters measured in biominerals than in geological analogues were attributed to the presence of intra-crystalline organic molecules in the biomineral lattice because when decomposed by heating, the removal of the intra-crystalline organic molecules results in similar lattice parameters in biominerals and geological minerals. Measuring lattice distortions at different temperatures is nowadays a strategy widely used in order to study the organic–inorganic interactions in biominerals and synthetic systems (Zolotoyabko 2017).

By using this approach, the study presented in this chapter (Albéric et al. 2018b) additionally addressed the role of ACC on lattice distortions. It was previously suggested that lattice distortions in biominerals may develop during the transformation of ACC into the crystalline polymorphs during biomineralization (Zolotoyabko 2017). When ACC crystallizes into calcite, a decrease of the total volume by ~13vol% is observed due to the different density of ACC and calcite (d(calcite)=2.7 $g.cm^{-3}$ and d(ACC)= 2.4 $g.cm^{-3}$ (Jensen et al. 2018).

It is further suggested that intra-crystalline organic molecules counterbalance this volume reduction and thereby induce stresses into biominerals (Zolotoyabko 2017). As previously mentioned, sea urchin skeletal elements maintain remnant ACC in their structure after biomineralization processes and are therefore excellent model systems to test this hypothesis by inducing ACC crystallization.

8.5.2. *Biogenic calcite lattice distortions*

Sea urchin spine powders were annealed from 50°C to 400°C before being measured at the ID22 beamline (Albéric et al. 2018b). ACC crystallization in sea urchin spines occurs at ~170–240°C and organics combustion (e.g. proteins, saccharides, lipids) starts at ~250°C (Albéric et al. 2018b). Figure 8.10(a) shows the (006) XRD reflection peaks of the native sample (in blue), and those at annealing temperatures of 200°C (in green) and 400°C (in red). The small peak shift of 0.01° towards lower 2Θ position, measured for the sample annealed at 200°C (in comparison to the native one) indicates that ACC crystallization leads to an expansion of the lattice along the c-axis as confirmed by the increase of the lattice distortions Δc/c(%) calculated with respect to geological calcite (Figure 8.10(b)). The peak back-shift, toward its original position, at 400°C reflects a lattice reduction due to the removal of organic molecules. Interestingly, the presence of remnant ACC in the spines leads to low initial lattice distortion (Δc/c=0.03%) when compared, for example, with the calcite prisms of *Pinna nobilis* shells (Δc/c=0.24%) that lack significant amounts of ACC (Pokroy et al. 2006; Albéric et al. 2018b). This study confirms that lattice distortions in biominerals develop during the crystallization of ACC when confined by organic molecules, which tend to counterbalance the volume reduction induced by the transformation.

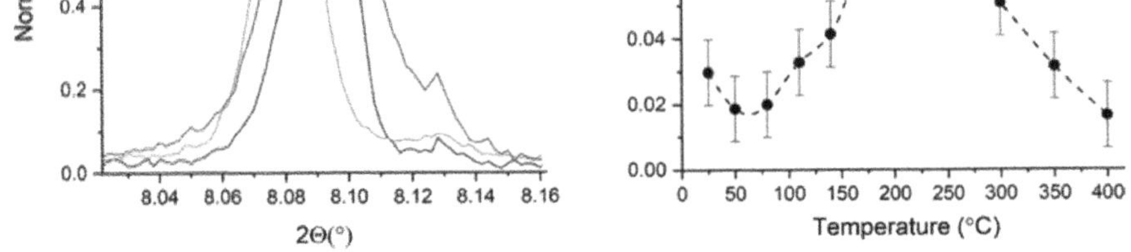

Figure 8.10. *A) (006) XRD reflection peaks of the native, 200°C and 400°C annealed sea urchin spines powder and B) lattice distortions Δc/c of the c-parameter as a function of temperatures. For a color version of this figure, see www.iste.co.uk/dejoie/synchrotron.zip*

8.5.3. *Why in situ Small Angle X-ray Scattering and how?*

Small Angle X-ray Scattering (SAXS) is a well-suited technique for analyzing internal interfaces having high electron density contrast, such as those of nanostructured bio-composites (e.g. bones and teeth as well as synthetic systems, for reviews, see Pabisch et al. (2013), Stawski and Benning (2013), Wagermaier et al. (2013) and DiMasi (2014)).

In situ heating SAXS measurements of intact sea urchin spines (Albéric et al. 2022) were performed at mySpot, the microfocus beamline of BESSY II (Helmholtz Zentrum, Berlin) (Paris et al. 2007). Spines were mounted vertically inside a Linkam heating stage where a heating ramp from 25 to 400°C at 2°C/min was applied. The crystallization of the remnant ACC and the destruction of organics in the spines were therefore induced at ~200°C and ≥ 250°C respectively. With a sample-detector distance of 830 mm, and taking into account the beamstop position, the SAXS q-range provided by this setup was 0.2–10 nm^{-1}, which allowed the measuring of scattering objects ranging from a few nanometers to 100 nm.

The calibration as well as the data evaluation were performed with the fit2D and DPDAK v1_2_0 software (Benecke et al. 2015). Briefly, the 2D SAXS signals were azimuthally integrated to obtain the corresponding SAXS profiles that represent the X-ray scattering intensity as a function of the scattering vector q (Figure 8.11(a)) (Fratzl et al. 1991). The T parameter analysis was also performed with the DPDAK software in order to determine the minimum dimension of the scattering objects (Fratzl et al. 1991).

8.5.4. *Quantification of the nanoporosity*

Figure 8.11 displays the SAXS profiles of an intact *P. lividus* sea urchin spine during ACC crystallization and organics removal. The SAXS profile of the native spine, measured at room temperature, shows a shoulder at ~2 nm^{-1} that most likely originates from the presence of nano-pores composed of hydrated organics with electron density differing from that of calcite (Albéric et al. 2022). Similar SAXS profiles were obtained when measuring *Anthocidaris crassispina* sea urchin spines (Seto et al. 2012) and in vitro calcite crystals with occluded polyelectrolytes and polymeric additives (Schenk et al. 2012; Kim et al. 2014).

Upon heating, the shoulder observed at ~2 nm^{-1} shifts to lower q (Figure 8.11(a)). This shift, which is better visualized in the Kratky plots (Figure 8.11(b)), suggests an increase in size of the nano-nanopores with increasing temperature. It is worth mentioning that after 275°C, the shoulder seems to disappear from the q-range window likely due to an increase in size of the nano-pores beyond the resolution of this specific SAXS setup. In order to quantify the size increase, the radius of gyration (Rg) of the nano-pores can be calculated by fitting the SAXS profiles with a Guinier–Porod model up to 275°C (Hammouda 2010). Figure 8.11(c) displays the Rg values from 25°C to 275°C and shows a most substantial increase of the nano-pore size above 200°C, due to ACC crystallization. Further data analyses suggest that the growth of the nano-pores in sea urchin spines upon heating occurs via a coarsening process governed by bulk diffusion (Albéric et al. 2022).

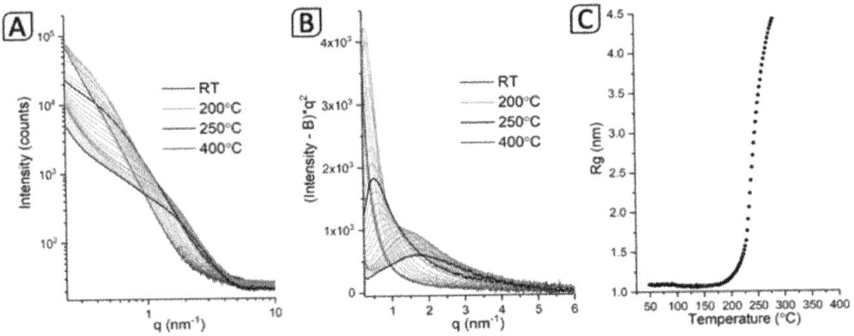

Figure 8.11. *A) Sea urchin spine SAXS intensity profiles for selected temperatures as a function of the scattering vector q, in log-log scale; B) Kratky plots, (Intensity-B)*q^2 as a function of q, with B, the Laue constant; and C) the radius of gyration (Rg) as a function of the temperatures. For a color version of this figure, see www.iste.co.uk/dejoie/synchrotron.zip*

8.6. Conclusion

The goal of this chapter was to outline a range of established SR-techniques in the field of biomineralization, which provided crucial information about the biomineralization processes in sea urchin skeletons. Finally, we would like to highlight that of SR-data analysis in SR-experiments, which can be time-consuming and tedious, but must be performed very carefully to ensure the correct interpretation of the results. In addition, SR-measurements may generate considerable amounts of data, especially tomography measurements. Moreover, analyzing SR-data often requires the development of custom-designed software, for example, involving machine learning methods allowing the quantitative descriptions of complex biomineralized structures (Beliav et al. 2020).

Finally, the study of heterogeneous and complex materials such as biominerals requires complementary approaches involving SR- and laboratory experiments together to draw final conclusions. In the studies presented in this chapter, for example, thermogravimetric analysis, coupled with differential scanning calorimetry allowed the quantification of ACC and organic molecules in sea urchin skeletal elements (Albéric et al. 2018b). Microgravimetry and isothermal calorimetry methods provided valuable insights into the mechanisms leading to the stability of synthetic ACC systems (Albéric et al. 2018a). Furthermore, focused ion beam-scanning electron microscopy (FIB-SEM) greatly complements SAXS measurements by bridging the gap between the nano- and the micro-scale (Albéric et al. 2022).

The constant evolution of synchrotron beamlines promises to offer more in situ measurements in the coming years that will allow the study of various mechanisms acting in vivo. This includes, for example, time-resolved experiments or cryo-setups for investigating the formation of biominerals closer to their native state and across developmental ages and species.

8.7. Acknowledgments

The authors thank Jonathan Perrin, Timm Weitkamp and Mario Schell for the CT data acquired at the ANATOMIX beamline of SOLEIL (Paris, France), which is an Equipment of Excellence (EQUIPEX) funded by the Investments for the Future program of the French National Research Agency (ANR), project NanoimagesX, grant no. ANR-11-EQPX-0031. The authors are grateful to Clemens Schmitt and Maryam Tadayon for the measurements performed at the Tomcat beamline of SLS (Zurich, Switzerland). Finally, the authors also thank Yael Politi and Luca Bertinetti from BCube and TU Dresden for their advices for writing the chapter.

8.8. References

Addadi, L., Raz, S., Weiner, S. (2003). Taking advantage of disorder: Amorphous calcium carbonate and its roles in biomineralization. *Advanced Materials*, 15(12), 959–970.

Aizenberg, J., Hanson, J., Koetzle, T.F., Leiserowitz, L., Weiner, S., Addadi, L. (1995). Biologically induced reduction in symmetry: A study of crystal texture of calcitic sponge spicules. *Chemistry - A European Journal*, 1(7), 414–422.

Aizenberg, J., Hanson, J., Koetzle, T.F., Weiner, S., Addadi, L. (1997). Control of macromolecule distribution within synthetic and biogenic single calcite crystals. *Journal American Chemical Society*, 119(5), 6.

Albéric, M., Bertinetti, L., Zou, Z.Y., Fratzl, P., Habraken, W., Politi, Y. (2018a). The crystallization of amorphous calcium carbonate is kinetically governed by ion impurities and water. *Advanced Science*, 5(5), 1701000.

Albéric, M., Caspi, E.N., Bennet, M., Ajili, W., Nassif, N., Azais, T., Berner, A., Fratzl, P., Zolotoyabko, E., Bertinetti, L. et al. (2018b). Interplay between calcite, amorphous calcium carbonate, and intracrystalline organics in sea urchin skeletal elements. *Crystal Growth & Design*, 18(4), 2189–2201.

Albéric, M., Stifler, C.A., Zou, Z., Sun, C.-Y., Killian, C.E., Valencia, S., Mawass, M.-A., Bertinetti, L., Gilbert, P.U.P.A., Politi, Y. (2019). Growth and regrowth of adult sea urchin spines involve hydrated and anhydrous amorphous calcium carbonate precursors. *Journal of Structural Biology: X*, 1(100004).

Albéric, M., Zolotoyabko, E., Spaeker, O., Chenghao, L., Tadayon, M., Schmitt, C.N.Z., Politi, Y., Bertinetti, L., Fratzl, P. (2022). Heat-mediated micro- and nano-pore evolution in sea urchin biominerals. *Crystal Growth & Design*, 22(6), 3727–3739.

Als-Nielsen, J. and McMorrow, D. (eds) (2011). Kinematical scattering II: Crystalline order. In *Elements of Modern X-ray Physics*. Wiley, Chichester, 147–205.

Amini, S. and Miserez, A. (2013). Wear and abrasion resistance selection maps of biological materials. *Acta Biomaterialia*, 9(8), 7895–7907.

Beliav, M., Zöllner, D., Pacureabu, A., Zaslansky, P., Bertinetti, L., Zlotnikov, I. (2020). Quantification of sheet nacre morphogenesis using X-ray nanotomography and deep learning. *Journal of Structural Biology*, 209, 107432.

Benecke, G., Flucke, G., Chenghao, L. (2015). DPDAK: Directly programmable data analyses kit for online and offline analysis of 2D scattering data [Online]. Available at: http://dpdak.desy.de/.

Beniash, E., Aizenberg, J., Addadi, L., Weiner, S. (1997). Amorphous calcium carbonate transforms into calcite during sea urchin larval spicule growth. *Proceedings of the Royal Society B-Biological Sciences*, 264(1380), 461–465.

Beniash, E., Addadi, L., Weiner, S. (1999). Cellular control over spicule formation in sea urchin embryos: A structural approach. *Journal of Structural Biology*, 125(1), 50–62.

Beniash, E., Benzerara, K., Blamart, D., Boskey, A.L., Buehler, M.J., Chung, P., Chung, S., Cosmidis, J., Cui, Q., Cusack, M. et al. (2014). *Biomineralization Sourcebook: Characterization of Biominerals and Biomimetic Materials*. CRC Press, Boca Raton.

Gorzelak, P., Stolarski, J., Dubois, P., Kopp, C., Meibom, A. (2011). 26Mg labeling of the sea urchin regenerating spine: Insights into echinoderm biomineralization process. *Journal of Structural Biology*, 176(8).

Gorzelak, P., Dery, A., Dubois, P., Stolarski, J. (2017). Sea urchin growth dynamics at microstructural length scale revealed by Mn-labeling and cathodoluminescence imaging. *Frontiers in Zoology*, 14(1), 42.

Hammersley, A.P., Svensson, S.O., Hanfland, M., Fitch, A.N., Hausermann, D. (1996). Two-dimensional detector software: From real detector to idealised image or two-theta scan. *High Pressure Research*, 14(4–6), 235–248.

Hammouda, B. (2010). A new Guinier-Porod model. *Journal of Applied Crystallography*, 43, 716–719.

Hasse, B., Ehrenberg, H., Marxen, J.C., Becker, W., Epple, M. (2000). Calcium carbonate modifications in the mineralized shell of the freshwater snail *Biomphalaria glabrata*. *Chemistry-a European Journal*, 6(20), 3679–3685.

Heatfield, B.M. (1971). Origin of calcified tissue in regenerating spines of sea urchin, *Strongylocentrotus-Purpuratus* (Stimpson): A quantitative radioautographic study with tritiated thymidine. *Journal of Experimental Zoology*, 178(2), 233–246.

Heatfield, B.M. and Travis, D.F. (1975). Ultrastructural studies of regenerating spines of the sea urchin. *Strongylocentrotus Purpuratus*. *Journal of Morphology*, 145(37), 13–72.

Ihli, J., Bots, P., Kulak, A., Benning, L.G., Meldrum, F.C. (2013). Elucidating mechanisms of diffusion-based calcium carbonate synthesis leads to controlled mesocrystal formation. *Advanced Functional Materials*, 23(15), 1965–1973.

Jensen, A.C.S., Imberti, S., Parker, S.F., Schneck, E., Politi, Y., Fratzl, P., Bertinetti, L., Habraken, W.J.E.M. (2018). Hydrogen bonding in amorphous calcium carbonate and molecular reorientation induced by dehydration. *The Journal of Physical Chemistry C*, 122(6), 3591–3598.

Juhas, P., Davis, T., Farrow, C.L., Billinge, S.J.L. (2013). PDFgetX3: A rapid and highly automatable program for processing powder diffraction data into total scattering pair distribution functions. *Journal of Applied Crystallography*, 46, 560–566.

Killian, C.E., Metzler, R.A., Gong, Y.U.T., Olson, I.C., Aizenberg, J., Politi, Y., Wilt, F.H., Scholl, A., Young, A., Doran, A. et al. (2009). Mechanism of calcite co-orientation in the sea urchin tooth. *Journal of the American Chemical Society*, 131(51), 18404–18409.

Kim, Y.-Y., Schenk, A.S., Ihli, J., Kulak, A.N., Hetherington, N.B.J., Tang, C.C., Schmahl, W.W., Griesshaber, E., Hyett, G., Meldrum, F.C. (2014). A critical analysis of calcium carbonate mesocrystals. *Nature Communications*, 5(1), 4341.

Ko, J.Y.P., Zhou, X.T., Heigl, F., Regier, T., Blyth, R., Sham, T.K. (2007). X-ray absorption near-edge structure (XANES) of calcium L3,2 edges of various calcium compounds and X-ray excited optical luminescence (XEOL) studies of luminescent calcium compounds. *AIP Conference Proceedings*, 882(1), 538–540.

Lam, R.S.K., Charnock, J.M., Lennie, A., Meldrum, F.C. (2007). Synthesis-dependant structural variations in amorphous calcium carbonate. *CrystEngComm*, 9(12), 1226–1236.

Larsson, A.-K. and Christy, A.G. (2008). On twinning and microstructures in calcite and dolomite. *American Mineralogist*, 93(1), 103–113.

Levi-Kalisman, Y., Raz, S., Weiner, S., Addadi, L., Sagi, I. (2002). Structural differences between biogenic amorphous calcium carbonate phases using X-ray absorption spectroscopy. *Advanced Functional Materials*, 12(1), 43–48.

Lowenstam, H.A. and Weiner, S. (1989). *On Biomineralization*. Oxford University Press, New York.

Markel, K. and Roser, U. (1983). The spine tissues in the echinoid Eucidaris - tribuloides. *Zoomorphology*, 103(1), 25–41.

Metzler, R.A., Kim, I.W., Delak, K., Evans, J.S., Zhou, D., Beniash, E., Wilt, F., Abrecht, M., Chiou, J.W., Guo, J.H. et al. (2008). Probing the organic–mineral interface at the molecular level in model biominerals. *Langmuir*, 24(6), 2680–2687.

Michel, F.M., Macdonald, J., Feng, J., Phillips, B.L., Ehm, L., Tarabrella, C., Parise, J.B., Reeder, R.J. (2008). Structural characteristics of synthetic amorphous calcium carbonate. *Chemistry of Materials*, 20(14), 4720–4728.

Moureaux, C., Pérez-Huerta, A., Compère, P., Zhu, W., Leloup, T., Cusack, M., Dubois, P. (2010). Structure, composition and mechanical relations to function in sea urchin spine. *Journal of Structural Biology*, 170, 41–49.

Pabisch, S., Wagermaier, W., Zander, T., Li, C., Fratzl, P. (2013). Imaging the nanostructure of bone and dentin through small- and wide-angle X-ray scattering. In *Methods in Enzymology*, De Yoreo, J.J. (ed.). Academic Press, San Diego, CA.

Paris, O., Li, C.H., Siegel, S., Weseloh, G., Emmerling, F., Riesemeier, H., Erko, A., Fratzl, P. (2007). A new experimental station for simultaneous X-ray microbeam scanning for small- and wide-angle scattering and fluorescence at BESSY II. *Journal of Applied Crystallography*, 40, S466–S470.

Pilkington, J.B. (1969). The organization of skeletal tissues in the spines of echinus esculentus. *Journal of the Marine Biological Association of the United Kingdom*, 49(25), 857–877.

Pokroy, B., Quintana, J.P., Caspi, E.N., Berner, A., Zolotoyabko, E. (2004). Anisotropic lattice distortions in biogenic aragonite. *Nature Materials*, 3(12), 900–902.

Pokroy, B., Fitch, A.N., Marin, F., Kapon, M., Adir, N., Zolotoyabko, E. (2006). Anisotropic lattice distortions in biogenic calcite induced by intra-crystalline organic molecules. *Journal of Structural Biology*, 155(8).

Pokroy, B., Kapon, M., Marin, F., Adir, N., Zolotoyabko, E. (2007). Protein-induced, previously unidentified twin form of calcite. *Proceedings of the National Academy of Sciences*, 104(18), 5.

Polishchuk, I., Bracha, A.A., Bloch, L., Levy, D., Kozachkevich, S., Etinger-Geller, Y., Kauffmann, Y., Burghammer, M., Giacobbe, C., Villanova, J. et al. (2017). Coherently aligned nanoparticles within a biogenic single crystal: A biological prestressing strategy. *Science*, 358(6368), 1294.

Politi, Y., Arad, T., Klein, E., Weiner, S., Addadi, L. (2004). Sea urchin spine calcite forms via a transient amorphous calcium carbonate phase. *Science*, 306(5699), 1161–1164.

Politi, Y., Levi-Kalisman, Y., Raz, S., Wilt, F., Addadi, L., Weiner, S., Sagi, I. (2006). Structural characterization of the transient amorphous calcium carbonate precursor phase in sea urchin embryos. *Advanced Functional Materials*, 16(10), 1289–1298.

Politi, Y., Metzler, R.A., Abrecht, M., Gilbert, B., Wilt, F.H., Sagi, I., Addadi, L., Weiner, S., Gilbert, P. (2008). Transformation mechanism of amorphous calcium carbonate into calcite in the sea urchin larval spicule. *Proceedings of the National Academy of Sciences of the United States of America*, 105(45), 17362–17366.

Radha, A.V., Fernandez-Martinez, A., Hu, Y.D., Jun, Y.S., Waychunas, G.A., Navrotsky, A. (2012). Energetic and structural studies of amorphous $Ca_{1-x}Mg_xCO_3.nH_2O$ ($0 <= x <= 1$). *Geochimica et Cosmochimica Acta*, 90, 83–95.

Raup, D.M. (1959). Crystallography of echinoid calcite. *Journal of Geology*, 67(6), 661–674.

Reeder, R.J., Tang, Y.Z., Schmidt, M.P., Kubista, L.M., Cowan, D.F., Phillips, B.L. (2013). Characterization of structure in biogenic amorphous calcium carbonate: Pair distribution function and nuclear magnetic resonance studies of lobster gastrolith. *Crystal Growth & Design*, 13(5), 1905–1914.

Rez, P. and Blackwell, A. (2011). Ca L23 spectrum in amorphous and crystalline phases of calcium carbonate. *The Journal of Physical Chemistry B*, 115(38), 11193–11198.

Rietveld, H.M. (1969). A profile refinement method for nuclear and magnetic structures. *Journal of Applied Crystallography*, 2(2), 65–71.

Rodriguez Carvajal, J. (1993). Recent advances in magnetic-structure determination by neutron powder diffraction. *Physica B*, 192(1–2), 55–69.

Saharay, M., Yazaydin, A.O., Kirkpatrick, R.J. (2013). Dehydration-induced amorphous phases of calcium carbonate. *Journal of Physical Chemistry B*, 117(12), 3328–3336.

Schenk, A.S., Zlotnikov, I., Pokroy, B., Gierlinger, N., Masic, A., Zaslansky, P., Fitch, A.N., Paris, O., Metzger, T.H., Cölfen, H. et al. (2012). Hierarchical calcite crystals with occlusions of a simple polyelectrolyte mimic complex biomineral structures. *Advanced Functional Materials*, 22(22), 4668–4676.

Schmidt, I., Lee, K., Zolotoyabko, E., Werner, P., Shim, T.S., Oh, Y.K., Fratzl, P., Wagermaier, W. (2014). Nanocrystalline calcitic lens arrays fabricated by self-assembly followed by amorphous-to-crystalline phase transformation. *Acs Nano*, 8(9), 9233–9238.

Scholl, A. (2014). PEEM Vision 2.001 [Online]. Available at: http://xraysweb.lbl.gov/peem2/webpage/Project/Resources/PEEMVision.pdf.

Seknazi, E., Kozachkevich, S., Polishchuk, I., Bianco Stein, N., Villanova, J., Suuronen, J.-P., Dejoie, C., Zaslansky, P., Katsman, A., Pokroy, B. (2019). From spinodal decomposition to alternating layered structure within single crystals of biogenic magnesium calcite. *Nature Communications*, 10(1), 4559.

Seto, J., Ma, Y.R., Davis, S.A., Meldrum, F., Gourrier, A., Kim, Y.Y., Schilde, U., Sztucki, M., Burghammer, M., Maltsev, S. et al. (2012). Structure–property relationships of a biological mesocrystal in the adult sea urchin spine. *Proceedings of the National Academy of Sciences of the United States of America*, 109(10), 3699–3704.

Smith, B.A. (1980). Stereom microstructure of the echinoid test. *Special Papers in Palaeontology*, 25, 1–324.

Stawski, T.M. and Benning, L.G. (2013). SAXS in inorganic and bioinspired research. In *Methods in Enzymology*, De Yoreo, J.J. (ed.). Academic Press, San Diego, CA.

Strathmann, R.R. (1981). The role of spines in preventing structural damage to echinoid tests. *Paleobiology*, 7(3), 400–406.

Taylor, M.G., Simkiss, K., Greaves, G.N., Okazaki, M., Mann, S. (1993). An X-ray absorption spectroscopy study of the structure and transformation of amorphous calcium carbonate from plant cystoliths. *Proceedings of the Royal Society of London. Series B: Biological Sciences*, 252(1333), 75–80.

Tsipursky, S.J. and Buseck, P.R. (1993). Structure of magnesian calcite from sea urchins. *American Mineralogist*, 78(7–8), 775–781.

Wagermaier, W., Gourrier, A., Aichmayer, B. (2013). Understanding hierarchy and functions of bone using scanning X-ray scattering methods. In *Materials Design Inspired by Nature: Function through Inner Architecture*, Fratzl, P., Dunlop, J.W.C., Weinkamer, R. (eds). Royal Society of Chemistry, Cambridge.

Weiner, S. and Addadi, L. (2011). Crystallization pathways in biomineralization. *Annual Reviews of Materials Research*, 41, 21–40.

Weiner, S., Levi-Kalisman, Y., Raz, S., Addadi, L. (2003). Biologically formed amorphous calcium carbonate. *Connective Tissue Research*, 44(1), 214–218.

Weitkamp, T., Scheel, M., Giorgetta, J.L., Joyet, V., Le Roux, V., Cauchon, G., Moreno, T., Polack, F., Thompson, A., Samama, J.P. (2017). The tomography beamline ANATOMIX at Synchrotron SOLEIL. *J. Phys.: Conf. Ser.*, 849, 012037.

Zizak, I., Paris, O., Roschger, P., Bernstorff, S., Amenitsch, H., Klaushofer, K., Fratzl, P. (2000). Investigation of bone and cartilage by synchrotron scanning-SAXS and -WAXD with micrometer spatial resolution. *Journal of Applied Crystallography*, 33(3-1), 820–823.

Zolotoyabko, E. (2017). Anisotropic lattice distortions in biogenic minerals originated from strong atomic interactions at organic/inorganic interfaces. *Advanced Materials Interfaces*, 4(1), 1–12.

Zou, Z.Y., Bertinetti, L., Politi, Y., Jensen, A.C.S., Weiner, S., Addadi, L., Fratzl, P., Habraken, W. (2015). Opposite particle size effect on amorphous calcium carbonate crystallization in water and during heating in air. *Chemistry of Materials*, 27(12), 4237–4246.

List of Authors

Marie ALBÉRIC
Laboratoire chimie de la matière condensée
Sorbonne University
Paris
France

Michèle ALVAREZ-MURGA
Institut Néel CNRS/UGA
Grenoble
France

Michel ANNE
Institut Néel CNRS/UGA
Grenoble
France

Pierre-Olivier AUTRAN
European Synchrotron Radiation Facility
Grenoble
France

Nils BLANC
Institut Néel CNRS/UGA
Grenoble
France

Pierre BLEUET
CEA-LETI
Grenoble
France

Pierre BORDET
Institut Néel CNRS/UGA
Grenoble
France

Nathalie BOUDET
Institut Néel CNRS/UGA
Grenoble
France

Chantal BROUCA-CABARRECQ
CEMES
Toulouse
France

Sophie CERSOY
Centre de Recherche sur la Conservation
Paris
France

Heng CHEN
Shandong Provincial Key Lab for the Preparation and Measurement of Building Materials
University of Jinan
China

Marine COTTE
European Synchrotron Radiation Facility
Grenoble
France

Catherine DEJOIE
European Synchrotron Radiation Facility
Grenoble
France

Philippe DILLMANN
LAPA NIMBE/IRAMAT
CEA/CNRS
Université Paris-Saclay
Gif-sur-Yvette
France

Eric DOORYHÉE
Institut Néel CNRS/UGA
Grenoble
France

Eddy FOY
LAPA NIMBE/IRAMAT
CEA/CNRS
Université Paris-Saclay
Gif-sur-Yvette
France

Jean-Louis HODEAU
Institut Néel CNRS/UGA
Grenoble
France

Marie D. JACKSON
Department of Geology and Geophysics
University of Utah
Salt Lake City
USA

Florian KERGOULAY
Institut Néel CNRS/UGA
Grenoble
France

Olivier LEYNAUD
Institut Néel CNRS/UGA
Grenoble
France

Pauline MARTINETTO
Institut Néel CNRS/UGA
Grenoble
France

Cagla MERAL AKGUL
Department of Civil Engineering
Middle East Technical University
Ankara
Turkey

François MIRAMBET
Centre de recherche et de restauration des musées de France
Paris
France

Letizia MONICO
CNR-SCITEC
Perugia
Italy

Judith MONNIER
ICMPE CNRS
Thiais
France

Delphine NEFF
LAPA NIMBE/IRAMAT
CEA/CNRS
Université Paris-Saclay
Gif-sur-Yvette
France

Jacob G. PETERSON
Department of Geology and Geologic Engineering
Colorado School of Mines
Golden
USA

Emeline POUYET
Laboratoire d'archéologie moléculaire structurale
Sorbonne University
Paris
France

Alain PRAT
Institut Néel CNRS/UGA
Grenoble
France

Solenn RÉGUER
Synchrotron Soleil
Saint-Aubin
France

Bryony RICHARDS
Energy and Geoscience Institute
University of Utah
Salt Lake City
USA

Philippe SCIAU
CEMES
Toulouse
France

Ronald SEIDEL
Center for Molecular and Cellular Bioengineering
Technische Universitat Dresden
Germany

Nobumichi TAMURA
Advanced Light Source
Lawrence Berkeley National Lab
USA

Philippe WALTER
Laboratoire d'archéologie moléculaire structurale
Sorbonne University
Paris
France

Index

C, D

E, F

G, H

I, J

K, L

M, N

P, Q

R, S

Printed and bound by CPI Group (UK) Ltd, Croydon, CR0 4YY
13/09/2026
14961072-0003